# ADVANCES IN ELECTRONICS AND ELECTRON PHYSICS

## VOLUME 36

## Contributors to This Volume

Aaron Barnes
J. S. Bakos
Henning F. Harmuth
Harold R. Kaufman
D. Midgley

# Advances in
# Electronics and
# Electron Physics

EDITED BY
## L. MARTON

*Smithsonian Institution, Washington, D.C.*

*Assistant Editor*
CLAIRE MARTON

EDITORIAL BOARD

## VOLUME 36

1974

**ACADEMIC PRESS**     New York   San Francisco   London
A Subsidiary of Harcourt Brace Jovanovich, Publishers

ACADEMIC PRESS, INC.
111 Fifth Avenue, New York, New York 10003

*United Kingdom Edition published by*
ACADEMIC PRESS, INC. (LONDON) LTD.
24/28 Oval Road, London NW1

LIBRARY OF CONGRESS CATALOG CARD NUMBER: 49-7504

ISBN 0–12–014536–7

PRINTED IN THE UNITED STATES OF AMERICA

# CONTENTS

### Theoretical Studies of the Large-Scale Behavior of the Solar Wind

#### AARON BARNES

### Multiphoton Ionization of Atoms

#### J. S. BAKOS

### Recent Advances in the Hall Effect: Research and Application

#### D. MIDGLEY

## Research and Development in the Field of Walsh Functions and Sequency Theory

HENNING F. HARMUTH

## Technology of Electron-Bombardment Ion Thrusters

HAROLD R. KAUFMAN

# CONTRIBUTORS TO VOLUME 36

AARON BARNES, Space Science Division, Ames Research Center-NASA, Moffett Field, California

J. S. BAKOS, Central Research Institute for Physics, Department of Quantum Electronics, Budapest, Hungary

HENNING F. HARMUTH, Department of Electrical Engineering, The Catholic University of America, Washington, D.C.

HAROLD R. KAUFMAN, Colorado State University, Fort Collins, Colorado

D. MIDGLEY, Department of Electronic Engineering, The University of Hull, Hull, United Kingdom

# FOREWORD

Five reviews on widely differing subjects constitute the present volume. The first, by Aaron Barnes considers the theoretical aspects of a large-scale electron physical phenomenon: the solar wind. The author discusses the basic assumptions of various solar wind models. Whatever model is proposed, it has to account for the heating and acceleration of the charged particles flowing from the sun, for their angular momentum, and for their penetration into interstellar space.

Multiphoton ionization of atoms is the subject of J. S. Bakos' review. Although the concept of two-photon processes may go back as far as 1925, serious investigation of multiphoton events is barely 20 years old. The review covers theoretical and experimental work in this field and the author points out that the investigation of multiphoton ionization provides information on such atomic processes as multiphoton absorption, multi-photon-induced emission and scattering with simultaneous absorption and emission of many quanta of different frequencies.

Six years ago various aspects of Hall-effect research and development were reviewed by S. Stricker in our Volume 25 (1968). D. Midgley takes up the subject from then on and shows how much has been done in the meantime in theoretical and experimental research, as well as in applications of the effect.

In the past the mathematical treatment of electrical communication problems was based essentially on the use of sinusoidal functions. In recent years transmitters and receivers were developed for use with nonsinusoidal waves. Walsh functions are eminently suited for their treatment and the review of H. F. Harmuth offers an introduction to these functions, as well as to their application to a number of problems.

We started this volume with a review on a space science subject. It is proper to finish it with a review on space technology. H. R. Kaufman examines the present status of electron-bombardment ion thrusters. It has been recognized for several years that, although at the present stage of technology chemical propulsion of space vehicles is more advantageous, electrical propulsion is sufficiently intriguing to warrant a close look. While emphasis in the review is on space applications, the author points out that technology developed in the space program has many applications in unrelated areas.

It has been our practice to list expected future contributions and their authors. In the next volumes we plan to publish the following reviews:

<table>
<tr><td>Intraband Magneto-Optical Studies of Semiconductors in the Far Infrared, I & II</td><td>B. D. McCombe and R. J. Wagner</td></tr>
<tr><td>The Gyrator in Electronic Systems</td><td>K. M. Adams, E. F. A Deprettere and J. O. Voorman</td></tr>
<tr><td>Image Sensors for Solid State Cameras</td><td>P. K. Weimer</td></tr>
<tr><td>Ion Implantation in Semiconductors</td><td>S. Namba and K. Masuda</td></tr>
<tr><td>The Impact of Solid State Microwave Devices: A Preliminary Technology Assessment</td><td>J. Frey and R. Bowers</td></tr>
<tr><td>The Future Possibilities for Neural Control</td><td>K. Frank and F. T. Hambrecht</td></tr>
<tr><td>Charged Pigment Xerography</td><td>F. W. Schmidlin and M. E. Scharfe</td></tr>
<tr><td>Nonlinear Electron Acoustic Waves, II</td><td>R. G. Fowler</td></tr>
<tr><td>The Effects of Radiation in MIS Structures</td><td>Karl Zaininger and R. J. Powell</td></tr>
<tr><td>The Photovoltaic Effect</td><td>Joseph J. Loferski</td></tr>
<tr><td>Semiconductor Microwave Power Devices, I & II</td><td>S. Teszner</td></tr>
<tr><td>The Excitation and Ionization of Ions by Electron Impact</td><td>John W. Hooper and R. K. Feeney</td></tr>
<tr><td>Whistlers and Echoes</td><td>Robert A. Helliwell</td></tr>
<tr><td>Experimental Studies of Acoustic Waves in Plasmas</td><td>J. L. Hirshfield</td></tr>
<tr><td>Auger Electron Spectroscopy</td><td>N. C. Macdonald and P. W. Palmberg</td></tr>
<tr><td>Interpretation of Electron Microscope Images of Defects in Crystals</td><td>M. J. Whelan</td></tr>
<tr><td>Time Measurements on Radiation Detector Signals</td><td>S. Cova</td></tr>
<tr><td>Energy Distribution of Electrons Emitted by a Thermionic Cathode</td><td>W. Franzen and J. Porter</td></tr>
<tr><td>In Situ Electron Microscopy of Thin Films</td><td>A. Barna, P. B. Barna, J. P. Pocza, and I. Pozsgai</td></tr>
<tr><td>Afterglow Phenomena in Rare Gas Plasmas between 0° and 300°K</td><td>J. F. Delpech</td></tr>
<tr><td>Physics and Technologies of Polycrystalline Si in Semiconductor Devices</td><td>I. Kobayashi</td></tr>
<tr><td>Advances in Molecular Beam Masers</td><td>D. C. Lainé</td></tr>
<tr><td>Electron Beam Microanalysis</td><td>D. R. Beaman</td></tr>
<tr><td>Charge Coupled Devices</td><td>M. F. Tompsett and C. H. Séquin</td></tr>
<tr><td>Development of Charge Control Concept</td><td>J. teWinkel</td></tr>
<tr><td>Charge Particles as a Tool for Surface Research</td><td>J. Vennik</td></tr>
<tr><td>Electron Micrograph Analysis by Optical Transforms</td><td>G. Donelli and L. Paoletti</td></tr>
</table>

Comments on the published papers and suggestions for future reviews are cordially invited. Our warmest thanks go to all who helped to put this volume together.

L. MARTON
CLAIRE MARTON

# ADVANCES IN ELECTRONICS AND ELECTRON PHYSICS

## VOLUME 36

# Theoretical Studies of the Large-Scale
# Behavior of the Solar Wind

## AARON BARNES

*Space Science Division,*
*Ames Research Center-NASA,*
*Moffett Field, California*

## I. Introduction

This review is a survey primarily of theoretical work aimed at understanding the large-scale dynamics of the solar wind. The emphasis is mainly on acceleration, heating, angular momentum transfer, and termination of the wind. Topics such as the interplanetary magnetic field, fluctuations of the solar wind, and the origin of the solar wind are discussed only insofar as they are related to the main topics of the review. We do not consider the interaction of the solar wind with planets, or the modulation of galactic cosmic rays, etc. Observations are reviewed only to the extent that they have direct relevance to the problems discussed in the text.

The most recent published reviews of solar wind phenomena may be found in the Proceedings of the Solar Wind Conference (Sonett *et al.*, 1972, esp. Chs. 3, 4, and 9), and the reviews of Parker (1972) (see also Parker, 1971) and Axford (1972) are particularly relevant to the present review. Earlier reviews of various aspects of the solar wind and interplanetary

1

magnetic field have been given by Parker (1965, 1967, 1969), Dessler (1967), Lüst (1967), Ness (1968), Axford (1968), Wilcox (1968), Hundhausen (1968, 1970), and Holzer and Axford (1970). Jokipii (1971) has recently reviewed interaction of the solar wind with cosmic rays. Spreiter and Alksne (1970) have reviewed the interaction of the solar wind with planetary obstacles. Monographs on the solar wind have been written by Parker (1963), Brandt (1970), and Hundhausen (1972).

The goals of the present review are to survey critically recent theoretical work in the field, and, at the same time, to provide a comprehensive description of the subjects for a broader audience than just specialists in the field. The references constitute a fairly complete bibliography of theoretical work in the period 1970–72. More complete bibliographies of observational studies and earlier theoretical work can be found in the earlier reviews cited above. The format of the review is as follows: the following section sets out the problem of the large-scale solar wind in a general way; Section III discusses the limitations of certain idealizations, such as spherical symmetry, steady flow, etc., which commonly are made in theoretical studies of the solar wind; Section IV is concerned with heating and radial acceleration of the solar wind; Section V discusses the outer solar wind and its transition to the interstellar medium; in Section VI we consider angular momentum transfer in the solar wind.

## II. PRELIMINARY REMARKS

The transition between the sun and nearby interstellar matter is largely governed by the steady expansion of the hot tenuous outer atmosphere (corona) of the sun. This expansion starts low in the corona, slowly at first, then accelerates outward. The flow passes continuously into a supersonic state a few solar radii ($R_\odot$) above the solar surface, and continues flowing outward as a supersonic "solar wind." The solar wind persists far from the sun, certainly beyond the orbit of the earth, and must eventually make a transition to the local interstellar medium. This picture of the solar environment is very likely applicable to many other stars that are similar to the sun.

This state of affairs was unsuspected twenty years ago. Although it had long been thought that occasional bursts of solar particles impacted the earth's magnetic field, it was not until the 1950's that the existence of *continuous* emission of solar particles was suggested by Biermann (1951, 1957), from his studies of ionic comet tails. Parker (1958a) pointed out that the very hot ($\geq 10^6$K) solar corona could be expected on theoretical grounds to expand continuously away from the sun. If so, this steadily expanding corona, or solar wind, could be identified with Biermann's "solar corpuscular radiation." This viewpoint, although not generally accepted when it was proposed, was shortly verified by *in situ* measurements of the solar wind from earth-orbiting spacecraft. For interesting, readable accounts of the ideas and discoveries that eventually led to the discovery of the solar wind, the reader is referred to the books of Brandt (1970) and Hundhausen (1972) and the review article of Dessler (1967).

Just as the existence of the solar wind was first established by spacecraft observation (Gringauz *et al.*, 1960; Shklovsky *et al.*, 1961), most of our current knowledge of the solar wind comes from earth- and solar-orbiting spacecraft. Reviews of these observations have been given (Axford, 1968; Hundhausen, 1968, 1970; Ness, 1968; Wilcox, 1968; Wolfe, 1972). The solar wind at 1 astronomical unit* (AU) from the sun is a completely ionized (but electrically neutral) plasma, whose composition is typically 96% hydrogen and 4% helium by number, which flows outward from the sun with a speed of about 400 km sec$^{-1}$. Typical parameters of the solar wind as measured by spacecraft-borne plasma probes and magnetometers are listed in Table I.

TABLE I

PROPERTIES OF THE INTERPLANETARY MEDIUM AT 1 AU

| | |
|---|---|
| Composition | ~ 96% H, 4% He, completely ionized |
| Density | ~ 6 protons cm$^{-3}$ |
| Flow velocity | ~ 400 km sec$^{-1}$, radially outward |
| Proton temperature | ~ 4 × 10$^4$K |
| Electron temperature | ~ 10–20 × 10$^4$K |
| Helium temperature | ~ 15 × 10$^4$K |
| Magnetic field strength | ~ 5 × 10$^{-5}$ G, oriented in the ecliptic plane at about 45° to the radial direction |

The parameters of Table I indicate that the solar wind at 1 AU is quite collisionless (the mean free path for Coulomb collisions is comparable to 1 AU), and is therefore not in thermal equilibrium; in fact, the kinetic temperatures of the various constituents are different. Also, the flow is hypersonic at 1 AU; the flow speed exceeds characteristic speeds of hydromagnetic waves by a factor of order ten. It should be emphasized that the above listed values of velocity, etc., are only averages, and fluctuate considerably on a time scale of days, or even hours (Fig. 1). The nature and possible origins of these fluctuations will be discussed later in this review.

Although spacecraft observations have been invaluable, they have so far been restricted to the ecliptic plane and to a relatively narrow range in heliocentric distance (0.7 to 5 AU), and so cannot give definitive data on

*The mean distance of the earth from the sun. 1 AU = 1.496 × 10$^8$ km.

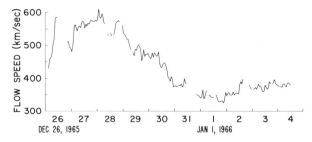

Fig. 1. Hourly average solar wind flow speed from December 26, 1965 to January 4, 1966 (ARC plasma experiment, Pioneer 6). Note the fluctuations on the scale of a few hours modulated by larger variations on the scale of days. Blank spaces are due to data gaps.

the solar wind taken as a whole. Observations at 1 AU are taken too far from the sun for optimum study of the origins of the wind, and too near the sun for optimum study of the transition to the interstellar medium. The range of spacecraft observation in the ecliptic plane should be significantly extended in the next decade; the planned Mariner Venus-Mercury and Helios missions will penetrate much nearer the sun than any past spacecraft, and Pioneers 10 and 11, and Mariner Jupiter-Saturn, may eventually leave the solar system. These missions are summarized in Table II. At this writing, Pioneer 10 is beginning to penetrate previously unexplored regions of the solar system (see Fig. 2).

TABLE II

SUMMARY OF SPACECRAFT MISSIONS[a]

| Spacecraft | Launch | Comments on orbit |
|---|---|---|
| Pioneer 10 | March 3, 1972 | Encountered Jupiter $(r \approx 5$ AU) December, 1973; survived Jovian radiation belts. Now leaving the solar system at speed of $\sim 3$ AU/yr |
| Pioneer 11 | April 6, 1973 | Encounters Jupiter in 1974–1975, and possibly Saturn $(r \sim 10$ AU) in about 1980 |
| Mariner Venus-Mercury | October, 1973 | Encounters Venus and Mercury, perihelion 0.4 AU |
| Helios A | Scheduled for August, 1974 | Perihelion 0.3 AU |
| Helios B | Scheduled for November, 1976 | Perihelion 0.3 AU |
| Mariner Jupiter-Saturn (proposed) | Proposed for 1977 | Dual launch. Would encounter Jupiter and Saturn, and eventually leave the solar system |

[a] This list should not be considered official, final, or necessarily complete.

Data from these missions will certainly clarify at least part of our picture of the large-scale solar wind. However, none of these missions will venture far from the ecliptic plane, or nearer the sun than 0.3 AU. Furthermore, although spacecraft passing through the outer solar system may eventually leave the solar system, they may not send back useful data from beyond 5–20 AU. Hence, it will be necessary for the foreseeable future to probe these important regions with other observational techniques. A variety of solar-astronomical methods, including reflection of terrestrial radar signals from the corona (James, 1970), are used to provide information about the corona near the sun. Some information about the character of the interplanetary

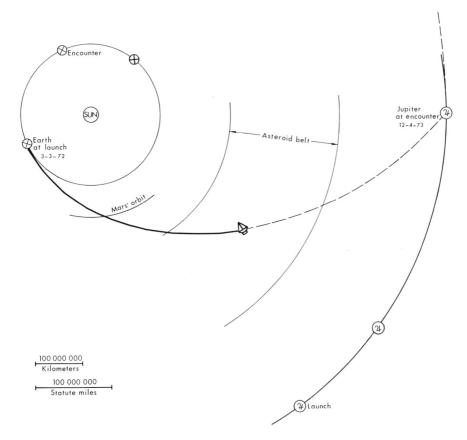

FIG. 2. Orbit of the Pioneer 10 spacecraft, showing its position on October 2, 1972. Pioneer 10 encountered Jupiter in December, 1973. It survived the Jovian radiation belts and continues to transmit solar wind measurements as it leaves the solar system at the speed of ∼3 AU/yr.

medium between the sun and 1 AU may be obtained from analysis of scattering and scintillation of signals from distant natural radio sources (e.g., see Hewish, 1972) and from the propagation of radio signals from solar-orbiting spacecraft when they happen to be far from earth (e.g., see Croft, 1972). Study of ionic comet tails has remained a useful tool for studying the solar wind at distances other than 1 AU and out of the ecliptic plane (Brandt, 1970). It has recently been established that the local interstellar medium may be observed in reradiated solar Lyman-α by the use of detectors borne by earth-orbiting spacecraft (e.g., see Thomas, 1972; Bertaux and Blamont, 1972). Certain phenomena associated with galactic or solar cosmic rays also provide at least tentative information on the state of the solar wind at various heliocentric distances (Jokipii, 1971). A summary of most of the above techniques, and their results up to 1968, has been given by Axford (1968); Chapter 4 of Brandt's (1970) book is also useful.

Altogether, our empirical knowledge of the solar wind is a mixture consisting partly of relatively unambiguous measurements, made *in situ* near 1 AU in the ecliptic plane, and partly of a number of kinds of indirect observations, whose interpretation is often difficult and sometimes ambiguous, but which cover a wide range of heliocentric distance and heliographic latitude. It is reasonable to expect that within the next decade or so our empirical knowledge of the large-scale character of the solar wind will be much more definite, largely because of space probes sent to both the outer and inner parts of the solar system, but also because of extension and refinement of the indirect observations and their interpretation. Despite these advances, one would be overly sanguine to expect that all the basic questions about the solar wind, its origin, and its termination will be finally resolved in the near future.

Therefore, any current attempt to construct a unified view of the large-scale solar environment must include some uncertain elements, which in turn usually consist of two or more alternate theories. These theories may be elaborate mathematical constructs, or they may be simple qualitative ideas. In any case, such a theory must be judged as to whether it is consistent with *those classes of observations which are most relevant to the essential features of the theory*. This last qualification is very important, because the relevancy of a given class of observation to a certain theory is often a matter of dispute and misunderstanding. For example, it has been fashionable to discuss the validity of theories of heating and acceleration of the solar wind primarily in terms of the agreement (or disagreement) of their predictions with spacecraft observations near 1 AU. This standard is convenient, because the direct spacecraft observations are the most reliable measurements we have. On the other hand, as we will discuss later, processes not directly related to the essential heating and acceleration mechanisms can strongly influence the detailed behavior of the solar wind at 1 AU. Accordingly, detailed comparison of a theory with observations at 1 AU may not always be the most suitable criterion for evaluating the validity of that theory with respect to its essential elements. To give a concrete example, we note that Parker's first models of the solar wind predicted flow velocity up to about $10^3$ km sec$^{-1}$ and density of $10^2$–$10^3$ protons cm$^{-3}$ at 1 AU. These predictions are not in quantitative agreement with observations (cf. Table I), but this disagreement does not invalidate the essential features of the theory. The basic prediction of that theory is that *the solar corona expands continuously and supersonically*. Judged by this standard, Parker's theory has been a brilliant success.

The point of these remarks is that since the solar wind is such a complicated physical system, one cannot reasonably expect any single model or theory to predict accurately the " weather " in the solar wind at all places and times. The most that should be expected of a theory is that it gives a new, qualitatively correct insight into the processes that govern the gross behavior of the solar wind. Although it would probably not be useful to attempt to set out general " rules of evidence " for evaluating theoretical ideas about the solar wind, in evaluating a given concept it is always important to consider carefully the relevance, as well as the reliability, of each class of observation.

### III. Critical Comments on the Basic Assumptions of Various Solar Wind Models

As mentioned previously, the solar wind near 1 AU is observed to behave in a complicated nonsteady fashion. Furthermore, at 1 AU the proton–proton collision time is comparable to the characteristic solar wind expansion time. The inner solar corona itself is normally rather asymmetric. In contrast, theoretical models of the solar wind often require simplifying assumptions such as that the wind behaves as a fluid, and that its flow is steady and spherically symmetric. Obviously such models cannot, and are not intended to predict the details of solar wind weather at 1 AU.

The purpose of such models is to provide insight as to what physical processes have major influences on the large-scale properties of the solar wind. On the other hand, it has often been argued that perhaps the observed behavior of the solar wind at 1 AU is so influenced by essentially nonsteady or nonfluid processes that the construction of steady, fluid models becomes a dubious and perhaps completely pointless exercise. In this section we attempt to assess the relative merits of these two viewpoints. Although we do not expect to attain a final answer to this question, it appears that we must reject the two extreme versions. It is equally incorrect to suppose either (1) that the complicated behavior of the solar wind at 1 AU implies that steady, fluid models cannot provide important insight into the large-scale dynamics of the solar wind, or (2) that it is justified to construct more and more steady, fluid models, differing in minor details, without extremely careful consideration of the relevance, reliability, and limitations of current observations.

The question of the validity of fluid models is logically distinct from that of the validity of steadiness and spherical symmetry; nonsteady fluid flows and steady noncollisional flows are quite possible. Accordingly, we discuss the two issues separately.

### A. Steady Flow versus Nonsteady Flow

The solar wind requires about 4–5 days to flow from the sun to 1 AU. Consequently, a steady model of this flow can be rigorously valid only if velocity, density, etc., of the solar wind at 1 AU change very little on a time scale long compared with 4 days. While this criterion is not rigorously satisfied at 1 AU, it has been argued that solar wind is organized into streams that are statistically quasi-steady, but which have large amplitude fluctuations superposed on them.

This point of view is illustrated in Fig. 3. Suppose that we idealized the solar wind as essentially consisting of a few streams of different speeds. As the sun rotates on its axis, the feet of the streams will also rotate, so that the faster streams will tend to overtake the slower ones eventually. As a fast stream catches up with a slower one, a noisy interaction region will be generated. If the source of the solar wind is steady, and corotates with the

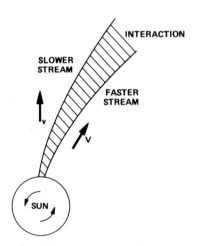

FIG. 3. Schematic illustration of the organization of the solar wind into streams of different speeds. As the sun rotates, so do the sources of the streams, and faster streams will eventually catch up with slower ones.

sun, the stream *structure* (but not the plasma itself) will also corotate with the sun. Hence, a fixed observer initially sitting in a slower stream would later be immersed in the interaction region, and finally in the faster stream. Except for the noise in the interaction region, variations of flow velocity seen by the observer would be due to rotation of the large-scale structure. In turn, if the large-scale streams are fairly homogeneous over a large enough range in solar longitude, they might be adequately approximated by a steady spherically symmetric model.

This notion that the solar wind is organized into quasi-steady streams with superposed fluctuations is often used as a working hypothesis. According to this viewpoint, steady spherically symmetric models may adequately describe the gross behavior of the large-scale streams, although they cannot describe the superposed fluctuations. In order to evaluate this view, we first consider in more detail the nature of the solar wind fluctuations at 1 AU.

We adopt the terminology of Burlaga and Ness (1968) in dividing solar wind fluctuations according to time scale into three classes:* microscale (time scale 1 hr or less); mesoscale ($\sim$ 10 hr); and macroscale ($\geq$ 100 hr). Because the solar wind is hypersonic at 1 AU, microfluctuations are due to convection of microsostructure (length scale $\lesssim$ 0.01 AU) past the observing spacecraft, independently of whether that microstructure is stationary or propagating relative to the local solar wind plasma. The origin and nature of the microstructure is an extremely interesting subject by itself (see reviews in Burlaga, 1971, 1972a; Scarf, 1970; Barnes, 1972). As we discuss later in this review, microfluctuations may play an essential role in the momentum, energy, and angular momentum budgets of the solar wind. Since steady models refer to average conditions over much longer times than 1 hr, the presence of microfluctuations does not vitiate the use of steady-state models.

The character of macroscale and, to a smaller degree, mesoscale variations has a direct bearing on the validity of steady-state models. Macroscale variations may originate in genuine temporal variations in the solar corona, in rotation of the spatially inhomogeneous corona beneath the point of observation, or both. Whatever their source, such variations will undergo dynamical changes in the interplanetary medium as they are convected to 1 AU. This dynamical evolution may distort the fluctuations so much that it becomes impossible to relate variations at 1 AU to source variations. Distortion of this kind may be considerable even if the fluctuations are of small amplitude; for example, theory indicates that small amplitude fluctuations in density at 0.1 AU may be strongly amplified by the time they reach 1 AU (Siscoe and Finley, 1969, 1970, 1972; Carovillano and Siscoe, 1969). The

---

* Hundhausen (1972) has recently proposed a similar, but somewhat more detailed, classification scheme.

dynamical evolution of larger amplitude fluctuations will be further distorted by nonlinear effects (Hundhausen, 1971, 1972, 1973; Gosling *et al.*, 1972a). These studies suggest that macroscale variations at the source evolve into a mixture of meso- and macrovariations at 1 AU. An example of meso-scale fluctuations produced in this way is the rapid density increase that occurs in the region of interaction between fast and slow streams. Such mesoscale fluctuations cannot be predicted by steady-state models. However, it seems reasonable that macroscale averages over a stream should not be grossly inconsistent with the predictions of steady-state models which contain the correct heating and acceleration mechanisms.

Observational studies of the persistence of large-scale solar wind structure give somewhat conflicting results. On the one hand, the sector pattern of interplanetary magnetic field polarity (Wilcox, 1968) and structure determined by radio scintillation measurements (Hewish, 1972; Cronyn, 1972) may persist in many solar rotations. On the other hand, correlation studies of the solar wind flow speed at 1 AU indicates a relatively weak persistence of velocity between one solar rotation and the next (Gosling and Bame, 1972; Gosling *et al.*, 1972a,b; Gosling, 1971, 1972).

It is fair to say that at present the difficult problem of establishing to what degree mesoscale and macroscale variations at 1 AU are due to variations at the source of the wind, as opposed to temporal evolution in the interplanetary medium, is not completely resolved. However, considerable progress in interpreting some of these features has recently been made. Linearized theory of small amplitude, corotating perturbations gives results that can be compared, at least qualitatively, with observed phase relations between (for example) density and flow speed in the stream structure at 1 AU (Siscoe and Finley, 1969, 1970, 1972; Carovillano and Siscoe, 1969). This theory gives results in best correspondence with observation when the source fluctuations are perturbations in temperature (Siscoe and Finley, 1972).

Another related problem which has been partially clarified recently is that of the correlations between various flow parameters measured at 1 AU. Since the flight of Mariner 2 it has been known that high speed solar wind streams tend to have relatively high proton temperatures and low densities (see Neugebauer and Snyder, 1966). A careful study of this relationship by Burlaga and Ogilvie (1970) showed that by far the strongest correlation is between flow speed and proton temperature. Burlaga and Ogilvie characterized the relation between proton temperature $T$ and flow speed $v$ by the relation

$$T^{1/2} = (0.036 \pm 0.003)v - (5.54 \pm 1.50),$$

where $T$ is measured in units of $10^3$K and $v$ is in km sec$^{-1}$. This result has been confirmed by the work of other observers (Hundhausen et al., 1970; Mihalov and Wolfe, 1971), although other empirical formulas may fit the data as well as the above equation.

Burlaga and Ogilvie interpreted the $T-v$ correlation as a large-scale effect which indicated extended heating in the solar wind. This interpretation has been adopted in some theoretical work on solar wind heating (Hartle and Barnes, 1970; Barnes et al., 1971, 1972) which showed that the variability in $T$ and $v$, and their relation, is easily understood in terms of variable heat sources for solar wind streams. This work is based on the assumption that the large-scale streams and their heating can be described adequately by steady-flow models.

In contrast, it had been suggested that possibly the $T-v$ correlation is not so directly related to extended heating of the solar wind streams, but rather is primarily an effect of temporal evolution of macroscale variations (Hundhausen, 1970; Montgomery, 1972b). To test this possibility, Hundhausen (1971, 1973) carried out numerical calculations of the nonlinear evolution of variations in a simple gasdynamical model of the solar wind. His approach was to require various time-dependent variations at the model base (near the sun, but beyond the distance of transition to supersonic flow), to compute the resulting variations in velocity, temperature, and density as functions of time at 1 AU, and to compare these predictions with observation. He found that *if the base perturbation is a suitable pure temperature perturbation*, the qualitative relations between the flow parameters at 1 AU agree well with the typical observed behavior of a solar wind stream. This conclusion therefore supports the similar conclusion of Siscoe and Finley (1972), based on linearized calculations.

In particular, Hundhausen's calculations predict the relation between $T$ and $v$ at 1 AU. A plot of their temporal variation in the $T-v$ plane indicates that $T(v)$ is a multivalued function, as one might intuitively expect. However, *on the average $T(v)$ is consistent with the empirical $T-v$ relation*.

Thus, Hundhausen's calculations show that the $T-v$ relation may be obtained from a time-dependent calculation, but only if the base perturbation includes temperature variation, which in turn implies extended heating below the model base. Hence it seems clear that the $T-v$ relation at 1 AU is due *essentially* to extended heating within about 0.1 AU, rather than to temporal evolution of macroscale variations in the interplanetary medium. Thus the conclusion of Burlaga and Ogilvie (1970, 1973) that the $T-v$ relation is an indicator of large-scale heating is supported. Furthermore, the steady-state models of Hartle and Barnes (1970) and others indicate that the $T-v$ relation is readily explained in terms of heating. Therefore, it appears that the essential features of solar wind heating and acceleration within 1 AU may be understood without considering the complexities associated with time-dependent processes.

At this point it may be worthwhile to reemphasize that, from the standpoint of solar wind acceleration and heating, considerations like the above place undue importance on the flow at 1 AU. After all, the main heating and acceleration processes that govern solar wind flow probably act within the first 0.1 AU or so. The time scale for flow in this region is less than a day, so that the assumption of steady flow is probably much better justified in this region than at 1 AU. From this viewpoint the question of whether the flow beyond the active heating and acceleration regions is steady or not is uninteresting.

Unfortunately, the most reliable observations at present are made near 1 AU, so that it is important to understand how the flow at 1 AU is modified by nonsteady phenomena. At the same time, it must be remembered that caution is appropriate when models are judged by their agreement with observations at 1 AU. We must not conclude that a model is correct in its essential features just because its agreement with observations at 1 AU is "perfect"; likewise, a model is not necessarily incorrect in its essential features if it predicts one or another parameter at 1 AU incorrectly by (say) a factor two.

### B. Fluid Flow and Noncollisional Flow

The terms "fluid" and "noncollisional" are somewhat ambiguous and, in fact, not mutually exclusive. These or closely related concepts have from time to time generated controversy which we see, with hindsight, to have been unnecessary. In this review we use the term "fluid flow" to mean flow of plasma characterized solely by macroscopic parameters such as flow velocity, density, stress tensor, and heat flow. We use the term "noncollisional flow" to mean plasma flow in which the mean *Coulomb* collision times of the plasma particles are long in comparison with some time that characterizes the large-scale behavior of the flow. It is possible for noncollisional flow also to be fluid flow in two senses: (1) the flow may be adequately described by a macroscopic theory, such as that of Chew–Goldberger–Low (Chew *et al.*, 1956), which is based on the collisionless kinetic (Vlasov) equation; or (2) noncollisional scattering of particles from fluctuations in the electromagnetic fields may reduce the effective particle collision time enough to make it much smaller than the characteristic time of the flow, so that the plasma behaves as a fluid.

In addition, collisional flows may be affected by noncollisional processes.

For example, consider a flow that is collisional in the sense that the Coulomb collision time is suitably small. If this flow contains microfluctuations whose periods are small compared with the Coulomb collision times, the plasma will appear noncollisional to the fluctuations. The fluctuations may dissipate by noncollisional processes such as Landau damping, thereby heating the plasma.

The issue of fluid flow versus noncollisional flow in the solar wind dates back to the beginning of the subject. Parker's (1958a) original model used fluid equations. Chamberlain (1961) argued for subsonic solutions on the basis of both fluid and exospheric models. Brandt and Cassinelli (1966) pointed out that in principle supersonic exospheric expansion is possible, because the energy dependence of the Coulomb cross section implies that higher energy particles can escape from deeper in the corona than lower energy particles. Jockers (1970) showed that, independent of this effect, a self-consistent treatment of the electric field in an expanding ion exosphere may require supersonic expansion.

Although such exospheric solutions are interesting, they are probably not relevant to the solar corona. The proton–proton collision time is probably much smaller than the expansion time of the wind for heliocentric distances smaller than about $10R_\odot$. Since the flow becomes supersonic well below this point, the plasma is collisional throughout the region of subsonic flow. Thus, exospheric expansion could begin only at some point where the plasma is already expanding outward supersonically. Exospheric models based on this premise have been studied by Hollweg (1970, 1971), Eyni and Kaufman (1971), and by Chen et al., (1972).

However, even this modified picture of laminar exospheric flow is probably not realistic for the solar wind. Plasma instabilities generated by distortion of the particle velocity distributions will tend to restore fluid-like properties to the collisionless solar wind (Eviatar and Schulz, 1970; Schulz and Eviatar, 1972; Perkins, 1973). In general, transport of momentum and energy in such a noncollisional fluid will be governed by "collisions" between particles and plasma fluctuations, rather than by Coulomb collisions. In this connection it should be remarked that such scattering does not require that the fluctuations be associated with instabilities in the plasma flow; fluctuations generated by large-scale motions near the solar surface, and convected outward in the wind, can have similar effects.

On the other hand, one might expect on physical grounds that the gross features of a solar wind model should not be very sensitive to whether a fluid or exospheric description is used. After all, velocity moments of the Vlasov equation give macroscopic equations that are closely analogous to the fluid equations. The main difference between the two approaches is manifested in the details of the stress tensor and temperature. But if exospheric flow begins in a region where the flow is already supersonic, the details of pressure and temperature do not strongly influence the large-scale flow.

If these statements are correct, it follows that apparent differences between fluid and exospheric models whose physical bases are otherwise similar may not be due to fluid or exospheric character of the flows. The differences may be caused by relatively subtle differences in boundary conditions, treatment of the equations, the details of the electron temperature distribution, etc. (cf. the discussion in Hollweg, 1971).

Leer and Holzer (1972) have examined this point in considerable detail. They constructed equivalent laminar exospheric and fluid models of the collisionless solar wind proton gas. The (kinetic) exospheric model was that of Hollweg (1970) modified to include a nonradial magnetic field (Chen *et al.*, 1972). The fluid model was based on the double-adiabatic version of the Chew–Goldberger–Low (Chew *et al.*, 1956) theory, which implies that pressure anisotropy is allowed, but proton thermal conduction and viscosity are not. Of course the exospheric model automatically contains the correct proton heat flow and viscous stress.

Leer and Holzer then compared the detailed velocity and temperature profiles of the two kinds of models for the same base conditions. They found the various profiles to be essentially the same for both classes of model. For example, at 1 AU the flow speeds predicted by the fluid equations are $\lesssim 8\%$ smaller than those given by the kinetic model; the proton temperature of the fluid model is a few percent higher than that of the exospheric model. From this comparison, they suggest that fluid models are adequate to describe the behavior of the solar wind within 1 AU. Furthermore, since their fluid model neglected viscosity and *proton* heat conduction, they conclude that these processes are not important in the solar wind between 0.1 AU and 1 AU (cf. discussion in Section IV).

The lesson to be drawn from the study of Leer and Holzer (1972) is that the gross description of the solar wind given by solution of the kinetic equations is not significantly different from the description given by an *appropriately chosen* fluid theory. The differences between one model and another are then differences in the physical mechanisms used or in base conditions. Since kinetic equations are difficult to handle, one normally will use a fluid description. On the other hand, one must not conclude that collisionless processes have negligible influence on the properties of the solar wind. As we shall see later, heating acceleration, and related transport phenomena may be profoundly affected by noncollisional processes, which must ultimately be described by plasma kinetic theory. However, such processes may usually be incorporated into a description of the large-scale solar wind by the introduction of a few macroscopic parameters. This kind of a procedure may not always be rigorous, but it will be adequate for the kind of coarse-grained description that is needed to understand the gross behavior of the solar wind. In the remainder of this section we discuss some of the noncollisional processes that may significantly affect the flow of the solar wind.

## 1. Pressure Anisotropy

One essential difference between collisional and noncollisional flow is that in noncollisional flows the stress (pressure) tensor may be quite anisotropic. In a collisionless, magnetized plasma whose spatial and temporal variations occur on scales long compared with the Larmor radii and gyroperiods of the electrons and ions, the pressure tensor is specified by two scalars $p_\perp$ and $p_\parallel$,

$$\mathbf{p} = p_\perp \mathbf{1} + (p_\parallel - p_\perp)\mathbf{ee} \tag{1}$$

where $\mathbf{e}$ is the unit vector parallel to the local magnetic field and $\mathbf{1}$ is the unit tensor (Chew et al., 1956). It has been verified observationally that solar wind protons at 1 AU are generally anisotropic and the anisotropy is oriented with respect to the magnetic field as prescribed by Eq. (1) (Hundhausen, 1970, and references therein). The pressure anisotropy is variable, but typically is of the order of $p_\parallel/p_\perp \sim 2$. Observations of solar wind electrons at 1 AU indicate that they are more isotropic than the protons, $p_\parallel/p_\perp \sim 1$–1.2 (Montgomery et al., 1968). It should be noted that it is more difficult to measure electrons than protons, because the energy of solar wind electrons at 1 AU ($\sim$ 10 eV) is not very different from that of photoelectrons produced by solar ultraviolet radiation impinging on the spacecraft.[*]

The pressure (and temperature) anisotropy of the solar wind is affected by a number of processes. If the flow is assumed to be steady and collisionless beyond some point, and the interplanetary magnetic field is assumed to be the ideal spiral[†] (Parker, 1963), then it is straightforward to compute the temperature anisotropy in the absence of interaction between particles and fluctuations in the plasma (Parker, 1958b). The result is (e.g., see Hollweg, 1971)

$$T_\perp r^2[1 + (r\Omega/v)^2]^{-1/2} = \text{const.}$$
$$T_\parallel v^2[1 + (r\Omega/v)^2] = \text{const.}$$

so that the anisotropy is

$$T_\perp/T_\parallel = (\text{const})(v^2/r^2)[1 + (r\Omega/v)^2]^{3/2}. \tag{2}$$

We expect $v(r)$ to increase slowly with increasing $r$, becoming essentially constant for $r > 1$ AU. Hence we expect $T_\perp/T_\parallel$ to decrease roughly as $r^{-2}$ out to about 1 AU, where $r\Omega/v \sim 1$, and to increase as $r$ far beyond 1 AU.

---

[*] By contrast, at 1 AU the energy of a solar wind proton is 0.5–1 keV.

[†] Radial component $B_r \propto r^{-2}$, azimuthal component $B_\phi \propto (r\Omega/v)B_r$, where $r$ is the heliocentric distance, $v$ is the radial flow velocity, and $\Omega$ is the angular velocity of the sun's rotation.

   The predicted value of the anisotropy at 1 AU depends on the constant in Eq. (2), which is determined by the distance at which the flow becomes collisionless. Reasonable assumptions about this distance imply enormous proton anisotropy at 1 AU$(T_{\parallel}/T_{\perp} \sim 50)$ if solar rotation is neglected $(\Omega = 0)$. Inclusion of solar rotation reduces the predicted anisotropy to $T_{\parallel}/T_{\perp} \sim 10$. The combination of solar rotation (as manifested in the spiral magnetic field) and occasional proton–proton collisions is sufficient to reduce the predicted anisotropy to observed values (Griffel and Davis, 1969; Hollweg, 1971; Leer and Axford, 1972). In addition, instabilities and other noncollisional mechanisms may play a role in limiting the thermal anisotropy of the wind.

   In short, although we understand qualitatively a number of physical mechanisms that may influence the thermal anisotropy of protons and electrons in the solar wind, we cannot sort out their relative importance at present. In any case, understanding the thermal anisotropy *per se* is not critical to understanding the large-scale behavior of the solar wind.* In the first place, the plasma is in fact fairly isotropic at 1 AU in the sense that $T_{\parallel}/T_{\perp}$ is nearer 1 than 10 even for the protons. Therefore it seems likely that the plasma is fairly isotropic throughout the region inside 1 AU. Secondly, even in the case of extreme anisotropy [Eq. (2) with $\Omega = 0$], the gross properties of the flow will not be very different from the corresponding isotropic case, because the details of the pressure tensor do not strongly affect supersonic flow.

## 2. Instability of Exospheric Flow

   We have seen above that if the solar wind were truly exospheric in the region of supersonic flow, it could be adequately described by a fluid theory which permits thermal anisotropy. However, an exospheric flow is likely to be unstable. For example, Parker (1958b) argued that laminar collisionless solar wind flow would probably pass the threshold for the firehose instability

$$P_{\parallel} > P_{\perp} + B^2/4\pi, \tag{3}$$

well within 1 AU (see also Kennel and Scarf, 1968; Hollweg and Völk, 1970; Scarf, 1970). Recently, Patterson (1971) has shown that nonlinear effects may prevent the expanding plasma from ever reaching this instability threshold. In any case, the pressure would be more isotropic than a naive exospheric picture would predict. If $P_{\perp} > P_{\parallel}$ far beyond 1 AU, the mirror

---

*This statement must be qualified when angular momentum transfer is considered (Section VI).

instability would tend to restore isotropy. Other anisotropy-driven instabilities may impose even more stringent requirements on isotropy. Such processes may affect the gross properties of the solar wind, not so much because they modify the pressure anisotropy, but because the fluctuations generated by instabilities can scatter particles and thereby modify the transport of energy and momentum. The most important instabilities in exospheric flow may be associated with distortions of the proton or electron velocity distributions not directly associated with thermal anisotropy.* The related anomalous transport of momentum and energy would probably be most important fairly near the sun ($r \lesssim 40R_{\odot}$), so that there are no relevant observations of the distortion of the velocity distributions. Unfortunately, the problem is very difficult to handle theoretically, largely because of the nonlinear character of the relevant wave–particle and wave–wave interactions. Eviatar and Schulz (1970) have addressed this problem to the extent of analyzing the stability of the velocity distribution of ions that are expanding in a collisionless manner.† They then construct a kinetic equation for the flow which includes the wave-particle scattering associated with the instabilities, as well as particle–particle collisions. Their work suggests that the solar wind ion flow out to 1 AU occurs in essentially three regions: (1) the collision-dominated outer corona; (2) a spherical shell of laminar exospheric expansion; (3) an outer region in which instabilities govern the expansion. As they point out, the existence and dimensions of region (2) are determined by many unknown parameters, including the flux of waves of solar origin. This work shows that plasma kinetic effects may influence proton transport properties in region (3).

The related problem for the electron component of the solar wind is even more complicated. In the first place, it is not clear how collisional the electrons are. Since the electron thermal speed is much higher than the flow speed, noncollisional effects can be important if the mean free path for electron–electron collisions is comparable to the scale height of the plasma, even though the collision time is small compared with the expansion time of the wind. The electron mean free path varies as $T_e^2$, and so is quite uncertain at heliocentric distances smaller than 1 AU. Secondly, if the electrons are collisionless, their velocity distribution is much more complicated than the corresponding velocity distribution for collisionless protons, because the electron thermal speed is much larger than the solar wind flow speed, and because the electrons tend to be quasi-trapped in a region whose size is

---

* Distortion of the velocity distribution by collisional transport processes may be unstable, as in the case of collisional thermal conduction (Forslund, 1970).

† Obviously, for this purpose the solar wind must be described by kinetic theory rather than fluid theory.

determined by the interplanetary magnetic and electrostatic fields (Eviatar and Schulz, 1970; Schulz and Eviatar, 1972; Perkins, 1973).

If it is assumed that the electron flow is noncollisional beyond some point, a number of plasma instabilities may arise (Schulz and Eviatar, 1972). Schulz and Eviatar have outlined some of the effects that such instabilities may have on the transport properties of the solar wind. One such instability has been suggested to account for the small-scale turbulence that produces the scintillations of radio sources (Perkins, 1973). In Perkins' model, like that of Schulz and Eviatar, the electrons execute quasi-trapped orbits between occasional collisions; the long orbits reduce the effective collisional mean free path, and hence reduce thermal conduction by the electrons. A related mechanism for reducing electron thermal conduction (but for very different, nontrapped electron orbits) had been proposed by Hollweg and Jokipii (1972).

To discuss these various mechanisms in detail would take us far beyond the scope of the present article. Present data does not justify attempting to incorporate *specific* mechanisms for (say) inhibition of thermal conduction into a detailed model of the solar wind. The importance of the studies of exospheric stability for solar wind dynamics has been to indicate that a purely laminar exospheric flow is unstable, and what qualitative effects the instability may have on solar wind flow. We will return to some of these points in the next section.

## IV. Acceleration and Heating of the Solar Wind

### A. Introductory Remarks

The question of heating has been fundamental to theoretical studies of the solar wind ever since Parker's original prediction. It had been known for many years that the sun's extended, tenuous corona is very hot ($\gtrsim 10^6$K). At such temperatures hydrogen is fully ionized and has very high thermal conductivity. On the other hand, the plasma density is so low ($\lesssim 10^9$ protons cm$^{-3}$) that the corona is not an efficient radiator or absorber of light. Hence it is conceivable that the energy transport of the corona is largely governed by thermal conduction, which in turn means that the plasma temperature will decline slowly with heliocentric distance. In the idealized case in which the energy transport is entirely due to thermal conduction, and the magnetic field is radial, the temperature $T$ is related to heliocentric distance $r$ by (Chapman, 1957)

$$T(r) = T(a)(a/r)^{2/7} \qquad (4)$$

(*a* is an arbitrary reference level). If the atmosphere is in static equilibrium, it satisfies the barometric equation

$$(d/dr)2nkT = -GM_\odot m_p n/r^2 \qquad (5)$$

whose solution is

$$n(r) = n(a)[T(a)/T(r)]\exp[-(GM_\odot m_p/2k)\int_a^r d\rho/\rho^2 T(\rho)], \qquad (6)$$

where $n$ is the proton number density, $M_\odot$ is the solar mass, $m_p$ is the proton mass, $k$ is Boltzmann's constant, and $G$ is the constant of gravitation. Equation (8) is physically unacceptable for many temperature profiles, because $n(\infty) \to \infty$. Furthermore, far from the sun the pressure would be

$$p_\infty = 2n(a)kT(a)\exp[-(GM_\odot m_p/2k)\int_a^\infty dr/r^2 T(r)]. \qquad (7)$$

Thus a nonzero confining pressure is required at infinity unless $T(r)$ declines faster than $1/r$ as $r \to \infty$. In particular, the Chapman conduction dominated atmosphere [Eq. (6)] gives*

$$p_{\infty cond.} = 2n(a)kT(a)\exp[-\tfrac{7}{10} GM_\odot m_p/kaT(a)]. \qquad (8)$$

For typical coronal parameters the confining pressure $p_{\infty cond.}$ would be much larger than any that could reasonably be expected for interstellar gas or magnetic fields. Hence a static equilibrium of the corona would not be consistent with the conductive temperature profile (4).

In the opposite extreme case of adiabatic temperature,

$$\left(\frac{n(r)}{n(a)}\right)^{2/3} = \frac{T(r)}{T(a)} = 1 + \frac{GM_\odot m_p}{5akT(a)}\left(\frac{a}{r} - 1\right) \qquad (9)$$

out to

$$r = a[1 - 5akT(a)/(GM_\odot m_p)]^{-1} \qquad (10)$$

and $n(r) = 0$ beyond, provided that $GM_\odot m_p/a > 5akT(a)$. If $GM_\odot m_p/a < 5akT(a)$, a finite external pressure is again required to maintain static equilibrium.

The problems with finite pressure as $r \to \infty$ led Parker to suggest that the solar corona is not static, but that it expands continuously throughout

*A more thorough treatment of this atmosphere shows that convective instability may modify the solution (Mestel, 1968).

the solar system until it is terminated by interstellar particles or fields. Steady spherically symmetric radial flow of velocity $v(r)$ must satisfy the equation of continuity

$$nvr^2 = J = \text{const.} \tag{11}$$

and the momentum equation

$$nm_p v \frac{dv}{dr} + \frac{d}{dr} 2nkT + \frac{GM_\odot m_p n}{r^2} = 0. \tag{12}$$

In Equation (12) we have assumed that the atmosphere is an inviscid fluid, with no external forces other than gravity and that the electron and proton temperatures are equal (these assumptions will be discussed critically later).

Parker (1963, 1965) analyzed such an expanding atmosphere for general $T(r)$ as follows. $n$ may be eliminated from Eq. (12) using Eq. (11), with the result

$$\begin{aligned} v(dv/dr)(1 - C^2/v^2) &= R(r) \\ &= -r^2(d/dr)(C^2/r^2) - W^2a/r^2, \end{aligned} \tag{13}$$

where $T$ is expressed in terms of the characteristic thermal speed

$$C(r) = [2kT(r)/m_p]^{1/2}, \tag{14}$$

and $W = [GM_\odot/a]^{1/2}$ is the characteristic escape velocity. Parker showed that in a tightly bound atmosphere $[W^2 \gg C^2(a) \gg V^2(a)]$ in which $T(r)$ declines less rapidly than $1/r$, $R(a) < 0$ and $R(\infty) > 0$, so that there exists an intermediate distance $r_c$ where $R(r_c) = 0$. It follows further that physically sensible solutions with $C(a) > V(a)$ must have $dv/dr > 0$ at $r < r_c$. Then at $r = r_c$, either (a) $dv/dr = 0$ and the flow remains subsonic beyond $r = r_c$, and $p(\infty)$ is finite, or (b) $v(r_c) = C(r_c)$, and the flow is supersonic (in the sense $v > C$) for $r > r_c$, and $p \to 0$ as $r \to \infty$. Hence for $T(r)$ declining more slowly than $1/r$, a flow described by Eqs. (11) and (12) must make a transition from subsonic to supersonic flow at $r = r_c$, and remain supersonic beyond $r_c$, in order to be consistent with small pressure far from the sun.

These considerations and other closely related ones have been discussed thoroughly by Parker (1963, 1965). Altogether, it is clear that the temperature profile $T(r)$, and therefore the heating, of an atmosphere determine to a large degree how, and even whether, the atmosphere will expand. Similarly, external forces due (for example) to magnetic fields or radiation pressure can have major effects on the flow. As an illustration, consider the addition of an outward radially directed velocity-independent external force to Eq. (12). If the temperature profile $T(r)$ is unchanged, the effect of the new force is to add a positive term to $R(r)$ in Eq. (13). In turn, this change means that the critical radius $r_c$ moves inward. In particular, a strong enough outward force can result in $r_c < a$, so that the supersonic flow is choked off. This point was first discussed by Marlborough and Roy (1970) in the context of whether stellar winds can be driven by radiation pressure.

### B. Possible Heating and Acceleration Mechanisms

In view of the preceding comments, we consider what physical processes might reasonably be expected to participate in the heating and acceleration of the solar wind, and then discuss how their relative importance can be sorted out. We have already seen that gravity and fluid pressure must be essential features of any model of the solar wind. In fact, for many purposes it is adequate to neglect other forces in modeling the solar wind. However, in general there are a number of other forces which must be considered, at least in principle. Similarly, it is generally agreed that heat conduction outward from the corona is an essential part of the heat budget of the solar wind. As in the case of momentum transfer, several other mechanisms may also be important, and should be considered.

The general steady-state momentum equation, which is valid even for exospheric flow, is

$$\mathbf{v} \cdot \nabla \mathbf{v} + (1/\rho)\nabla \cdot \mathbf{p} + (GM_\odot/r^2)\mathbf{e}_r = \mathscr{F}. \tag{15}$$

Here $\mathbf{v}$ is the bulk velocity of the ions,[*] $\rho$ is the mass density, $\mathbf{p}$ is the stress tensor (possibly including viscous stress), $\mathbf{e}_r$ is the unit vector in the outward radial direction, and $\rho\mathscr{F}$ is the sum of all nongravitational forces (per unit volume). We will discuss in some detail the momentum and energy transfer that may be associated with magnetic fields, hydromagnetic waves, and viscosity. We also note for completeness that the outer corona and solar wind are so transparent to light that radiation plays no role in their energy and momentum balances.

### 1. Magnetic Forces

In the outer corona the magnetic pressure $B^2/8\pi$ exceeds the fluid pressure $p(= \frac{1}{3} \mathrm{Tr}\, \mathbf{p})$. Hence it might appear that magnetic forces dominate solar wind flow near the sun. However, it turns out that the main contribution of

---

[*] If the sun is to remain electrically neutral, the outward electron charge flux must match the outward ion charge flux. Furthermore, any plasma must remain electrically neutral, on the average, over scales much longer than the Debye length ($\lesssim 10$ m in the solar wind). Since the solar wind ions are mostly protons, it follows to an excellent approximation that the electron number density equals that of the protons, and the electron bulk speed is $\mathbf{v}$.

magnetic forces is to channel the flow along the magnetic field lines. This is so because the very electrical conductivity of the coronal plasma is so high that the magnetic field is "frozen" into the plasma (e.g., Cowling, 1957), and hence the plasma flows parallel to the local magnetic field $\mathbf{B}$, i.e. $\mathbf{v} = v\mathbf{e} = v\mathbf{B}/B$. Now the magnetic force is

$$\rho \mathscr{F}_{mag} = (1/4\pi)(\text{curl } \mathbf{B}) \times \mathbf{B}$$
$$= (1/4\pi)\mathbf{B} \cdot \nabla \mathbf{B} - (1/8\pi)\nabla B^2, \tag{16}$$

and $\mathbf{e} \cdot \mathscr{F}_{mag} = 0$. Therefore if $\mathscr{F}$ is purely magnetic, the component of Eq. (15) along the magnetic field lines is

$$v(dv/ds) + (1/\rho)\mathbf{e} \cdot (\nabla \cdot \mathbf{p}) + (GM_\odot/r^2)\mathbf{e}_r \cdot \mathbf{e} = 0, \tag{17}$$

where $ds$ is the increment of distance along a field line. From the equation of continuity

$$\nabla \cdot (\rho \mathbf{v}) = 0 \tag{18}$$

and Maxwell's equation

$$\nabla \cdot \mathbf{B} = 0 \tag{19}$$

we obtain

$$\rho(s)v(s)/B(s) = \text{const.} \tag{20}$$

Equations (17) and (20) are actually (11) and (12) generalized to a geometry determined by the magnetic field. Hence the coronal magnetic field, to the extent that it is nonradial, modifies solar wind flow by channeling it along the magnetic field lines.

The geometry of the coronal magnetic field is, of course, determined by its source currents; these may be due, in part, to the flow of the solar wind itself. It has not so far been possible to measure coronal magnetic fields. Hence our ideas about these fields are based on extrapolation of the photospheric field with appropriate assumptions about coronal electric currents. Techniques of making such extrapolations have advanced to the stage where they seem to be reasonably reliable. A detailed discussion of these matters is far beyond the scope of this article. The interested reader is referred to the review articles of Howard (1972), Newkirk (1972), Schatten (1972), and Davis (1972).

Detailed analysis of the effect of the complicated coronal magnetic field on the plasma flow is not needed to understand the gross properties of the solar wind. However, it is important to understand in a general way the kinds of effects that the coronal magnetic field may have on the flow. We shall return to this question later.

## 2. *Hydromagnetic Waves*

Hydromagnetic waves are associated with large-scale disturbances in a magnetized plasma, as sound waves are associated with large-scale distur- bances in a neutral gas. However, hydromagnetic waves are more com- plicated than sound waves, having several modes of propagation. In this article we do not attempt to cover theory and observation of these waves in detail (see reviews in Burlaga, 1972a; Barnes, 1972). Nevertheless, hydro- magnetic waves may play a central role in both heating and accelerating the solar wind, and it will be necessary to consider certain of the properties of these waves.

One of the hydromagnetic wave modes, the Alfvén mode, is observed to dominate the microstructure of the interplanetary medium at 1 AU 30–50% of the time (Belcher and Davis, 1971). The other wave modes, the magneto- acoustic modes, have never been observed at 1 AU. Presumably the nonAlf- vénic components of the microstructure are stationary equilibria being convected with the wind (Burlaga, 1972a; Barnes, 1972). Plane Alfvén waves are associated with fluctuations in direction, but not magnitude, of the mag- netic field, and in velocity, but not density or pressure, of the plasma. They propagate relative to the plasma at the speed $B_\perp/(4\pi\rho)^{1/2}$, where $B_\perp$ is the component of the magnetic field transverse to the planes of constant phase of the wave. The phase relation between fluctuations in velocity and mag- netic field implies the sense of propagation relative to the mean magnetic field, in a frame of reference at rest with respect to the plasma.

Belcher and Davis used these criteria to identify Alfvén waves, and the sense of their propagation. They found the remarkable fact that, in the plasma frame, all the Alfvén waves are propagating outward from the sun. Belcher and Davis interpreted this observation as indicating that the waves probably originated very near the sun. Magneto-acoustic waves originating in the same region would not be able to reach 1 AU because they undergo strong Landau damping (Barnes, 1966, 1971). Hollweg (1972) has suggested that the Alfvén waves are generated by fluid motions in the chromospheric supergranulation.

Whatever their source, these Alfvén waves will exert pressure, analogous to radiation pressure, with important effects on the solar wind (Belcher, 1971; Alazraki and Couturier, 1971). This mechanism appears to provide a natural explanation for low density, high velocity streams (in which Alfvén waves are especially abundant at 1 AU). Furthermore, Alfvén wave pressure can, in principle, drive a supersonic flow, which would be subsonic in the absence of the waves.

The Alfvén radiation pressure is easily calculated in the simple case of Alfvén waves propagating radially outward. Waves whose wavelength is small compared with the scale height of the plasma may be described by the WKB approximation. A standard theorem of plasma physics states that steady propagation of a wave of frequency $\omega$ and wavevector $\mathbf{k}$ in a nondissipative medium is characterized by conservation of the flux $\mathbf{G} = W\, \partial\omega/\partial\mathbf{k}$, where $W$ is a pseudo-energy density.* For Alfvén waves, propagating outward in a radial magnetic field, it may be shown that

$$\mathbf{G} = \mathbf{e}_r[(v + C_A)^2/C_A]\langle\delta\mathbf{B}^2\rangle/4\pi \qquad (P_\perp = P_\parallel), \qquad (21)$$

where $\delta\mathbf{B}$ is the (real) fluctuation in the magnetic field, $C_A = B/(4\pi\rho)^{1/2}$ is the Alfvén speed, and $\langle\ \rangle$ denotes an average over the wave period. Conservation of $\mathbf{G}$ along the (radial) rays implies that $G r^2 = \text{const.}$, so that

$$\langle\delta B^2(r)\rangle^{1/2} = \langle\delta B^2(a)\rangle^{1/2}\left(\frac{\rho(r)}{\rho(a)}\right)^{1/4}\left(\frac{v(a)}{C_A(a)} + 1\right)\bigg/\left(\frac{v(r)}{C_A(r)} + 1\right). \qquad (22)$$

We have used the fact that a radial magnetic field declines as $r^{-2}$ (since $\nabla \cdot \mathbf{B} = 0$) in deriving Eq. (22).† This expression was first derived by Parker (1965), who used somewhat different arguments.

However, the conserved flux $\mathbf{G}$ is not the physical energy flux unless $v = 0$. The true mean energy flux to second order in the wave amplitude is

$$\mathbf{F} = (c/4\pi)\langle(\delta\mathbf{E} \times \delta\mathbf{B})\rangle + \tfrac{1}{2}\rho\mathbf{V}\langle\delta V^2\rangle$$
$$= [(\tfrac{3}{2}V + C_A)\langle\delta\mathbf{B}^2\rangle/4\pi]\mathbf{e}_r \qquad (23)$$

(recall that $\delta\rho = \delta T = 0$ for Alfvén waves). This expression for the energy flux was derived by Belcher (1971); see also Scholer and Belcher (1971). Comparing (21) and (23), and noting that $r^2 G_r = \text{const.}$, we find that

$$r^2 F_r(r) = a^2 F_r(a)\frac{\tfrac{3}{2}v/C_A + 1}{(v/C_A + 1)^2}\frac{[V(a)/C_A(a) + 1]^2}{\tfrac{3}{2}V(a)/C_A(a) + 1}. \qquad (24)$$

* $W$ is generally defined in terms of complex electric and magnetic field amplitudes $\delta\mathbf{E}_0$ and $\delta\mathbf{B}_0$ by

$$W = (1/16\pi)[\,|\delta\mathbf{B}_0|^2 + \delta\mathbf{E}_0^* \cdot (\partial/\partial\omega)(\omega\mathbf{K}^h) \cdot \delta\mathbf{E}_0],$$

where $\mathbf{K}^h$ is the hermitian part of the dielectric tensor (e.g., see Stix, 1962; Kadomtsev, 1965). In this notation $\langle\delta\mathbf{B}^2\rangle = \tfrac{1}{2}|\delta\mathbf{B}_0|^2$, etc.

† The general expression for Alfvén waves may also be written in the convenient form

$$(d/dt)\ln(\langle\delta\mathbf{B}^2\rangle/\rho^{1/2}) = -\text{div } \mathbf{v} \qquad (P_\perp = P_\parallel),$$

where $d/dt$ refers to the time derivative along an Alfvén ray (Jeffrey and Taniuti, 1964; Völk and Alpers, 1973). For radial rays and spherical symmetry, $d/dt \to (v + C_A)\, d/dr$. Since $B \propto r^{-2}$ and $\rho v r^2 = \text{const.}$,

$$\text{div } \mathbf{v} = 2(dv/dr) - v(d/dr)\ln C_A^2$$
$$= (v + C_A)(d/dr)\ln[(v + C_A)^2/C_A^2]$$

and Eq. (21) follows immediately.

If we then require that the sum of the energy effluxes in the plasma and waves be constant, it is easily shown that the force on the plasma by the waves is

$$-(1/r^2 v)(d/dr)(r^2 F_r),$$

i.e. we must substitute

$$\mathscr{F} = -[1/r^2 v(r)](d/dr)(r^2 F_r(r))\mathbf{e}_r \tag{25}$$

in Eq. (15) with $F_r$ given by Eq. (24). Parker (1965) first pointed out that this force exists, although he did not obtain the correct expression for $F_r$. Later, Belcher (1971) and Alazraki and Couturier (1971) independently obtained solutions of Eq. (15) with $\mathscr{F}$ given by (25).

The qualitative effects of this Alfvén radiation pressure are easily seen. Near the sun, where $v \ll C_A$, $r^2 F_r$ is nearly constant and $\mathscr{F}$ is small. Far beyond the Alfvénic distance, where $v \gg C_A$, $v$ is essentially constant and $C_A \propto 1/r$. Then $\mathscr{F}_r \simeq F(r)/vr$, so that in order of magnitude

$$\mathscr{F}_r/(dp/dr) \sim F(r)/p(r) v(r).$$

Hence $\mathscr{F}_r$ is the dominant force in super-Alfvénic flow if the Alfvén wave flux is greater than the thermal energy flux of the wind. Altogether, it appears that the Alfvén radiation pressure does not strongly influence the flow within a few $R_\odot$ of the sun, but can be a strong outward force in regions where $v \gtrsim C_A$ ($r \gtrsim 10$–$20 \, R_\odot$). Hence this force may account for the high velocity, low density streams (Belcher, 1971). This suggestion is supported by the observed abundance of Alfvén waves in the high velocity streams (Belcher and Davis, 1971).

It is reasonable to expect nonAlfvénic hydromagnetic wave modes near the sun, as well. These magnetoacoustic modes are associated with fluctuations in density, pressure, and magnetic field strength, unlike Alfvén waves. Moreover, magnetoacoustic waves whose periods are short compared with the particle Coulomb collision times are subject to strong collisionless (Landau) damping (Stepanov, 1958; Barnes, 1966; Tajiri, 1967). This Landau damping accounts for the fact the magnetoacoustic waves have not been observed in the solar wind at 1 AU (Unti and Neugebauer, 1968; Belcher and Davis, 1971).

Physically, the Landau damping is caused by resonant interaction between waves (frequency $\omega$, wavevector $\mathbf{k}$) and particles whose velocity $v_\parallel = \mathbf{v} \cdot \mathbf{B}/B$ along the magnetic field lines matches that of a plane of constant phase, $v_\parallel = \omega/k_\parallel$. The wave tends to pick up resonant particles of mass $m$ and magnetic moment $\mu = \frac{1}{2}mv_\perp^2/B$ because of its gradient in magnetic field strength,

$$m \, dv_\parallel/dt = -\mu(\nabla B)_\parallel. \tag{26}$$

(This process does not occur for Alfvén waves where $\mathbf{V}B = 0$.) The process by which this resonant acceleration leads to exchange of energy between waves and particles has been analyzed in detail by Barnes (1967). For an isotropic velocity distribution the energy exchange is proportional to the derivative $\partial f/\partial v_\parallel$ of the velocity distribution evaluated at $v_\parallel = \omega/k_\parallel$; this energy exchange may involve either electrons or ions. It turns out that for conditions expected in the solar wind between the sun and 1 AU, most directions of propagation will be associated with heating of ions by the waves. The least damped magnetoacoustic mode will be damped out in 1–4 × $\exp(1/\beta_p)$ wavelengths, where $\beta_p = 8\pi nkT_p/B^2$ is the ratio of the proton pressure to the magnetic pressure. At 1 AU $\beta_p$ is typically 0.4 so that magnetoacoustic waves of period $\lesssim$ 3 hr could not reach 1 AU from the sun. These points are discussed in more detail in the review of Barnes (1972).

Like Alfvén waves, magnetoacoustic waves might exert a force on the solar wind. However, a more important effect of magnetoacoustic waves is that their dissipation can be a very efficient mechanism of heating the wind. Magnetoacoustic waves of five minute period (the period of maximum photospheric noise) propagating outward from the lower corona would damp out in 10–20 $R_\odot$, thereby providing an extended heat source which leads to considerable acceleration of the wind (Barnes *et al.*, 1971). Magnetoacoustic waves generated by transient events could provide efficient local heating farther out in the wind (Jokipii and Davis, 1969).

## 3. *Viscosity*

The role of viscosity in solar wind expansion has been a topic of some controversy, which has not yet been finally resolved [e.g., see Section 7 of Parker's (1965) review, and Chapter 3.4 of Brandt's (1970) book]. There seems to be general agreement that near the sun viscosity plays some role in angular momentum transport, but has a minor effect on the radial flow. However, divergent opinions have arisen concerning the importance of viscosity for the radial flow far from the sun. It has been variously stated that the inclusion of viscosity strongly reduces the flow velocity near 1 AU, and that as $r \to \infty$ the solar wind velocity increases without bound! Neither assertion accords with physical intuition or with observation. The problem arises with use of the collisional coefficient of viscosity, $\eta \sim nm_p v_{th} \lambda$ ($v_{th}$ is the proton thermal speed and $\lambda$ the mean free path for Coulomb collisions), to describe momentum transport in the outer solar wind. In this region $\lambda \gtrsim r$, whereas the collisional transport coefficients such as viscosity are derived for systems in which $\lambda$ is small compared with any macroscopic scale height.

At present there is no reliable theory for handling viscosity in the region of noncollisional flow. On the one hand, the large mean free path suggests that microscopic momentum transport might be fairly efficient. On the other hand, experiments with dilute neutral gases indicate that the viscosity becomes small when the mean free path is large (Chapman and Cowling, 1961, p. 103). Moreover, the previously discussed comparison of kinetic exospheric models (with exact microscopic momentum transport) and fluid models (with zero viscosity) indicates that viscosity may not strongly affect the gross properties of the solar wind (Leer and Holzer, 1972). The Leer–Holzer calculations give a quiet, cool wind, however, and it is not certain that their conclusions remain valid for hotter winds.

From a theoretical viewpoint, the question of viscosity is exceptionally difficult. This would not be so if it were realistic either to use a laminar exospheric description of momentum transport, or the ordinary collisional viscosity. However, it is more probable that viscous transport in the region of noncollisional flow is governed by " collisions " between ions and fluctuations in the electromagnetic fields. The result of the wave–particle interactions would be to reduce the effective mean free path and thus reduce the viscosity from the classical collisional value. Such an anomalous viscosity has been suggested as governing the viscous interaction between the solar wind and the earth's magnetosphere (Eviatar and Wolf, 1968).

It has recently been shown that it is possible to construct reasonable models of the quiet solar wind in which viscosity is important* (Wolff et al., 1971). They find that collisional viscosity, as modified by the interplanetary magnetic field, has a relatively minor effect on the velocity profile, but viscous heating can greatly enhance the proton temperature at 1 AU. They recognized that use of the mean free path for Coulomb collisions tends to exaggerate the viscous heating, but suggested that a somewhat smaller mean free path due to wave–particle interactions may cause viscous effects which are qualitatively similar to those they described. On the other hand, the heating models of Hartle and Barnes (1970) show that collisional viscous heating is formally significant only when the Coulomb collision time is comparable to the macroscopic expansion time of the wind. On the basis of these calculations and those of Leer and Holzer (1972), it seems unlikely that viscous heating is a major effect in the solar wind. Nevertheless, the physics of viscous transport by wave–particle interaction is uncertain enough that the question of viscous heating cannot be regarded as closed.†

---

* See also the more recent paper of Brandt et al. (1973).

† This last point is underlined by the recent exchange of letters in the Journal of Geophysical Research, from Brandt and Wolff (1973) on the one hand, and Holzer and Leer (1973) on the other.

## C. Recent Models of Solar Wind Heating and Acceleration

We now consider how far one can go in sorting out the relative influences of the heating and acceleration mechanisms discussed above. We hope that these issues will eventually be decided by definitive observations, which very likely will require sending spacecraft much nearer the sun than has been possible up to now. At present, the best one can do is to calculate models based on various assumptions about the physics of the solar wind, and to compare these models with whatever observations happen to be available.

Currently, models must be judged primarily by comparing their predictions against data from spacecraft near 1 AU, since these measurements are the most reliable by far. However, as we emphasized earlier, the most reliable measurements are not necessarily the most relevant ones. The values of flow speed, density, etc., measured at 1 AU fluctuate over a fairly wide range, so that the criteria for agreement between models and observation are necessarily somewhat subjective. In particular, as we pointed out in Section III, it is not completely clear to what extent the observed variability is due to the spacecraft's sampling different large-scale states of the solar wind, and to what extent this variability is due to temporal evolution of fluctuations in the interplanetary medium.

Nevertheless, most models of the solar wind treat it as consisting of large, steady streams, and regard macroscale variability in the observations as indicating the sampling of different streams. Many of the models based on this assumption give good agreement with a large class of direct observations at 1 AU, and with indirect observations at other heliocentric distances. Hence it appears that the idealization of large, steady streams does not grossly distort our picture of the large-scale physics of the solar wind. At any rate, in the ensuing discussion we will adopt as a working hypothesis the idea that various heating and acceleration mechanisms can be evaluated in the framework of steady, spherically symmetric flow.

If this assumption is accepted, there are essentially two views about what should be required of a solar wind model. One is that models can be expected at most to predict the most quiet state of the wind, and the other is that models should predict a *range* of relatively active states, with the quiet wind as a special case.

The quiet solar wind is attractive for theoretical study because of the hope that it is simpler than more active states of the wind. The quiet wind fluctuates relatively little, and therefore the flow at 1 AU may be less distorted by evolution of inhomogeneities in the interplanetary medium. Moreover, perhaps the physics of the quiet state is less complicated than that of the more active wind, and therefore the quiet state may be easier to model. The principal difficulty of focusing on the quiet wind is that it is less typical than the more active states. For example, a commonly used operational definition of the quiet wind is that its flow speed should be less than 350 km sec$^{-1}$ at 1 AU. However, the solar wind meets this criterion only about 30% of the time, even in a period of relatively low solar activity (Hundhausen et al., 1970). Moreover, fluctuations are not negligible even in this velocity range, so that very quiet states of the wind are rare.

The other alternative is to consider the quiet wind as a special case of the normally active wind. Steady models based on this viewpoint strain the hypothesis of steady streams a bit more, particularly with respect to predictions as far from the sun as 1 AU. Nevertheless, results of models based on this assumption seem to justify its use. If the quiet wind is regarded as a special case of the normally active wind, then a satisfactory model of the wind must actually be a series of models, giving a range of flow velocity, etc., for a given heliocentric distance. One test of such a series of models is that the predicted flow parameters at 1 AU should vary from model to model in a way that is consistent with observed large-scale correlations among the flow parameters. By far the strongest correlation turns out to be that between flow speed and proton temperature (Burlaga and Ogilvie, 1970; Hundhausen et al., 1970; Mihalov and Wolfe, 1971). Why the measured flow velocity correlates with the proton temperature (whose thermal energy flux is only 2% of the total energy flux) more strongly than with other fluxes of energy is not fully clear. Nevertheless, it is now apparent that the $T_p-v$ correlation is associated with the large-scale heating of the wind (cf. Section III; see also Burlaga, 1972b, Burlaga and Ogilvie, 1973; and Hundhausen, 1973).

## 1. Basic Two-Fluid Model of the Quiet Solar Wind

The basic two-fluid model (Sturrock and Hartle, 1966; Hartle and Sturrock, 1968) is a reasonable starting point for a detailed review of acceleration and heating mechanisms. This relatively simple model is based on the assumptions of radial, steady, spherically symmetric flow, that the acceleration is governed by fluid pressure and gravity, and that the heating is due to thermal conduction. The transport coefficients are taken to be due to Coulomb collisions, but viscosity is neglected. Then the flow is described by the equation of continuity,

$$nvr^2 = \text{const.,} \tag{27}$$

the momentum equation,

$$nm_p v(dv/dr) + (d/dr)[nk(T_e + T_p)] + (GM_\odot m_p n/r^2) = 0, \tag{28}$$

and the electron and proton heat equations,

$$\frac{3}{2}nvk\frac{dT_e}{dr} - vkT_e\frac{dn}{dr} - \frac{1}{r^2}\frac{d}{dr}\left(r^2K_e\frac{dT_e}{dr}\right) = -\frac{3}{2}v_E nk(T_e - T_p), \quad (29)$$

$$\frac{3}{2}nvk\frac{dT_p}{dr} - vkT_p\frac{dn}{dr} - \frac{1}{r^2}\frac{d}{dr}\left(r^2K_p\frac{dT_p}{dr}\right) = \frac{3}{2}v_E nk(T_e - T_p). \quad (30)$$

The expressions for the electron and proton thermal conductivities, $K_e$ and $K_p$, and the energy-exchange rate $v_E$ are those given by Braginskii (1965). In order of magnitude

$$\begin{aligned}
K_e &\sim 6 \times 10^{-7}T_e^{5/2} \text{ erg cm}^{-1} \text{ sec}^{-1}\,^\circ\text{K} \\
K_p &\sim (m_e/m_p)^{1/2}(T_p/T_e)^{5/2}K_e \\
v_E &\sim 10^{-1}nT_e^{-3/2} \text{ sec}^{-1}.
\end{aligned} \quad (31)$$

The proton thermal conductivity is usually small compared with the electron thermal conductivity. As Sturrock and Hartle (1966) first pointed out, the collisional energy exchange rate at 1 AU is considerably smaller than the expansion rate $2v/r$. Hence in the basic two-fluid model the protons and electrons are thermally decoupled. The electrons are strongly heated by thermal conduction throughout the flow, and the protons are unheated except for small contributions from proton thermal conduction and collisional energy exchange with electrons.

Hartle and Sturrock (1968) numerically solved Eqs. (27)–(30), subject to appropriate boundary conditions. The results were presented as profiles of velocity, density, and temperatures as functions of heliocentric distance. Their predictions were compared against observations at 1 AU (and, in the case of density $n$, also in the range $1 R_\odot < r \lesssim 10 R_\odot$). The results are very grossly consistent with the observations, but there are significant discrepancies at 1 AU. The actual numbers predicted at 1 AU depend on the choice of boundary conditions, but for any reasonable choice, the following qualitative discrepancies always showed up: (1) the predicted density profile is too flat, in the sense that either the density is too high at 1 AU, or too low near the sun; (2) the predicted flow speed ($\sim 250$ km sec$^{-1}$) is lower than observed even at the quietest times; (3) the predicted proton temperature ($\sim 4 \times 10^3$K) is too low at 1 AU by about a factor 10.

In addition, the basic two-fluid model predicts electron temperatures at 1 AU in the range $2.4$–$3.5 \times 10^5$K, depending on the base conditions. This value is somewhat higher than the value $T_e = 1.5 \pm 0.5$ typically reported by observers (Montgomery, 1972a; Montgomery et al., 1968). This discrepancy would not be troublesome except for that fact that the energy flux due to thermal conduction depends on temperature to a sensitive degree, $F_{\text{cond.}} \propto T_e^{7/2}$. For the highest of Hartle and Sturrock's predicted electron temperatures, the electron thermal conduction flux is comparable to the kinetic energy flux of the protons, contrary to observation (Hundhausen, 1970). Moreover, such a high conduction flux exceeds the product of the electron thermal energy density and electron thermal speed, and therefore is not physically reasonable. By suitably choosing the base conditions (e.g., see Barnes et al., 1971, Table 1), $T_e$ can be reduced enough so that the predicted thermal conduction flux is much smaller than the kinetic energy flux, although it is still larger than reported observations of the conduction flux.

Obviously these two-fluid models are highly idealized, so that the next step has been to consider which of the idealizations produce the severest discrepancies. Several classes of possible remedies have been suggested, as we shall discuss below.

## 2. Modified Two-Fluid Models of the Quiet Solar Wind with External Heating

Sturrock and Hartle (1966; also Hartle and Sturrock, 1968) suggested that the most probable remedy for the ills of the basic two-fluid model is heating by dissipation of nonthermal energy far out into the solar wind. Parker (1965) reached a similar conclusion on the basis of his earlier studies of the " one-fluid " solar wind ($T_e = T_p$). It seems to be generally agreed that dissipation of some sort of wave motion accounts for the high temperature of the inner corona (see review of Kuperus, 1969). It therefore seems plausible that some wave energy passes beyond the inner corona, heating or accelerating the wind far from the sun (Barnes, 1969; Fredricks, 1969; Barnes et al., 1971; Belcher, 1971; Hollweg, 1972). The two-fluid model can be greatly changed by dissipating $2-5 \times 10^{26}$ erg sec$^{-1}$ beyond $2R_\odot$. Extended heating of this magnitude appears modest in contrast with $\sim 10^{28}$ erg/sec required to maintain the inner corona.

Studies of the effects of arbitrary hypothetical heating profiles $P_p(r)$ added to the right-hand side of the proton energy equation (30) have been carried out by Hartle and Barnes (1970) and Leer and Axford (1972). It is fairly obvious that external heating of the protons results in both higher flow velocities and higher proton temperatures at 1 AU. Moreover, it turns out that heating tends to steepen the density profile, and, if the heating is limited to ions, the electrons will cool somewhat by expansion. All these effects modify the two-fluid model in the direction of better agreement with observation.

However, the spatial distribution of the heating turns out to be quite important. Hartle and Barnes (1970) used the hypothetical heating profile

$$P_p(r) = D_0(n/n_0)\exp[-(r/R_\odot - a)^2/b^2] \qquad (32)$$

which permits variation of the heating strength and geometry by varying the parameters $D_0$, $a$, and $b$. They found that heating spread out over a region $\sim 20\,R_\odot$ in radius can modify the basic two-fluid model to give a reasonably good representation of the observed quiet solar wind. If the heating region is much more compact that $\sim 20\,R_\odot$, the flow velocity is raised but the proton temperature at 1 AU remains low. If the heating region is much larger, the temperature increases but the velocity remains low. In fact, if the heating region is extended enough, the flow can choke off and become subsonic.

Another possible remedy for the basic two-fluid model is to add forces besides pressure and gravity to accelerate the wind. There is good evidence that Alfvén waves in the solar wind originate at or near the sun (Belcher and Davis, 1971). Theoretical studies carried out independently by Belcher (1971) and Alazraki and Couturier (1971) show that the force exerted on the plasma by Alfvén waves can produce high velocity, low density streams. It is possible that these waves also provide some heat to the wind by nonlinear dissipation (Hollweg, 1973a), or by scattering from other fluctuations (Valley, 1971). Magnetoacoustic waves of solar origin would also accelerate and heat the wind.

Stationary magnetic fields near the sun will also affect the flow, by channeling it along the magnetic field lines. Models incorporating this mechanism indicate that its main effect is to steepen the density profile without strongly increasing the velocity far from the sun (Pneuman and Kopp, 1970). The suggestion of Whang (1971) that the interplanetary magnetic field significantly accelerates the radial flow of the wind appears to be incorrect, since it would require an appreciable systematic azimuthal component of the interplanetary magnetic field at the base of the corona.

Altogether, these several heating and acceleration mechanisms seem to modify the basic two-fluid model toward agreement with observations. However, it has been pointed out that most of these mechanisms involve the addition of external energy to the solar wind, and that in fact for some choices of boundary conditions the basic two-fluid model gives the correct *total* (kinetic plus conductive) energy flux at 1 AU (Hundhausen, 1969, 1970). The low predicted velocity is compensated by high density and high electron thermal conduction, so that the predicted total energy flux agrees with observation. Hence, Hundhausen has suggested that the deficiencies in the basic two-fluid model need not be remedied by the addition of energy. In particular, cutoff of thermal conduction by one of several mechanisms may convert the conduction flux into kinetic energy, thereby increasing the velocity at 1 AU. Some other noncollisional mechanism might exchange thermal energy between electrons and protons, thus lowering the electron temperature and increasing the proton temperature. These alternatives deserve serious consideration, especially since there are theoretical reasons to expect noncollisional modification of thermal conduction beyond $10$–$20\,R_\odot$ (Forslund, 1970; Schulz and Eviatar, 1972; Hollweg and Jokipii, 1972; Perkins, 1973).

Recently a number of workers have developed detailed models of the quiet solar wind aimed at sorting out various remedies to the basic two-fluid model (Cuperman and Harten, 1971 (CH); Wolff *et al.*, 1971 (WBS); Barnes *et al.*, 1971 (BHB); see also Hartle and Barnes, 1972; Barnes and Hartle, 1972). These models are compared in the paper by Barnes *et al.* (1972; see also Cuperman, 1973). All these models have the common features of azimuthally symmetric, steady flow. The models of CH and BHB employ the simplified momentum equation (28), while the WBS models include the collisional viscous force, magnetic stress, and related nonradial flow. However, these last effects do not strongly influence radial motion, and the important differences among the models remain in the energy equations. CH consider the effect of various hypothetical reductions of the electron thermal conduction coefficient, together with an enhancement of proton–electron energy exchange or proton thermal conduction. WBS also reduce thermal conduction by including both the spiral interplanetary magnetic field and an additional hypothetical reduction factor. In addition, WBS include heating as predicted for viscosity due to Coulomb collisions. BHB used the unmodified collisional thermal conductivity, but included heating due to dissipation of hydromagnetic waves generated at the sun. They neglected viscous heating on the grounds that it is formally significant only if the Coulomb mean free path is so large that the collisional viscous coefficients are invalid (however, see our previous discussion of viscosity). Hence the BHB model differs from the basic two-fluid models only through collisionless dissipation of magnetoacoustic waves. In this model the state of the wind is determined by the wave energy efflux at the coronal base. The quiet state of the wind then corresponds to a relatively small, but nonzero, wave efflux.

The three kinds of models described above may be compared with observations at 1 AU. The only other available observations that are relevant to models of the quiet wind are the density profile of the corona (e.g., see Brandt, 1970; Newkirk, 1967) and the temperature near the coronal base. A comparison of the CH, WBS, and BHB models with some of these observations is given in Table III (Barnes *et al.*, 1972).

All these models are in reasonable agreement with observation. In each case the agreement is not perfect, however. All these models suffer from a too flat density profile; the density is either too low near the sun or too high near 1 AU. This problem arises because the predicted density variations in the range 2–10 $R_\odot$ are not steep enough. It may be that this difficulty cannot be resolved with radial flow models. Nonradial channeling of the flow by coronal magnetic fields naturally results in a steeper density profile, as indicated by the studies of Pneuman and Kopp (1970).

Except for the base coronal temperatures, which are comparable to the

TABLE III

MODELS OF THE QUIET SOLAR WIND

| | Observed quiet wind | WBS | BHB[a] | CH[b] |
|---|---|---|---|---|
| $v_E$ (km sec$^{-1}$)[c] | 300–350 | 303 | 330 | 270 |
| $n$: At 1 AU (cm$^{-3}$) | 8.7 ± 4.6 | 9 | 15 | 6 |
|    At 3 $R_\odot$ (10$^5$ cm$^{-3}$) | 2–8 | 1.7 | 4.2 | — |
| $T_{eE}$ (10$^4$K) | 14.0 ± 5.0 | 20 | 22 | 16 |
| $T_{pE}$ (10$^4$K) | 4.4 ± 1.8 | 4 | 3.2 | 4.3 |
| $B_E$ (10$^{-5}$G) | 4.6 ± 1.4 | 8 | 2.6 | — |

[a] Taken from BHB paper, wave flux $F_0 = 3.4 \times 10^3$ erg cm$^{-2}$ sec$^{-1}$.

[b] Model X of CH paper (Cuperman and Harten, 1971), reduced electron and increased proton thermal conductivity.

[c] The subscript E stands for an observation made near 1 AU. The sources of the measurements and their uncertainties are discussed in the paper by Barnes *et al.* (1972).

observed $1$–$2 \times 10^6$K in all models, good measurements of parameters other than density are presently available only near 1 AU. All three models predict reasonable values for the temperatures and bulk velocity, although the CH velocity is somewhat low, and the electron temperatures of WBS and BHB are marginally high. The magnetic field is too high in WBS, but could be lowered without grossly changing the radial flow. The magnetic field is too low in BHB. Since the magnetic field sensitively controls the rate of hydromagnetic wave damping (Barnes, 1972), its low value may be of some concern.

One other measurement at 1 AU, that of the electron heat flux, is sometimes compared against various models (Cuperman and Harten, 1971; Hundhausen, 1970). Since this heat flux is observed to be small with the kinetic energy flux (Montgomery *et al.*, 1968), this comparison is probably not critical unless the model in question predicts a conduction flux that is comparable to the kinetic energy flux at 1 AU. As mentioned above, some choices of conditions at the coronal base can give such high conduction fluxes in the basic two-fluid model. This problem persists, *for the same base conditions,* in models with external heating. However, for somewhat different base conditions, such as those used in the BHB models, the basic two-fluid model gives a smaller electron temperature at 1 AU, hence greatly reducing the predicted conductive heat flux ($\propto T_e^{7/2}$).

Altogether, it appears that the observed quiet solar wind is adequately approximated at 1 AU by the two-fluid model modified to include either (a) dissipation of hydromagnetic waves of solar origin, or (b) noncollisional cutoff of thermal conduction plus some ion-heating mechanism such as viscosity or noncollisional proton–electron energy exchange. The relative importance of these mechanisms cannot be determined on the basis of quiet wind observations at 1 AU. Definitive observations on the question may come from future spacecraft observations nearer the sun, since the different models give different profiles of the flow parameters, especially proton temperature (see Burlaga, 1971, for a more detailed discussion of this point).

However, at present we are not limited to models of the quiet wind. Studies of the more normally active solar wind may provide additional circumstantial evidence as to what mechanisms govern the acceleration and heating of the solar wind.

## 3. *Heating and Acceleration of the Normally Active Wind*

The BHB models (Barnes *et al.*, 1971) which were discussed in the preceding section are primarily aimed at describing the normally active solar wind, of which the quiet wind is a special case. The BHB model of Table IV is one of a series of models generated by varying one parameter, $F_0$, which is the energy flux in the form of magneto-acoustic waves at the base of the model (taken at $r = 2 R_\odot$). The density, temperatures, and magnetic field are held fixed at the base of all models of the series. For definiteness, the wave spectrum was taken to be monochromatic, but the results are not sensitive to this assumption or to the choice of frequency. The wave period was chosen to be about five minutes, which corresponds to the peak of the observed acoustic and magnetic noise in the solar chromosphere and photosphere (Leighton *et al.*, 1962; Tanenbaum *et al.*, 1969).

These models also included the usual assumptions of spherically symmetric, steady flow. The dynamical equations of the flow are Eqs. (27)–(30), with the exception that the heating of ions due to Landau damping of the waves is added to Eq. (30). The amplitudes of the waves are computed as functions of $r$ in the WKB approximation, taking account of losses due to Landau damping. The simultaneous solution of the dynamical equations and the equations for the wave amplitudes then permit the generation of a series of models parameterized by the single quantity $F_0$. The predictions of such a series for $r = 1$ AU is given in Table IV (Barnes *et al.*, 1971).

In these models the base radius was fixed at $2R_\odot$, and the base number density, temperatures, and magnetic field strength were $n_0 = 1.4 \times 10^6$ cm$^{-3}$, $T_{e0} = 1.3 \times 10^6$ K, $T_{p0} = 1.7 \times 10^6$ K, and $B_0 = 0.18$ G. The assumption that $T_{p0} > T_{e0}$ implies preferential heating of the protons somewhere below $r = 2R_\odot$, and is consistent with at least one reasonable model of heating the inner corona (d'Angelo, 1968, 1969; for a discussion of the relevant coronal observations, see Newkirk, 1967).

The first column of Table IV lists the single variable parameter of the

TABLE IV

NUMBER DENSITY, FLOW SPEED, ELECTRON AND PROTON TEMPERATURES, AND MAGNETIC
FIELD STRENGTH AT 1 AU

| $F_o$ (erg cm$^{-2}$ sec$^{-1}$) | $n_E$ (cm$^{-3}$) | $v_E$ (km sec$^{-1}$) | $T_{eE}$ (°K) | $T_{pE}$ (°K) | $B_E$ ($\gamma = 10^{-5}$ G) |
|---|---|---|---|---|---|
| 0 | 20 | 210 | $2.4 \times 10^5$ | $8.5 \times 10^3$ | 3.5 |
| $1.4 \times 10^3$ | 18 | 270 | $2.3 \times 10^3$ | $1.3 \times 10^4$ | 2.9 |
| $2.2 \times 10^3$ | 16 | 290 | $2.3 \times 10^5$ | $2.0 \times 10^4$ | 2.8 |
| $3.4 \times 10^3$ | 15 | 330 | $2.2 \times 10^5$ | $3.2 \times 10^4$ | 2.6 |
| $4.5 \times 10^3$ | 14 | 360 | $2.2 \times 10^5$ | $5.0 \times 10^4$ | 2.4 |
| $5.2 \times 10^3$ | 14 | 370 | $2.2 \times 10^5$ | $6.2 \times 10^4$ | 2.4 |
| $5.8 \times 10^3$ | 14 | 380 | $2.1 \times 10^5$ | $7.9 \times 10^4$ | 2.4 |
| $6.5 \times 10^3$ | 14 | 390 | $2.1 \times 10^5$ | $9.6 \times 10^4$ | 2.3 |
| $1.2 \times 10^4$ | 13 | 430 | $2.1 \times 10^5$ | $1.7 \times 10^5$ | 2.2 |

series, the energy flux in magnetoacoustic waves at $r = 2R_\odot$. It is clear that the addition of enough wave energy raises the velocity predicted at 1 AU by any amount one chooses. As the velocity is raised, there are corresponding changes in some of the other parameters, most notably in the proton temperature. As discussed in Section III, the correlation between velocity and proton temperature at 1 AU is due to large-scale extended heating. Therefore, an important test of any model of the normally active solar wind is whether it can explain the correlation between $v_E$ and $T_{pE}$.

Consider for the moment the hypothesis that macroscale velocity variations in the solar wind at 1 AU are due only to variations in the flux of magnetoacoustic waves at the coronal base. Then the series of models of Table IV should be able to predict the $T_p$–$v$ correlation. Figure 4 shows a scatter plot of daily averages of $T_p$ and $v$ from the Pioneer 6 space probe, together with the theoretical $T_p$–$v$ correlation line from Table IV. The agreement between theory and observation is excellent for $v \leq 425$–450 km sec$^{-1}$, which corresponds to most of the observations. Theory and observation do not agree well at higher velocities. However, dispersion in the scatter plot suggests that the $T_p$–$v$ correlation is not strong at the higher velocities; this may indicate that nonsteady effects strongly influence the temperature in high velocity streams. Figure 5 (Barnes et al., 1971) gives another comparison of the series of models with the observed $T_p$–$v$ correlation. In this case $T_p^{1/2}$ is plotted against $v$ in order to compare with the empirical formula given by Burlaga and Ogilvie (1970).

The success of these models in predicting the $T_p$–$v$ correlation shows that modeling the normally active wind may be more instructive than just modeling the quiet wind. Moreover, there are other kinds of circumstantial

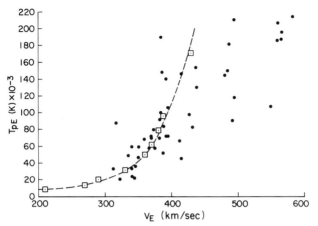

FIG. 4. Relation between proton temperature and flow speed. Points in the scatter plot correspond to daily averages of $T_{pE}$ and $v_E$ as measured by the ARC plasma probe on Pioneer 6, December 18, 1965 to February 3, 1966. Squares connected by the dashed curve correspond to values of $T_{pE}$ and $v_E$ predicted by the BHB models.

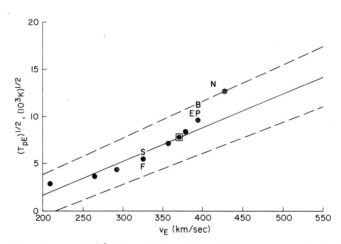

FIG. 5. Correlation of $T_{pE}^{1/2}$ with $v_E$. Solid and dashed lines are, respectively, the average and variances of the Burlaga–Ogilvie empirical formula. Filled circles correspond to the models listed in Table IV (Barnes *et al.*, 1971). The letters N, E, P, B correspond to models with slightly different base conditions, and F, S, correspond to somewhat different wave frequencies (Barnes *et al.*, 1971).

evidence that may be used to evaluate such models. For example, the rate of generation of magnetoacoustic waves used in the BHB models is quite reasonable, being comparable to the energy transport by thermal conduction in the corona and rather small compared with the wave energy required to maintain the high temperature of the inner corona. Furthermore, the BHB models imply a shell of turbulence in the solar wind over a fairly extended region. About half the wave energy is dissipated within $r = 6\,R_\odot$, but an appreciable amount survives out to $r - 15$–$20\,R_\odot$. The existence of a turbulent shell of roughly this size is supported by several lines of observational evidence. Large fluctuations in the solar wind velocity for $r \lesssim 20\,R_\odot$ are inferred from scintillations of radio "stars" (Ekers and Little, 1971). In addition, delay and anisotropy of certain solar cosmic ray events may be explained by the existence of such a turbulent envelope (see discussion in Burlaga, 1971). The solar wind appears from radar observations to be quite turbulent very near the sun (James, 1970). Finally, the Alfvén waves of solar origin in the interplanetary medium strongly support the idea that the solar wind is turbulent throughout an envelope greater than $20\,R_\odot$ in radius.

Altogether, this series of solar wind models, which is based on collisional thermal conduction with additional heating due to Landau damping of hydromagnetic waves, unifies a great variety of observations. It gives reasonable values of density, temperature, and flow speed at the orbit of the earth and near the sun. It very easily accounts for the correlation of velocity and proton temperature at 1 AU. It predicts a turbulent shell, about $20\,R_\odot$ in radius, about the sun, whose existence may be inferred from several kinds of observations. It predicts little or no correlation between velocity and density, electron temperature, and magnetic field strength at 1 AU; this prediction is also consistent with spacecraft observations. Finally, the series is generated by varying only one simple parameter, the wave energy flux at the base.

Clearly there is a strong case that dissipation of magnetoacoustic waves plays a major role in driving the solar wind. On the other hand, the picture given in the above models is not complete. In the first place, the above models do not include the acceleration that would automatically accompany the hydromagnetic wave dissipation; this acceleration is not large in the region of subsonic flow, but can be appreciable in the region of supersonic flow. Furthermore, there is no reason to suppose that magnetoacoustic waves should be stronger than Alfvén waves in the turbulent envelope. In fact, roughly half the magnetoacoustic wave energy is dissipated in the region of supersonic flow, which suggests that acceleration due to Alfvén waves may be stronger than that due to magnetoacoustic waves. However, Alfvén waves are not subject to Landau damping (Barnes, 1966; Barnes and Suffolk, 1971), and so will not heat the plasma as will magnetoacoustic waves. However, it is conceivable that Alfvén waves of sufficiently large amplitude may be dissipated by nonlinear processes (Hollweg, 1973a), and thus supply some heat to the wind. Whether or not heating of this sort can account for the $T_p$–$v$ correlation depends on its spatial distribution (Hartle and Barnes, 1970), and thus remains an open question at present.

In addition, the BHB series described above has certain minor difficulties found in all modified two-fluid models (see previous section). The density profile is somewhat too flat; this problem seems to be common to all radial flow models, and is quite probably due to nonradial flow near the sun. The predicted heat conduction flux at 1 AU is higher than observed, but much smaller than the proton kinetic energy flux; this problem is probably indicative of noncollisional reduction in thermal conduction (see previous section). These minor difficulties show that the BHB models are somewhat incomplete, but do not, by themselves, provide serious objections to the basic notion that hydromagnetic wave dissipation is the dominant process for heating solar wind ions.

In summary, models based on thermal conduction with the addition of heating by magnetoacoustic wave dissipation and acceleration, or Alfvén wave acceleration, or both, can account for most gross solar wind properties presently observable at 1 AU and elsewhere. These processes can account for both quiet and normally active states of the wind. However, noncollisional cutoff of thermal conduction, viscosity, and other noncollisional transport effects can be used to calculate reasonable models at least of the quiet wind. There is no reason why a variety of these mechanisms cannot be acting simultaneously. In any case, the importance of various mechanisms for heating and acceleration of the solar wind cannot be finally sorted out until definitive observations of the interplanetary medium over a wide range of heliocentric distances have been made. In this connection, we must not forget the possibility that some of our current theoretical ideas about what drives the solar wind may be quite far from reality. As Parker (1971) has said, "I am willing to bet that quantitative observation of the wind over the regions inside and outside the orbit of earth will show novel features (and inconsistencies in some of our present ideas) that could not possibly have been anticipated from theory and from analysis of observations near the orbit of earth."

## V. The Outer Solar Wind and Its Transition to the Interstellar Medium

The previous section on the acceleration and heating of the solar wind neglected any influence that the interstellar medium might have on solar wind dynamics. In the first approximation this neglect is certainly justified;

in fact, elementary arguments requiring the existence of a solar wind are based on the fact that the interstellar pressure is negligible compared with the pressure at the coronal base. It is perhaps less clear that the interstellar medium does not have some effect on the wind near the orbit of earth. It has been suggested that charge exchange by neutral interstellar hydrogen streaming into the solar system may detectably affect the proton temperature near 1 AU. However, at present most workers believe that the relatively low local density of interstellar neutral hydrogen means that the interstellar medium cannot sensibly affect the solar wind until it has flowed well beyond 1 AU (Chao and d'Angelo, 1972).

Therefore it is probably realistic, as well as conceptually useful, to divide solar wind flow into three regions. The inner region, which extends perhaps out to $25$–$50\ R_{\odot}$, is the region of primary acceleration and heating. The intermediate region ($0.25\ \mathrm{AU} \lesssim r \lesssim 5\ \mathrm{AU}$) is a region of turbulent and dynamically interesting flow, in which, however, primary heating and acceleration due to the sun or the interstellar medium does not occur. The outer region, beyond 5 AU, is where interaction between the solar wind and the interstellar medium, magnetic field, or cosmic rays may modify and eventually terminate the solar wind flow. Obviously the boundaries of these regions are not well known at present. Even the position of the orbit of earth in this scheme is not certain, although it is almost certainly in the intermediate region, probably nearer the inner than the outer boundary.

In the previous section on acceleration and heating we concentrated on the dynamics of the inner region. We noted that most of our data about the solar wind comes from near 1 AU, outside the region of interest, and that therefore definitive observations about the acceleration and heating of the wind await the acquisition of data from spacecraft sent much nearer the sun. Similarly, the orbit of earth is not a good location for observing the interaction between the sun and the interstellar medium, and definitive observations on this matter will probably require spacecraft to be sent well beyond 5 AU.

Nevertheless, solar wind conditions at 1 AU can be interpreted as the direct results of dynamical processes in the inner region, although the interpretation may sometimes be ambiguous. By contrast, the solar wind at 1 AU is insensitive to interstellar conditions. Accordingly, our ideas about the outer region of the solar wind must be highly speculative, and one must be cautious about constructing elaborate theories for this region. Nevertheless, progress has been made in delineating what dynamical processes are likely to be important in the outer solar wind. Axford (1972) has recently written an excellent, detailed review article on this subject, which we do not attempt to duplicate here. Rather, we briefly summarize the alternatives, with emphasis on the most recent theories and the prospects for answering these basic questions in the next decade.

## A. The Solar Wind Beyond 1 AU

It is probable that the solar wind is not strongly influenced by the interstellar medium for some distance beyond the orbit of earth (out to 5 AU or so). For a qualitative understanding of this region it may be appropriate to extrapolate models of the inner solar wind beyond 1 AU. Profiles of solar wind density and temperature beyond 1 AU have been calculated in the models of Hartle and Sturrock (1968), Hartle and Barnes (1970), Leer and Axford (1972), and Cuperman *et al.* (1972); Leer and Axford (1971) have also given attention to variations in heliographic latitude. However, care should be exercised in making such extrapolations, since processes that are negligible inside 1 AU may be of great importance far beyond 1 AU, and vice versa.

In any event, considerable progress can be made in analyzing the gross behavior of the solar wind far from the sun, without carrying out detailed calculations. For example, if the flow is steady, radial, and spherically symmetric,

$$nvr^2 = J = \text{const.} \tag{33}$$

If no energy is added to the flow from outside, the velocity $v$ is approximately constant, and the density declines as

$$n(r) \simeq n(a)(a/r)^2, \tag{34}$$

where $a$ is the base distance (1 AU, say). In reality, the mixing of streams of different velocities will probably increase beyond 1 AU, so that the velocity, density, etc., will fluctuate as much as or more than at 1 AU. Eventually the streams will become so mixed that the velocity variations of the streams will be randomized. This process would have the effect of heating the wind, but on the average the density profile would still satisfy Eq. (34). Departures from Eq. (34) (on the average) could be caused by systematic nonradial flow, addition of mass and deceleration of the wind by photoionization of the interstellar medium, or deceleration of the wind by charge exchange with neutral interstellar hydrogen atoms.

The behavior of the ion and electron temperatures beyond 1 AU is more difficult to predict than is the behavior of the density and velocity. Since the heat flux due to collisional thermal conduction is parallel to the magnetic field, the outward conductive heat flux will become small as the interplanetary magnetic field becomes azimuthally directed with increasing $r(B_r/B_\phi \propto 1/r)$. Hence, at first sight, it appears that the heat supply of the wind will be cut off and it will coast outward adiabatically. If so, the (electron or ion) temperature $T \propto n^{2/3}$ and so from Eq. (34)

$$T(r) \simeq T(a)(a/r)^{4/3}. \tag{35}$$

However, interstream mixing may convert a sizable fraction of the energy difference between streams into thermal energy (Jokipii and Davis, 1969). This process seems only to produce very local heating near 1 AU (Burlaga and Ogilvie, 1970; Burlaga et al., 1971), but may heat the wind more generally beyond 1 AU. Since velocity variations of 200 km sec$^{-1}$ or more are not uncommon at 1 AU, heating by interstream mixing may cause the ion temperature to decline much more slowly than Eq. (35) indicates.

If thermal conduction is cut off, however, it is quite possible that the electron temperature will decline as in Eq. (35). The observational study of interstream heating at 1 AU by Burlaga et al. (1971) suggests that electrons are negligibly heated. This observation is consistent with the suggestion of Barnes and Hung (1973) that the apparent ion heating in interstream regions is due to pitch angle scattering of ions by hydromagnetic waves in the interaction region (see also Goldstein and Eviatar, 1973). On the other hand, if interstream heating involves Landau damping of these waves (Jokipii and Davis, 1969), then some electron heating as well as ion heating should occur (Barnes, 1966, 1967).

Other effects may heat the solar wind plasma beyond 1 AU. In the equatorial plane the interplanetary magnetic field is typically divided into several sectors of opposite polarity (Wilcox and Ness, 1965; Wilcox, 1968). It has been suggested that the sector boundaries might become unstable somewhere beyond 1 AU (Davis, 1970; Axford, 1972). If so, recombination of magnetic field lines could release magnetic energy which would reappear in the plasma as heat. However, Fisk and Van Hollebeke (1972) have inferred from cosmic ray electron observations that the interplanetary sector pattern persists out to $\sim 30$ AU. Another source of heating could be interaction with the interstellar medium. Ionization of an interstellar neutral atom either by solar ultraviolet radiation or by charge exchange with a solar wind ion results in a new ion that is initially nearly at rest with respect to the sun. The new ion is quickly swept up by the solar wind, but maintains a high peculiar velocity relative to the solar wind plasma. The net result of this process is heating and deceleration of the solar wind, with addition of mass in the case of photoionization. Finally, it is possible that the interaction of the solar wind with galactic cosmic rays contributes to the deceleration of the wind (Jokipii, 1971; Axford, 1972).

It is of interest to consider the profiles of a few other parameters in the outer solar wind. Since the density $n \propto r^{-2}$, and the (azimuthal) magnetic field $B \propto r^{-1}$, the Alfvén speed $B/(4\pi m_p n)^{1/2}$ is constant beyond a few astronomical units. Hence the important parameters

$$\beta_p = 8\pi n k T_p / B^2$$
$$\beta_e = 8\pi n k T_e / B^2$$

vary in proportion to $T_p$ and $T_e$, respectively. The parameter $\beta_p$ governs ion Landau damping of magnetoacoustic waves; such a wave is damped out in roughly $\exp(1/\beta_p)$ wavelengths. If $\beta_p = 0.5$ at 1 AU, the ion Landau damping distance at heliocentric distance $r$ will be of the order of $\exp[2T_p(a)/T_p(r)]$ wavelengths. Therefore, once the plasma has cooled well below its temperature at 1 AU, its ions no longer will damp fast-mode hydromagnetic waves. Electron Landau damping, which is much less sensitive to $\beta_e$, will still occur, however.

Also, note that the Coulomb collision times vary as $T^{3/2}/n \propto r^2 T^{3/2}$. Hence the ratio of the collision times to the characteristic expansion time $r/v$ varies as $rT^{3/2}$, and the ratio of the electron mean free path to the scale height $r$ varies as $rT_e^2$. Hence the solar wind plasma tends to become more collisional with increasing heliocentric distance unless the temperatures decline less rapidly than $r^{-2/3}$. In particular, adiabatic expansion $(T \propto r^{-4/3})$ means that the outer solar wind is more collisional than the wind at 1 AU. Moreover, Coulomb collisions will tend to keep the outer solar wind thermally isotropic if the plasma cools adiabatically. If the plasma remains warm enough that the infrequent collisions do not produce isotropy, the plasma may tend to become anisotropic in the sense $T_\perp > T_\parallel$; however, this anisotropy will be limited by microinstabilities. In either case, we do not expect the thermal and anisotropy in the outer solar wind to be much greater than at 1 AU.

It is reasonable to expect that within the next decade the average radial profiles of density, velocity, proton temperature, magnetic field, and possibly other parameters will be observationally determined out to $r \sim 10$ AU, at least in the ecliptic plane. Such observations, especially of the proton temperature, may provide definitive evidence as to the termination of the solar wind, even though the terminal shock is too distant to be directly observed in foreseeable spacecraft missions.

## B. Transition to the Interstellar Medium

We expect the sun to be typical of many, perhaps most, stars in having a supersonically expanding corona. Thus, stellar winds rather than classical Strömgren spheres may govern the interaction between the interstellar medium and stars of the middle and late spectral type (Parker, 1963). Hence the character of the termination of the solar wind by the local interstellar medium is a topic of great astrophysical importance. This subject has become both exciting and timely since recent measurements of Lyman $\alpha$ emission by local interstellar hydrogen became available (Thomas and Krassa, 1971; Thomas, 1972; Bertaux and Blamont, 1971, 1972), and since space probe missions to the outer solar system are planned for this decade.

At present we do not know what mechanism terminates the solar wind. Whatever the mechanism is, however, the transition should be located, roughly speaking, at the distance $R$ at which the ram pressure $m_p nv^2$ of the wind is matched by the interstellar pressure. The interstellar pressure is due to some combination of pressures from the interstellar magnetic field, galactic cosmic rays, and (possibly) ionized interstellar gas. Parker (1969) has noted that even if the solar wind is primarily decelerated by charge exchange between solar wind ions and neutral interstellar hydrogen, complete degradation of the flow cannot occur in a distance smaller than $R$. The value of $R$ is not known at present; values in the range $30\,\text{AU} \le R \le 100\,\text{AU}$ are commonly quoted.

The hypothetical situation of the unmagnetized solar wind expanding symmetrically into a stationary ionized interstellar plasma with no magnetic field was analyzed early on by Parker (1963). In this case the solar wind would pass through a strong shock to subsonic flow near the distance $R$. Beyond $R$ the flow would continue to decelerate; on a scale large compared with $R$ the sun would appear as a point source of incompressible fluid. If this slowly expanding plasma is not swept away by motions of the interstellar medium, it could fill a sizable region ($\sim 3$ pc) in the lifetime of the sun (Parker, 1963), or possibly be terminated sooner in a recombination front (Newman and Axford, 1968). The above picture must be considerably modified if the interstellar plasma is moving (a velocity of 20 km sec$^{-1}$ relative to the sun is quite reasonable). In this case the shocked interplanetary plasma would be swept back into a long wake (Parker, 1963). Moreover, the transition shock would be deformed away from spherical symmetry.

If the termination of the solar wind is dominated by the galactic magnetic field, the flow will pass through a shock into subsonic flow, just as in the previous case. The flow of the shocked plasma will be asymmetric. The plasma pressure will tend to push local galactic field lines away from the sun, but the plasma is constrained to flow along the field lines. Therefore the solar wind plasma might escape freely along the local galactic magnetic field, but not transverse to it.

If the interplanetary magnetic field continues out to the transition shock in an ordered manner, it can significantly modify the picture presented above. The classical spiral model (Parker, 1963) predicts that the field will be essentially transverse to the flow direction far beyond the orbit of earth. Since the magnetic energy density is small compared to the kinetic energy density in the region of supersonic flow, the interplanetary field will not be of great importance in determining the position of the transitional shock. At the shock the magnetic energy density cannot increase by more than a factor 16 (Tidman and Krall, 1971), so that most of the upstream kinetic energy will be converted into thermal energy just downstream from the shock. Hence the shocked plasma would have a high value of $\beta = 8\pi nk(T_e + T_p)/B^2$. This in turn would mean that the shocked plasma would very efficiently dissipate much of the hydromagnetic turbulence generated at the shock (Barnes, 1966). Hence most of the turbulence generated at the shock would probably be confined to a fairly narrow region just beyond the shock. This turbulent shell, and the shock itself, might produce appreciable effects on low energy cosmic rays entering the solar system (see also Jokipii, 1968).

However, the region of high $\beta$ would not persist far beyond the terminal shock. Steady radial flow would imply that $n \propto 1/vr^2$ and $B \propto 1/vr$, so that $n \propto B/r$. If the flow were incompressible, $B$ would increase in proportion to $r$. Hence the idealization of incompressible flow would be untenable beyond the terminal shock (Axford, 1972). If the plasma expands adiabatically, the fluid pressure $p \propto n^{5/3} \propto (B/r)^{5/3}$. Hence $\beta \propto p/B^2 \propto B^{-1/3}r^{-5/3}$, and $\beta$ must decrease on going outward unless $B$ decreases at least as fast as $r^{-5}$. Solutions of the adiabatic flow equations by Cranfill (1971; see also Axford, 1972) indicate that $B$ does not decline so fast (and in fact, may increase with $r$). Hence $\beta$ should decline rapidly with increasing $r$ after attaining a high value in the post-shock plasma.

It now seems quite likely that the local interstellar medium is un-ionized, with a density of the order of 0.1 H atoms cm$^{-3}$ (for a thorough discussion of the observations, see Axford, 1972). It is reasonable to suppose that this gas is streaming past the solar system with a relative velocity $\sim 20$ km sec$^{-1}$. This gas will interact with the solar environment, being ionized either by solar ultraviolet radiation or by charge exchange with the solar wind ions, or by both processes. Some atoms may penetrate far into the solar wind; interstellar hydrogen may penetrate into $r \sim 4$ AU, and helium into $r < 1$ AU (Axford, 1972, Table 1), before being photoionized.

Ionization of the interstellar medium may have a significant effect on the deceleration and eventual termination of the solar wind. Holzer (1972) and Wallis (1971a,b) have calculated detailed models of the deceleration of the solar wind by such interaction with the interstellar medium. The neutral gas decelerates and heats the supersonic solar wind. This heating could be directly observed by future spacecraft plasma measurements beyond 5 AU. In principle, such deceleration can lead to a shock-free transition from supersonic to subsonic flow. However, such a transition is not necessary; in fact, Holzer (1972) concludes that a transitional shock should be expected on the basis of our current knowledge of interplanetary and interstellar parameters. The location of the shock sensitively depends on the density of interstellar hydrogen; for $N_H \sim 0.1$ cm$^{-3}$ Holzer finds $r \lesssim 50$ AU to be a reasonable estimate of the minimum shock distance.

Holzer (1972) has also calculated the effects of interstellar atomic hydrogen gas on the shocked subsonic flow. He finds that the interstellar gas, as well as the interplanetary magnetic field, produce significant cooling of the hot, subsonic plasma. This cooling might be inhibited by several mechanisms, however. It is possible that the rapid cooling will produce a super-adiabatic temperature gradient, and thus become convectively unstable. The net effect of this process would be to transfer energy convectively and reduce the temperature gradient. However, as long as the gas is fairly hot (thermal energy $\sim$ magnetic energy), Landau damping may inhibit the generation of turbulence. If the plasma becomes too cool for effective Landau damping, magnetic field line reconnection might occur and heat the plasma.

Holzer also found that the cooling of the plasma produces a population of hot neutral hydrogen atoms which can penetrate back through the shock into the solar wind. He considered whether these atoms could be a significant component of the interplanetary atomic hydrogen gas, and concluded that they are not important unless the shock transition is located in the inner solar system.

One cannot determine the details of the shape of the heliosphere from calculations of the type outlined above. However, it appears clear that charge exchange with interstellar hydrogen would tend to make the subsonic heliosphere quite asymmetric and elongated in the direction of flow of the interstellar gas, and to make the plasma in the distant heliosphere flow at the speed of the interstellar gas. The downwind heliosphere will eventually dissipate, probably into a feeble wake, by a variety of mechanisms. The extent of the downwind heliosphere is not known; Axford (1972) suggests a few hundred astronomical units as the appropriate dimension.

In summary, the outer solar wind will be decelerated by one or more of several mechanisms; fluid interaction with the local galactic magnetic field and/or ionized interstellar medium, or interaction with the neutral interstellar gas by charge exchange and photoionization. In all cases, a terminal shock whose minimum heliocentric distance lies in the range 25–100 AU is likely. The character of the subsonic flow of the plasma beyond the shock is complicated, and strongly depends on the mechanism of its interaction with the interstellar medium. The extent and shape of the heliosphere depend on the mechanism of interaction and on the state of motion of the interstellar gas. At present there is very little observational data that has direct bearing on these questions. It is expected that plasma, magnetic, ultraviolet radiation, and cosmic ray measurements from spacecraft missions to the outer solar system will go far in resolving at least some of these issues. Even if the terminal shock is too distant to be observed in the near future, direct observations of the radial profiles of various parameters may provide definitive information about the mechanism of transition to the interstellar medium.

## VI. Angular Momentum Transfer

Solar rotation, which has been of little concern in our considerations of heating and radial acceleration, is of primary importance in studies of angular momentum transfer in the wind. This transfer of angular momentum is fairly efficient, and in fact may have reduced appreciably the rotation rate of at least the outer solar layers during the sun's lifetime (see Brandt et al., 1972; Brandt, 1970; Parker, 1969; and references therein). Theoretical studies have succeeded in defining a number of mechanisms which may strongly influence this angular momentum transport. Unfortunately, the difficulty in measuring the azimuthal component $v_\phi$ of the solar wind velocity near 1 AU ($v_\phi \sim 0.01\, v_r$) makes it impossible to make a definitive comparison between theories and observations at present. The observational situation may improve during the next few years, when space probes penetrate the inner solar wind where $v_\phi$ is probably larger, and therefore more easily measured, than at 1 AU.

The solar wind is expected to influence the sun's angular momentum budget more than its mass budget, because the corona probably corotates with the sun far above the photosphere. Thus the solar wind should contain more angular momentum per unit mass than the sun. If the solar wind plasma corotates with the sun out to a certain distance $R_A$, and beyond $R_A$, each element of the plasma experiences no torque, then the azimuthal component of velocity is

$$v_\phi = \Omega r \qquad (r < R_A)$$
$$v_\phi = \Omega R_A^2/r \qquad (r > R_A) \tag{36}$$

where $\Omega$ is the angular velocity of the sun. Values of $v_\phi$ from spacecraft-borne plasma probes near 1 AU tend to lie in the range 0–10 km sec$^{-1}$ [e.g., see Chapter 4 of Solar Wind (Sonett et al., 1972)]. Comet tails and the tail of the earth's magnetosphere also give a measure of $v_\phi$ (Brandt, 1970; Brandt et al., 1972). If we take $v_\phi \approx 5$ km sec$^{-1}$, then according to (36) we must have $R_A \approx 0.1$ AU. Although the crude picture given by (36) should not be taken too seriously, it is clear that an azimuthal velocity of several kilometers per second at the orbit of earth implies significant rotational coupling between the sun and the solar wind over a distance of at least $\sim 0.1$ AU.

The forces which produce this rotational coupling are generally different from those which dominate radial acceleration. If the solar wind is steady and azimuthally symmetric, the momentum equation may be written in the form

$$\mathbf{V} \cdot (m_p n \mathbf{vv} - \mathbf{T}) = -(GM_0 m_p n/r^2)\mathbf{e}_r \tag{37}$$

where the stress tensor $\mathbf{T}$ describes the nongravitational forces. Contributions to $\mathbf{T}$ come from the pressure tensor, magnetic stress, viscous stress, and hydromagnetic waves. Since the pressure tensor is

$$\mathbf{p} = p_\perp \mathbf{1} + (p_\| - p_\perp)\mathbf{BB}/B^2 \qquad (38)$$

(Chew, Goldberger, and Low, 1956), we have

$$\mathbf{T} = (1/4\pi)\mathbf{BB}[1 + (4\pi/B^2)(p_\perp - p_\|)] - [p_\perp + (B^2/8\pi)]\mathbf{1} + \boldsymbol{\sigma} \qquad (39)$$

where $\boldsymbol{\sigma}$ represents stress due to viscosity and hydromagnetic waves. For the purpose of illustration, neglect the contribution from waves and suppose the viscous stress is given by the usual expression from fluid mechanics:

$$\sigma_{ij} = \eta[(\partial v_i/\partial x_j) + (\partial v_j/\partial x_i) - \delta_{ij}\mathbf{V} \cdot \mathbf{v}] + (\zeta + \tfrac{1}{3}\eta)\,\delta_{ij}\mathbf{V} \cdot \mathbf{v} \qquad (40)$$

where $\eta$ and $\zeta$ are (scalar) coefficients of viscosity. Let us further restrict our attention to the solar equatorial plane, and assume that $\partial v_\phi/\partial \vartheta = 0$ in this plane, so that $\sigma_{\theta\phi} = 0$. Thus $T_{\theta\phi} = 0$ so that the external torque $L_Z$ per unit solid angle in the direction of the sun's rotation axis, acting on the solar wind inside radial distance $r$, is

$$L_Z = r^3 T_{r\phi} . \qquad (41)$$

The angular momentum per unit solid angle convected outward at the same position is

$$K_Z = r^3 m_p n v_r v_\phi . \qquad (42)$$

For a steady state, conservation of angular momentum requires that $K_Z - L_Z$ is constant. Then, since $n v_r r^2$ is constant,

$$r v_\phi = (r T_{r\phi}/m_p n v_r) + J, \qquad (43)$$

where the constant $J$ is the angular momentum per unit mass carried away by the solar wind (cf. Weber and Davis, 1970). Equation (43) may also be derived by taking the $\phi$ component of Eq. (37). On substituting (39) and (40) into (43), we have

$$r v_\phi = (r B_r B_\phi/4\pi m_p n v_r)[1 + (4\pi/B^2)(p_\perp - p_\|)]$$
$$+ [r^2 \eta(r)/m_p n v_r](d/dr)(v_\phi/r) + J. \qquad (44)$$

The firehose stability criterion requires that $1 + 4\pi(p_\perp - p_\|)/B^2 > 0$. Thus the first term on the right-hand side of (44) is negative if $B_r B_\phi < 0$, as in the spiral model magnetic field. Also the second term on the right-hand side of (44) is negative if the angular velocity $v_\phi/r$ decreases with increasing $r$. Thus both viscous and magnetic torques will oppose the sun's rotation.

The magnetic field must also satisfy Maxwell's equations div $\mathbf{B} = 0$, $c$ curl $\mathbf{E} = -$curl$(\mathbf{v} \times \mathbf{B}) = 0$. Thus if $|V_r| \ll |V_\phi|$ at the coronal base,

$$r(V_r B_\phi - V_\phi B_r) = -\Omega r^2 B_r = \text{const.} \tag{45}$$

On solving (45) for $B_\phi$ and substituting the result into (44), we have

$$v_\phi\left(1 - \frac{A}{M_A^2}\right) = -\frac{r\Omega A}{M_A^2} + \frac{J}{r} + \frac{r\eta}{m_p n v_r} \frac{d}{dr}\left(\frac{v_\phi}{r}\right), \tag{46}$$

where $M_A = (4\pi m_p n v_r^2 / B_r^2)^{1/2}$ is the (radial) Alfvén number and $A = 1 + 4\pi(p_\perp - p_\parallel)/B^2$ is the anisotropy factor. Since $A \sim 0(1)$ near the sun, $M_A^2 \ll A$ near the sun and $M_A^2 \gg A$ far from the sun. Thus there is a critical distance $r_A$ such that

$$M_A^2(r_A) = A(r_A). \tag{47}$$

Since $v_\phi$ must be finite at $r_A$, (46) implies that

$$J = r_A^2\Omega - [r_A^4 \eta(r_A)/m_p(n v_r r^2)](d/dr)(v_\phi/r)|_{r = r_A}. \tag{48}$$

Then $\Omega r_A^2$ is a lower bound on the angular momentum per unit mass carried by the wind. This point was first discussed by Weber and Davis (1967) and, in a somewhat different way, by Modisette (1967). A closely related theory, but without trans-Alfvénic flow, was given earlier by Mestel (1961).

Thus, in the absence of viscosity, the angular momentum carried away from the sun by the solar wind is the same as if the wind corotated with the sun out to $r = r_A$, and coasted freely beyond [Eq. (36)]; cf. also the discussion by Brandt (1970). $J$ is sensitive to the radial magnetic field strength and the flow profile, since $r_A$ is defined by $B_r^2/4\pi + p_\perp - p_\parallel = m_p n v_r^2$. Hence the details of the radial flow, governed by radial acceleration and heating, strongly influence the angular momentum transport and the azimuthal velocity profile. The different predictions of $v_\phi$ at 1 AU by various models may be due as much to their different assumptions about radial acceleration and heating as to their different assumptions about the magnetic and viscous torques. We therefore will limit our remarks to a brief review of the qualitative effects of the different kinds of torques.

*Magnetic Stress.* If we neglect viscosity Eq. (46) can be rewritten as

$$v_\phi\left(1 - \frac{A(r)}{M_A^2}\right) = \frac{\Omega r_A^2}{r}\left(1 - \frac{A(r)}{A(r_A)} \frac{v_r(r_A)}{v_r(r)}\right). \tag{49}$$

Computations show that $r_A \approx 20$–$30\, R_\odot$ (Weber and Davis, 1967). Then near the sun $M_A^2 \ll 1$, $v_r(r) \ll v_r(r_A)$, $A(r) \sim A(r_A) \sim 1$ in order of magnitude, so that (49) implies

$$v_\phi \simeq (\Omega r_A^2/r)[v_r(r_A)/v_r(r)][M_A^2/A(r_A)] = r\Omega|_{(r \ll r_A)}, \tag{50}$$

i.e., the magnetic field forces the plasma to corotate with the sun if $r \ll r_A$. Qualitatively, the magnetic field "stiffens" the rotating plasma inside the Alfvénic point. Beyond the Alfvénic point the plasma kinetic energy exceeds the magnetic energy, and the behavior of the magnetic field is determined by the plasma flow. As $r \to \infty$, $M_A^2 \gg 1$, $A(r) \sim 0(1)$, so that (49) gives

$$rv_\phi \simeq \Omega r_A^2 \{1 - [A(r)/A(r_A)][v_r(r_A)/v_r(r)]\}_{(r \gg r_A)}, \qquad (51)$$

and at 1 AU the plasma flow and magnetic torque contribute to an angular momentum loss in the ratio

$$rv_\phi/(-rB_r B_\phi/4\pi m_p n v_r) = [A(r_A)/A(r)][v_r(r)/v_r(r_A)] - 1. \qquad (52)$$

Except for small effects due to pressure anisotropy, the above ratio is essentially constant with increasing $r$ beyond 1 AU. Hence the transfer of angular momentum between the sun and the interstellar medium is partly due to transport of mechanical angular momentum by the solar wind, and partly due to direct magnetic coupling to the interstellar medium. The two contributions are of comparable magnitude (Weber, 1972).

*Pressure Anisotropy.* We have argued earlier that pressure anisotropy does not strongly modify radial flow. If this is so, and if the solar wind remains fairly isotropic at least out to $r = r_A$, then according to Eq. (48) pressure anisotropy does not have a major influence on the *total* angular momentum carried by the solar wind. However, pressure anisotropy may have a sizable effect on the distribution of the angular momentum transport between the magnetic and flow terms. Since $A < 1$ at 1 AU, Eq. (51) and (52) show that anisotropy increases $v_\phi$ and its relative contribution to the angular momentum efflux at 1 AU over their values in a similar system with isotropic pressure. More detailed calculations have shown that models with anisotropic pressure do in fact give higher values of $v_\phi$ than isotropic models (Weber, 1970; Weber and Davis, 1970; Tan and Abraham-Shrauner, 1972).

*Viscosity.* As mentioned in our discussion of acceleration and heating, there is no firm theoretical basis for treating viscosity in regions beyond $r \sim 10\,R_\odot$, where the Coulomb collision time is not small in comparison with the expansion time of the wind. Although it is possible to integrate Eq. (46) for any specified viscous coefficient $\eta(r)$, we shall not do so, in view of the uncertainties as to the proper form of $\eta(r)$. Moreover, the anisotropy in viscous stress due to the magnetic field means that the form of the viscous term in Eq. (46) is not rigorously correct. Nevertheless, Eq. (46) is probably good enough to give reasonable order-of-magnitude estimates of the influence of viscosity on angular momentum transport, provided that we can estimate $\eta(r)$.

From (44) we see that viscous stresses produce appreciable torque on the plasma if $\eta(r) \gtrsim m_p n v_r r$. Near the sun, where the usual collisional viscosity coefficient should be correct, $\eta < m_p n v_r r$. Since the magnetic field is strong enough to enforce corotation near the sun, it is clear the magnetic stresses dominate viscous stresses for $r \ll r_A$. Beyond $r \gtrsim 10\,R_\odot$ the use of the collisional viscous coefficient is questionable; therefore let us write $\eta \sim \frac{1}{3}m_r n v_{th}\,\lambda$, where $\lambda$ is the proton mean free path due either to Coulomb collisions or to wave–particle collisions, and $v_{th}$ is the proton speed. Hence viscous stress is significant only if $v_{th}\,\lambda/(3v_r r) \gtrsim 1$. Therefore beyond $r \sim 5\,R_\odot$, where the flow is supersonic, the above inequality obtains only if $\lambda/r > 1$, which means that the viscous stress is important only if the mean free path is so large that the usual kinetic theory treatment of viscosity is incorrect. It will be recalled that a similar situation arises in the analysis of viscous heating in the solar wind.

Thus it is not clear whether viscosity is ever as important as magnetic stresses in governing the angular momentum transport. Models using conventional collisional viscosity, which should be considered an upper limit on the true viscosity, often find a large viscous effect at a few tens of $R_\odot$ and beyond (Pneuman, 1966; Weber and Davis, 1970). However, these models are one-fluid models, and the collisional viscosity depends sensitively on the proton temperature ($\eta \propto T_p^{5/2}$). Since the proton temperature is normally smaller than the electron temperature, one-fluid models tend to overestimate the collisional viscous coefficient. Inclusion of collisional viscosity in two-fluid models indicates that viscosity influences angular momentum transport much less than one-fluid models suggest (Wolff et al. 1971).

*Hydromagnetic Waves.* Preferential emission of hydromagnetic waves in certain directions, by (say) tilted sunspot regions, could result in a net transfer of angular momentum by the waves (Schubert and Coleman, 1968). The angular momentum would eventually be transferred to the plasma by mechanisms discussed earlier in the context of heating and radial acceleration by waves. Hollweg (1973b) recently studied the role of Alfvén waves in angular momentum transport, finding that such waves tend to reduce $v_\phi$ in the presence of thermal anisotropy.

At present it seems that there are more questions than answers about angular momentum transfer in the solar wind. A few things seem clear, however. Measurements of the interplanetary magnetic field and azimuthal plasma flow near 1 AU indicate that the solar wind and its magnetic field carry enough angular momentum away from the sun to have been significant in spinning down the sun in $5 \times 10^9$ yr (Brandt et al., 1972). Moreover, the solar magnetic field probably dominates angular momentum transfer near the sun, enforcing approximate corotation out to $r \gtrsim 15\,R_\odot$. The relative importances of other processes, such as viscous stress, pressure anisotropy, hydromagnetic waves, or preferentially guided flow near the sun are not well understood at present.

We may expect advances in this field when observations from spacecraft that penetrate to $r < 0.5$ AU become available. The azimuthal velocity should increase as $1/r$ going inward (until magnetic or viscous torques become significant), and will therefore be more easily measured than at 1 AU. Moreover, measurement of radial flow profiles may give us a better idea of how viscous the wind really is, and what are the important heating and radial acceleration mechanisms. These facts, in turn, may allow construction of more realistic models of angular momentum transport.

### ACKNOWLEDGMENTS

I wish to thank C. P. Sonett for critically reading the manuscript, and John H. Wolfe and John D. Mihalov for providing unpublished data from the Ames Research Center Pioneer 6 plasma probe.

### REFERENCES

Alazraki, G., and Couturier, P. (1971). *Astron. Astrophys.* **13**, 380.
Axford, W. I. (1968). *Space Sci. Rev.* **8**, 331.
Axford, W. I. (1972). *In* " Solar Wind " (C. P. Sonett, P. J. Coleman, Jr., and J. M. Wilcox, eds.), p. 609. NASA SP-308. U.S. Gov. Printing Office, Washington, D.C.
Barnes, A. (1966). *Phys. Fluids* **9**, 1483.
Barnes, A. (1967). *Phys. Fluids* **10**, 2427.
Barnes, A. (1969). *Astrophys. J.* **155**, 311.
Barnes, A. (1971). *J. Geophys. Res.* **76**, 7522.
Barnes, A. (1972). *In* " Solar Wind " (C. P. Sonett, P. J. Coleman, Jr., and J. M. Wilcox, eds.), p. 333. NASA SP-308. U.S. Gov. Printing Office, Washington, D.C.
Barnes, A., and Hartle, R. E. (1972). *In* " Solar Wind " (C. P. Sonett, P. J. Coleman, Jr., and J. M. Wilcox, eds.), p. 219. NASA SP-308. U.S. Gov. Printing Office, Washington, D.C.
Barnes, A., and Hung, R. J. (1973). *Cosmic Electrodyn.* **3**, 416.
Barnes, A., and Suffolk, G. C. J. (1971). *J. Plasma Phys.* **5**, 315.
Barnes, A., Hartle, R. E., and Bredekamp, J. H. (1971). *Astrophys. J.* **166**, L53.
Barnes, A., Brandt, J. C., Hartle, R. E., and Wolff, C. L. (1972). *Cosmic Electrodyn.* **3**, 254.
Belcher, J. W. (1971). *Astrophys. J.* **168**, 509.
Belcher, J. W., and Davis, L., Jr. (1971). *J. Geophys. Res.* **76**, 3534.
Bertaux, J. L., and Blamont, J. E. (1971). *Astron. Astrophys.* **11**, 200.
Bertaux, J. L., and Blamont, J. E. (1972). *In* " Solar Wind " (C. P. Sonett, P. J. Coleman, Jr., and J. M. Wilcox, eds.), p. 661. NASA SP-308. U.S. Gov. Printing Office, Washington, D.C.
Biermann, L. (1951). *Z. Astrophys.* **29**, 274.
Biermann, L. (1957). *Observatory* **107**, 109.
Braginskii, S. T. (1965). *In* " Reviews of Plasma Physics " (M. A. Leontovich, ed.), Vol. 1, p. 205. Consultants Bureau, New York.
Brandt, J. C. (1970). " Introduction to the Solar Wind." Freeman, San Francisco, California.
Brandt, J. C., and Cassinelli, J. P. (1966). *Icarus* **5**, 47.
Brandt, J. C., and Wolff, C. L. (1973). *J. Geophys. Res.* **78**, 3197.
Brandt, J. C., Roosen, R. G., and Harrington, R. S. (1972). *Astrophys. J.* **177**, 277.
Brandt, J. C., Thayer, N. N., Wolff, C. L., and Hundhausen, A. J. (1973). *Astrophys. J.* **183**, 1037.

Burlaga, L. F. (1971). *Space Sci. Rev.* **12**, 600.

Burlaga, L. F. (1972a). *In* "Solar Wind" (C. P. Sonett, P. J. Coleman, Jr., and J. M. Wilcox, eds.), p. 309. NASA SP-308. U.S. Gov. Printing Office, Washington, D.C.

Burlaga, L. F. (1972b). *In* "Cosmic Plasma Physics" (K. Schindler, ed.), p. 73. Plenum, New York. (See also Goddard Prepr. NASA-GSFC X692-71-400.)

Burlaga, L. F., and Ogilvie, K. W. (1973). *J. Geophys. Res.* **78**, 2028.

Burlaga, L. F., and Ness, N. F. (1968). *Can. J. Phys.* **46**, S962.

Burlaga, L. F., and Ogilvie, K. W. (1970). *Astrophys. J.* **159**, 659.

Burlaga, L. F., Ogilvie, K. W., Fairfield, D. H., Montgomery, M. D., and Bame, S. J. (1971). *Astrophys. J.* **164**, 137.

Carovillano, R. L., and Siscoe, G. L. (1969). *Solar Phys.* **8**, 401.

Chamberlain, J. W. (1961). *Astrophys. J.* **133**, 675.

Chao, J. K., and d'Angelo, N. (1972). *J. Geophys. Res.* **77**, 6226.

Chapman, S. (1957). *Smithson. Contrib. Astrophys.* **2**, 1.

Chapman, S., and Cowling, T. G. (1961). "The Mathematical Theory of Non-Uniform Gases." Cambridge Univ. Press, London and New York.

Chen, W. M., Lai, C. S., Lin, H. E., and Lin, W. C. (1972). *J. Geophys. Res.* **77**, 1.

Chew, G. F., Goldberger, M. L., and Low, F. L. (1956). *Proc. Roy. Soc. Ser. A* **236**, 112.

Cowling, T. G. (1957). "Magnetohydrodynamics." Wiley (Interscience), New York.

Cranfill, C. (1971). Ph.D. Thesis, Univ. of California, San Diego.

Croft, T. A. (1972). *In* "Solar Wind" (C. P. Sonett, P. J. Coleman, Jr., and J. M. Wilcox, eds.), p. 521. NASA SP-308. U.S. Gov. Printing Office, Washington, D.C.

Cronyn, W. M. (1972). *EOS Trans. AGU* **53**, 477.

Cuperman, S. (1973). *Astrophys. Space Sci.* **20**, 519.

Cuperman, S., and Harten, A. (1971). *Astrophys. J.* **173**, 383.

Cuperman, S., Harten, A., and Dryer, M. (1972). *Astrophys. J.* **177**, 555.

d'Angelo, N. (1968). *Astrophys. J.* **154**, 401.

d'Angelo, N. (1969). *Solar Phys.* **7**, 321.

Davis, L., Jr. (1970). *Leningrad STP Symp.* Pap. No. II-2.

Davis, L., Jr. (1972). *In* "Solar Wind" (C. P. Sonett, P. J. Coleman, Jr., and J. M. Wilcox, eds.), p. 93. NASA SP-308. U.S. Gov. Printing Office, Washington, D.C.

Dessler, A. J. (1967). *Rev. Geophys.* **5**, 1.

Ekers, R. D., and Little, L. T. (1971). *Astron. Astrophys.* **10**, 310.

Eviatar, A., and Schulz, M. (1970). *Planet. Space Sci.* **18**, 321.

Eviatar, A., and Wolf, R. A. (1968). *J. Geophys. Res.* **73**, 5561.

Eyni, M., and Kaufman, A. S. (1971). *Planet. Space Sci.* **19**, 1609.

Fisk, L. A., and Van Hollebeke, M. (1972). *J. Geophys. Res.* **77**, 2232.

Forslund, D. W. (1970). *J. Geophys. Res.* **75**, 17.

Fredricks, R. W. (1969). *J. Geophys. Res.* **74**, 2919.

Goldstein, M. and Eviatar, A. (1973). *Astrophys. J.* **179**, 627.

Gosling, J. T. (1971). *Solar Phys.* **17**, 499.

Gosling, J. T. (1972). *In* "Solar Wind" (C. P. Sonett, P. J. Coleman, Jr., and J. M. Wilcox, eds.), p. 202. NASA SP-308. U.S. Gov. Printing Office, Washington, D.C.

Gosling, J. T., and Bame, S. J. (1972). *J. Geophys. Res.* **77**, 12.

Gosling, J. T., Hundhausen, A. J., Pizzo, V., and Asbridge, J. R. (1972a). *J. Geophys. Res.* **77**, 5442.

Gosling, J. T., Pizzo, V., Neugebauer, M., and Snyder, C. W. (1972b). *J. Geophys. Res.* **77**, 2744.

Griffel, D. H., and Davis, L., Jr. (1969). *Planet. Space Sci.* **17**, 1009.

Gringauz, K. I., Bezruvkikh, V. V., Ozerov, V. D., and Rybchinskiy, R. E. (1960). *Sov. Phys.— Dokl.* **5**, 361.

Hartle, R. E., and Barnes, A. (1970). *J. Geophys. Res.* **75**, 6915.
Hartle, R. E., and Barnes, A. (1972). *In* " Solar Wind " (C. P. Sonett, P. J. Coleman, Jr., and J. M. Wilcox, eds.), p. 248. NASA SP-308. U.S. Gov. Printing Office, Washington, D.C.
Hartle, R. E., and Sturrock, P. A. (1968). *Astrophys. J.* **151**, 1155.
Hewish, A. (1972). *In* " Solar Wind " (C. P. Sonett, P. J. Coleman, Jr., and J. M. Wilcox, eds.), p. 477. NASA SP-308. U.S. Gov. Printing Office, Washington, D.C.
Hollweg, J. V. (1970). *J. Geophys. Res.* **75**, 2403.
Hollweg, J. V. (1971). *J. Geophys. Res.* **76**, 7491.
Hollweg, J. V. (1972). *Cosmic Electrodyn.* **2**, 423.
Hollweg, J. V. (1973a). *Astrophys. J.* **181**, 547.
Hollweg, J. V. (1973b). *J. Geophys. Res.* **78**, 3643.
Hollweg, J. V., and Jokipii, J. R. (1972). *J. Geophys. Res.* **77**, 3311.
Hollweg, J. V., and Völk, H. J. (1970). *J. Geophys. Res.* **75**, 5297.
Holzer, T. E. (1972). *J. Geophys. Res.* **77**, 5407.
Holzer, T. E., and Axford, W. I. (1970). *Annu. Rev. Astron. Astrophys.* **8**, 31.
Holzer, T. E., and Leer, E. (1973). *J. Geophys. Res.* **78**, 3199.
Howard, R. (1972). *In* " Solar Wind " (C. P. Sonett, P. J. Coleman, Jr., and J. M. Wilcox, eds.), p. 3. NASA SP-308. U.S. Gov. Printing Office, Washington, D.C.
Hundhausen, A. J. (1968). *Space Sci. Rev.* **8**, 690.
Hundhausen, A. J. (1969). *J. Geophys. Res.* **74**, 5810.
Hundhausen, A. J. (1970). *Rev. Geophys.* **8**, 729.
Hundhausen, A. J. (1971). *EOS Trans. AGU* **52**, 915.
Hundhausen, A. J. (1972). "Coronal Expansion and the Solar Wind." Springer, New York.
Hundhausen, A. J. (1973). *J. Geophys. Res.* **78**, 1528.
Hundhausen, A. J., Bame, S. J., Asbridge, J. R., and Sydoriak, S. J. (1970). *J. Geophys. Res.* **75**, 4643.
James, J. C. (1970). *Solar Phys.* **12**, 143.
Jeffrey, A., and Taniuti, T. (1964). "Non-Linear Wave Propagation." Academic Press, New York.
Jockers, K. (1970). *Astron. Astrophys.* **6**, 219.
Jokipii, J. R. (1968). *Astrophys. J.* **152**, 799.
Jokipii, J. R. (1971). *Rev. Geophys.* **9**, 27.
Jokipii, J. R., and Davis, L., Jr. (1969). *Astrophys. J.* **156**, 1101.
Kadomtsev, B. B. (1965). "Plasma Turbulence." Academic Press, New York.
Kennel, C. F., and Scarf, F. L. (1968). *J. Geophys. Res.* **73**, 6149.
Kuperus, M. (1969). *Space Sci. Rev.* **9**, 713.
Leer, E., and Axford, W. I. (1972). *Solar Phys.* **23**, 238.
Leer, E., and Holzer, T. E. (1972). *J. Geophys. Res.* **77**, 4035.
Leighton, R. B., Noyes, R. W., and Simon, G. W. (1962). *Astrophys. J.* **135**, 474.
Lüst, R. (1967). *In* "Solar-Terrestrial Physics" (J. W. King and W. S. Newman, eds.), p. 1. Academic Press, New York.
Marlborough, J. M., and Roy, J. R. (1970). *Astrophys. J.* **160**, 221.
Mestel, L. (1961). *Mon. Notic. Roy. Astron. Soc.* **122**, 473.
Mestel, L. (1968). *Mon. Notic. Roy. Astron. Soc.* **138**, 359.
Mihalov, J. D., and Wolfe, J. H. (1971). *Cosmic Electrodyn.* **2**, 326.
Modisette, J. L. (1967). *J. Geophys. Res.* **72**, 1521.
Montgomery, M. D. (1972a). *In* "Solar Wind" (C. P. Sonett, P. J. Coleman, Jr., and J. M. Wilcox, eds.), p. 208. NASA SP-308. U.S. Gov. Printing Office, Washington, D.C.
Montgomery, M. D. (1972b). *In* "Cosmic Plasma Physics" (K. Schindler, ed.), p. 61. Plenum, New York.
Montgomery, M. D., Bame, S. J., and Hundhausen, A. J. (1968). *J. Geophys. Res.* **73**, 4999.

Ness, N. F. (1968). *Annu. Rev. Astron. Astrophys.* **6**, 79.

Neugebauer, M., and Snyder, C. W. (1966). *In* "The Solar Wind" (R. J. Mackin, Jr. and M. Neugebauer, eds.), p. 3. Pergamon, Oxford.

Newkirk, G., Jr. (1967). *Annu. Rev. Astron. Astrophys.* **5**, 213.

Newkirk, G., Jr. (1972). *In* "Solar Wind" (C. P. Sonett, P. J. Coleman, Jr., and J. M. Wilcox, eds.), p. 11. NASA SP-308. U.S. Gov. Printing Office, Washington, D.C.

Newman, R. C., and Axford, W. I. (1968). *Astrophys. J.* **151**, 1145.

Parker, E. N. (1958a). *Astrophys. J.* **128**, 664.

Parker, E. N. (1958b). *Phys. Rev.* **109**, 1874.

Parker, E. N. (1963). "Interplanetary Dynamical Processes." Wiley (Interscience), New York.

Parker, E. N. (1965). *Space Sci. Rev.* **4**, 666.

Parker, E. N. (1967). *In* "Solar-Terrestrial Physics" (J. W. King and W. S. Newman, eds.), p. 45. Academic Press, New York.

Parker, E. N. (1969). *Space Sci. Rev.* **9**, 325.

Parker, E. N. (1971). *Rev. Geophys.* **9**, 825.

Parker, E. N. (1972). *In* "Solar Wind" (C. P. Sonett, P. J. Coleman, Jr., and J. M. Wilcox, eds.), p. 161. NASA SP-308. U.S. Gov. Printing Office, Washington, D.C.

Patterson, B. R. (1971). Lawrence Berkeley Lab. Prepr. LBL-45.

Perkins, F. W. (1973). *Astrophys. J.* **179**, 637.

Pneuman, G. W. (1966). *Astrophys. J.* **145**, 800.

Pneuman, G. W., and Kopp, R. A. (1970). *Solar Phys.* **13**, 176.

Scarf, F. L. (1970). *Space Sci. Rev.* **11**, 234.

Schatten, K. H. (1972). *In* "Solar Wind" (C. P. Sonett, P. J. Coleman, Jr., and J. M. Wilcox, eds.), p. 65. NASA SP-308. U.S. Gov. Printing Office, Washington, D.C.

Scholer, M., and Belcher, J. W. (1971). *Solar Phys.* **16**, 472.

Schubert, G., and Coleman, P. J., Jr. (1968). *Astrophys. J.* **153**, 943.

Schulz, M., and Eviatar, A. (1972). *Cosmic Electrodyn.* **2**, 402.

Shklovsky, I. S., Moroz, V. I., and Kurt, V. G. (1961). *Sov. Astron.—AJ* **4**, 871.

Siscoe, G. L., and Finley, L. T. (1969). *Solar Phys.* **9**, 452.

Siscoe, G. L., and Finley, L. T. (1970). *J. Geophys. Res.* **75**, 1817.

Siscoe, G. L., and Finley, L. T. (1972). *J. Geophys. Res.* **77**, 35.

Sonett, C. P., Coleman, P. J., Jr., and Wilcox, J. M., eds. (1972). "Solar Wind," Proc. 1971 Solar Wind Conf. NASA SP-308. U.S. Gov. Printing Office, Washington, D.C.

Spreiter, J. R., and Alksne, A. Y. (1970). *Annu. Rev. Fluid Mech.* **2**, 313.

Stepanov, K. N. (1958). *Sov. Phys.—JETP* **7**, 892.

Stix, T. H. (1962). "The Theory of Plasma Waves." McGraw-Hill, New York.

Sturrock, P. A., and Hartle, R. E. (1966). *Phys. Rev. Lett.* **16**, 628.

Tajiri, M. (1967). *J. Phys. Soc. Jap.* **22**, 1482.

Tan, M., and Abraham-Shrauner, B. (1972). *Cosmic Electrodyn.* **3**, 71.

Tanenbaum, A. S., Wilcox, J. M., Frazier, E. N., and Howard, R. (1969). *Solar Phys.* **9**, 328.

Thomas, G. E. (1972). *In* "Solar Wind" (C. P. Sonett, P. J. Coleman, Jr., and J. M. Wilcox, eds.), p. 668. NASA SP-308. U.S. Gov. Printing Office, Washington, D.C.

Thomas, G. E., and Krassa, R. F. (1971). *Astron. Astrophys.* **11**, 218.

Tidman, D. A., and Krall, N. A. (1971). "Shock Waves in Collisionless Plasmas." Wiley (Interscience), New York.

Unti, T. W. J., and Neugebauer, M. (1968). *Phys. Fluids* **11**, 563.

Valley, G. C. (1971). *Astrophys. J.* **168**, 251.

Völk, H. J., and Alpers, W. (1973). *Astrophys. Space Sci.* **20**, 267.

Wallis, M. (1971a). Unnumbered Rep. Roy. Inst. Technol., Stockholm.

Wallis, M. (1971b). *Nature (London)* **233**, 23.

Weber, E. J. (1970). *Solar Phys.* **13**, 240.

Weber, E. J. (1972). *In* "Solar Wind" (C. P. Sonett, P. J. Coleman, Jr., and J. M. Wilcox, eds.), p. 268. NASA SP-308. U.S. Gov. Printing Office, Washington, D.C.

Weber, E. J., and Davis, L., Jr. (1967). *Astrophys. J.* **148**, 217.

Weber, E. J., and Davis, L., Jr. (1970). *J. Geophys. Res.* **75**, 2419.

Whang, Y. C. (1971). *Astrophys. J.* **169**, 369.

Wilcox, J. M. (1968). *Space Sci. Rev.* **8**, 258.

Wilcox, J. M., and Ness, N. F. (1965). *J. Geophys. Res.* **75**, 6366.

Wolfe, J. H. (1972). *In* "Solar Wind" (C. P. Sonett, P. J. Coleman, Jr., and J. M. Wilcox, eds.), p. 170. NASA SP-308. U.S. Gov. Printing Office, Washington, D.C.

Wolff, C. L., Brandt, J. C., and Southwick, R. G. (1971). *Astrophys. J.* **165**, 181.

# Multiphoton Ionization of Atoms

## J. S. BAKOS

*Central Research Institute for Physics*
*Department of Quantum Electronics*
*Budapest, Hungary*

# I. HISTORICAL BACKGROUND

## A. Multiphoton Elementary Process

One of the multiphoton elementary processes has been known since the beginning of this century and has been used to investigate the structure of matter. This is the Raman scattering of light process, and it was first interpreted as a two-photon process by Kramers and Heisenberg (1925) in their quantum mechanical dispersion theory: one photon of frequency $\omega$ is absorbed, and the second photon of frequency $\omega' \neq \omega$ is emitted by the atomic system. The resonance scattering of the radiation (the special case of Raman scattering, $\omega = \omega'$), i.e. the second of the multiphoton processes other than Raman scattering, is explained on the same theoretical basis as the dispersion.

In addition to these multiphoton processes attention has been drawn to a new two-photon process (Kramers and Heisenberg, 1925)—the stimulated two-photon emission—which had not been observed before. Here, the presence of radiation of frequency $\omega$ stimulates the emission of a photon of the same frequency, together with the simultaneous emission of a second photon of frequency $\omega'$ by the atomic system as a result of its being in an excited state, so that $\omega + \omega' = \omega_{mn}$ which is the frequency of the transition.

The transition probabilities of the Raman and resonance scattering and stimulated two-photon emission can be calculated using the same order of approximation of the perturbation theory, and these probabilities were found to be of the same orders of magnitude.

Göppert-Mayer (1929, 1931) investigated the last process in more detail and simultaneously the inverse one, i.e. the two-photon excitation or absorption process, in which two photons of frequency $\omega$ and $\omega'$ are absorbed. The probability of this last process is also of the same order of magnitude as the probabilities of the previously mentioned processes but this probability depends quadratically on the light power density in contrast with the linear dependence on the light power density of the probability of Raman scattering, resonance scattering, and the stimulated two-photon emission.

In the visible region of the spectrum the photon-induced emission and the two-photon absorption were not observed until the advent of the laser. The Raman scattering was used for the investigation of the structure of molecules in the face of great experimental difficulties by using as intense an exciting light as possible in the case of classical light sources.

Multiphoton absorption was first observed by Brossel *et al.* (1953) in the radio frequency region of the spectrum. Using the method of optical pumping (Cohen-Tannoudji and Kastler, 1966), population differences were accomplished between the Zeeman sublevels of an atomic state. The simultaneously present radio frequency field induces multiphoton transitions between these sublevels.

In the visible region of the spectrum the multiphoton process was first observed by Franken et al. (1961) as the second harmonic generation in quartz, where two photons of frequency $\omega_1$ are absorbed and one photon of frequency $\omega_2 = 2\omega_1$ is emitted.

With the rapid development of laser techniques the number of invented multiphoton processes has increased sharply. Two-photon absorption was observed in $CaF_2 : Eu^{2+}$ by Kaiser and Garrett (1961); in cesium vapor by Abella (1962); in organic crystals by Peticolas et al. (1963); in $Cs_2$ by Giordmaine and Howe (1963) etc.

Maker et al. (1964) reported the appearance of a spark at the focal point of a focusing lens using a $Q$-switched ruby laser. The ignition process of the spark is a new type of elementary multiphoton process—the multiphoton photoemission from atoms, viz. multiphoton ionization. This new process, theoretically predicted by Bunkin and Prokhorov (1964), showed that every multiphoton process taking place in the atom will be surpassed by the multiphoton ionization if the light intensity increases.

## B. Multiphoton Ionization

The first rigorous theoretical treatment of multiphoton ionization was made by Keldish (1964). He demonstrated the intrinsic unity of the processes of multiphoton ionization and of the field emission in an alternating field by generalizing Oppenheimer's (1928) theory of field emission in a static field.

The observation of this new multiphoton process—multiphoton ionization—was first reported by Voronov and Delone (1965) while investigating the multiphoton ionization of Xe atoms.

Work on the ionization of negative ions of iodine $I^-$, on the multiphoton photodetachment, was published almost simultaneously by Hall et al. (1965) and had been theoretically predicted by Geltman (1963).

The first experimental observation of multiphoton ionization of rare gases showed the influence of the distortion of the atomic structure—the shifting and broadening of the excited atomic levels in the ionizing strong laser field on the multiphoton ionization. Therefore, following many thorough experimental investigations of the ionization of noble gases (Voronov et al., 1966a,b; Voronov and Delone, 1966; Bistrova et al., 1967; Agostini et al., 1968a,b; see details in Section III), works were published on the multiphoton ionization of alkaline atoms (Delone et al., 1969, 1971a; Fox et al.,

1971; Held *et al.*, 1972). The ionization potential of these atoms is considerably smaller than that of the rare gases and therefore the number of ionizing quanta is also smaller, and the light field needed for the ionization is consequently smaller.

Multiphoton ionization could be investigated without distortion of the atomic structure in this case and the results (see Section II) compared with the simplest, first, nonvanishing approximation of the perturbation theory of the process (Bebb, 1966, 1967; Morton, 1967; Rapoport, 1973; Delone, 1973).

As has been mentioned the distortion of the atomic structure plays an especially important role in the multiphoton ionization of rare gases. The influence of distortion on multiphoton ionization can be investigated in the case of multiphoton resonance (see Section IV). This means that the energy of one atomic state coincides with the energy of some quanta. The multiphoton ionization rate, as Keldish (1964) theoretically demonstrated, can be expressed as the product of the rate of multiphoton excitation and that of subsequent ionization of the resonant level (two-step process). The rate of the multiphoton excitation is especially sensitive to the shift of the resonance level, i.e. the distortion of the atomic structure. Much experimental work has been devoted to revealing the role which multiphoton resonance plays in the multiphoton ionization of rare gases (Baravian *et al.*, 1970; Bakos *et al.*, 1972a,c; Delone and Delone, 1969; Held *et al.*, 1973).

In a very intense field the distortion of the atomic structure is expected. It is reflected by the complicated dependence of the ionization rate on light intensity as theoretically demonstrated by Reiss. Experimental verification of the theory has not been carried out yet (Section V).

Though the strong relation between multiphoton ionization and field emission had already been revealed in the first theoretical paper dealing with these problems (Keldish, 1964) experimental verification is still lacking (see Section V).

The characteristics of multiphoton ionization reflect not only the feature of the atomic structure but the feature of the ionizing field. So the rate of multiphoton ionization depends not only on the field strength (which is trivial) and the frequency (resonance ionization), but also on the polarization and the correlation functions of high order of the ionizing light (see Section VI). Apart from the numerous theoretical papers (Mollow, 1968; Lambrapoulos, 1968, 1972a,b, 1973; Agarwal, 1970; Reiss, 1972a) there is only one piece of experimental work that has measured the dependence of the multiphoton ionization rate of cesium atoms on the polarization state of the ionizing radiation, viz. Fox *et al.* (1971) and Kogan *et al.* (1971).

Experimental investigation of the role of the light field statistics has been absent until now.

## II. Lowest Order Perturbation Theory of Multiphoton Ionization

### A. Formal Theory

Multiphoton ionization—the special kind of multiphoton transition—can be described at a moderately light power density using the formalism of the time-dependent perturbation theory (Messiah, 1965). In the following it is supposed that the light power density is so small that for the calculation of the transition probability of the multiphoton process it is sufficient to take into account the approximation of lowest nonvanishing order of the perturbation theory.

The Hamiltonian of the system (atom and radiation field) can be divided into the interaction Hamiltonian $V$ and the Hamiltonian $H_0$ of separate systems of field and atom without any interaction between them

$$H = H_0 + V. \tag{2.1}$$

The Schrödinger equation of motion is represented as

$$i(\partial/\partial t)U_I(t) = V_I \cdot U_I(t) \tag{2.2}$$

in the interaction representation (Heitler, 1954). $U(t)$ is the time development operator of the system so that

$$U_I(t) = e^{iH_0 t}U(t) \tag{2.3}$$

and

$$V_I = e^{iH_0 t} \cdot V \cdot e^{-iH_0 t}. \tag{2.4}$$

The initial condition is

$$U_I(0) = 1. \tag{2.5}$$

In this section the Hartree atomic units are used (Bethe and Salpeter, 1957). At time $t$ the state $|t\rangle$ of the system can be expressed as

$$|t\rangle = U(t)|g\rangle, \tag{2.6}$$

where $|g\rangle$ is the initial state. The solution of the Schrödinger equation (2.2) is the following

$$U_I(t) = 1 - i\int_0^t V_I(t')U_I(t')\,dt', \tag{2.7}$$

where the initial condition (2.5) was taken into account. The integral equation (2.7) can be solved by interation which leads to the result

$$U_I(t) = 1 - \sum_{k=1}^{\infty} U_I^{(k)}(t), \tag{2.8}$$

where

$$U_I^{(k)}(t) = (-i)^k \int_0^t dt_k \int_0^{t_k} dt_{k-1} \cdots \int_0^{t_2} dt_1 \, V_I(t_k) \cdots V_I(t_1). \tag{2.9}$$

In the nonrelativistic case of one electron the interaction Hamiltonian $V$ can be expressed as

$$V = (1/c)\mathbf{p}\mathbf{A} + (1/2c^2)\mathbf{A}^2, \tag{2.10}$$

where $\mathbf{p}$ is the momentum of the electron, $\mathbf{A}$ is the vector potential of the light wave, and $c$ is the velocity of light. Suppose that the vector potential is uniform in the volume of the atom then we can use the electric dipole approximation, and the interaction Hamiltonian is represented by the familiar form (Göppert-Mayer, 1931)

$$V = \mathbf{r}\mathbf{E}. \tag{2.11}$$

$\mathbf{E}$ is the electric field of the light wave which is expressed in the form of the usual series of the orthogonal eigenwaves, $\mathbf{A}(\mathbf{r})$ and $\mathbf{A}^*(\mathbf{r})$ (Heitler, 1954), i.e.

$$\mathbf{E} = (i/c)\sum_{\lambda} (\omega_{\lambda} q_{\lambda} \mathbf{A}_{\lambda} + \omega_{\lambda} q_{\lambda}^+ \mathbf{A}_{\lambda}^*); \tag{2.12}$$

$q, q^+$ are the photon annihilation and creation operators

$$q_{\lambda} | n_{\lambda} \rangle = (n_{\lambda}/2\omega_{\lambda})^{1/2} | n_{\lambda} - 1 \rangle,$$
$$q_{\lambda}^+ | n_{\lambda} \rangle = [(n_{\lambda} + 1)/2\omega_{\lambda}]^{1/2} | n_{\lambda} + 1 \rangle, \tag{2.13}$$

$n_{\lambda}$ is the photon number of the eigenwave

$$\mathbf{A}_{\lambda} = (4\pi c^2)^{1/2} \exp(i\mathbf{K}_{\lambda} \mathbf{r}) \cdot \mathbf{e}_{\lambda}, \tag{2.14}$$

$\mathbf{K}_{\lambda}$ is the wave vector, and $\mathbf{e}_{\lambda}$ is the polarization unit vector.

The $k$th-order approximation of the time development operator $[U_I^{(k)}(t)]$ contains the product of the creation and annihilation operators. The number of these operators in the products is $k$. This approximation of the time development operator describes the development of the system when the number of emitted or absorbed photons is $k$.

In the following we are interested in the process of the absorption of $k_0$ photons and it is supposed that the transition from the initial state $|g\rangle$ to the final one $|f\rangle$ is not possible by absorption of fewer than $k_0$ photons due to the conservation of energy. $|f\rangle$ and $|g\rangle$ are the energy eigenstates of the unperturbed separate systems of atom and field. Then the time development of the system is described in the lowest nonvanishing order by the $k_0$th order approximation of the time development operator which contains the product of $k_0$ annihilation operators. The probability that the system is in the final state $|f\rangle$ at the time $t$ after switching on the perturbation is presented as the square of the projection of the state $|t\rangle$ to the final state, i.e.

$$P_{f,g}^{(k_0)}(t) = |\langle f | U_I^{(k_0)}(t) | g \rangle|^2. \tag{2.15}$$

Using the unperturbed energy representation $|m\rangle$, that is the energy eigenfunctions of the isolated systems, atom and field, as the basic vectors, the expression (2.9) will have the form (Bebb and Gold, 1966)

$$U_I^{(k_0)}(t) = (-i)^{k_0} \sum_{m_{k(0)-1}} \cdots \sum_{m_1} \int_0^t dt_{k_0} \int_0^{t_{k(0)}} dt_{k_0-1} \cdots \int_0^{t_2} dt_1$$
$$\cdot V_I(t_{k_0}) | m_{k_0-1} \rangle \langle m_{k_0-1} | V_I(t_{k_0-1}) \cdots \langle m_1 | V_I(t_1). \tag{2.16}$$

It should be noticed here that in the case of continuous eigenvalues of the energy the summations have to be replaced by integrals.

The integrations are performed by taking minus infinity as the lower limits of the integrals except the last one (Heitler, 1954; Bebb and Gold, 1966; also see Section VI), and the probability that the system is in the final state is

$$P_{f,g}^{(k_0)}(t) = f(\omega_{f,g}) | V_{f,g}^{(k_0)} |^2, \tag{2.17}$$

where

$$f(\omega_{f,g}) = (e^{-i\omega_{f,g}t} - 1)/\omega_{f,g}. \tag{2.18}$$

$f(\omega_{f,g})$ is the function expressing the conservation of energy and the matrix element of the $k_0$th order of the interaction Hamiltonian is

$$V_{f,g}^{(k_0)} = \sum_{m_{k(0)-1}} \cdots \sum_{m_1} \langle f | V | m_{k_0-1} \rangle \langle m_{k_0-1} | V | \cdots | V | g \rangle / \omega_{k_0-1,g} \cdots \omega_{1,g}; \tag{2.19}$$

$\omega_{i,k} = \omega_i - \omega_k$ where $\omega_i = \mathscr{E}_i/\hbar$ is the frequency and $\mathscr{E}_i = \mathscr{E}_{a_i} + (n_p - i)\omega_p$ is the energy of the state $i$. $\mathscr{E}_{a_i}$ is the energy of the atom in the $i$ state, and $n_p$ the number of photons of frequency $\omega_p$ of the ionizing radiation. It is supposed that only one mode of the field ($p$) is excited, that the light is linearly polarized, and, using expression (2.11) that the time derivative of the probability $P(t)$, i.e. the transition rate, is given as

$$W_{f,g}^{(k_0)} = 2\pi(F\omega_p)^{k_0} | r_{a_f, a_g}^{(k_0)} |^2 \delta(\omega_{f,g}), \tag{2.20}$$

where $F$ is the photon flux in atomic units $[F_0 = (c/2\pi)(1/a_0^3)$, where $a_0$ is the Bohr radius] and the transition matrix element of $k_0$th order is

$$r^{(k_0)}_{a_f, a_g} = \sum_{a_{k(0)-1}} \cdots \sum_{a_1} \langle a_f | r | a_{k_0-1} \rangle$$

$$\cdot \frac{\langle a_{k_0-1} | r | a_{k_0-2} \rangle \cdots \langle a_1 | r | a_g \rangle}{[\omega_{a_{k(0)-1}, a_g} - (k_0 - 1)\omega_p] \cdots [\omega_{a_1, a_g} - \omega_p]}. \quad (2.21)$$

$|a_i\rangle$ is the wavefunction of the atom when in the state $i$. The letter "$a$" referring to the atom will be dropped in the following if no confusion arises. For instance $|a_i\rangle$ will be replaced simply by $|i\rangle$.

The diagrammatic explanation by Fig. 1 of the process gives a clear idea about the mechanism of the multiphoton absorption. With the absorption of the first photon the atom makes a transition from the initial state to the first virtual state.

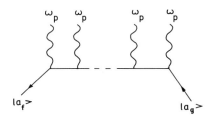

FIG. 1. Diagram of the multiphoton process.

The denominator $\omega_{a_1, a_g} - \omega_p$ in the first sum does not have to be equal to zero. Consequently the energy is not conserved in this transition. The second absorbed photon transfers the atom from the first virtual state to the second one, etc. After $k_0$ photon absorptions the atom will be in the final state. The energy is conserved for the process as a whole (Bonch-Bruevich and Khodovoi, 1965).

Let us specialize in the following, the general expression (2.20) for the $k_0$-photon transition rate of the bound-free transition, for the $k_0$-photon ionization of the atom. The final state vector of the ionization is considered to be the solution of the Schrödinger equation in the Coulomb field of the nucleus (Baz et al., 1971) with given wavenumber $\kappa$ (i.e. momentum) of the outgoing electron at infinity. (The solution is the superposition of incoming spherical and outgoing plane waves.) The wavefunction is normalized to unity in the $\kappa$ space. Then the density of states of given energy $(\mathscr{E}_\kappa)$ of the system is

$$\rho(\mathscr{E}_\kappa) = [\kappa/(2\pi)^3] \, d\Omega_\kappa . \quad (2.22)$$

With this density of states the expression (2.20), for the ionization rate which takes place in the final state of a given momentum, has to be multiplied in order to get the ionization rate in the state of given energy. The resultant expression for the differential cross section of the $k_0$-photon ionization is

$$(d\sigma_{f,g}^{(k_0)}/d\Omega_\kappa)(\theta_\kappa, \varphi_\kappa) = (\alpha/2\pi)(F\omega_p)^{k_0-1} |r_{f,g}^{(k_0)}|^2 \kappa\omega_p. \qquad (2.23)$$

$\alpha$ is the fine structure constant. The rate of the $k_0$-photon ionization in the state of given energy is

$$W_{f,g}^{(k_0)} = \sigma_{f,g}^{(k_0)} \cdot F, \qquad (2.24)$$

where $\sigma_{f,g}^{(k_0)}$ is the total cross section of the ionization. $d\Omega_\kappa$ is the solid angle and $\theta_\kappa$, $\varphi_\kappa$ are the colatitude and the asimuth of the direction of the wave vector of the emitted electron. The wavenumber $|\kappa|$ of the emitted electron can be calculated by the equation

$$\mathscr{E}_{f,g} = \mathscr{E}_0 - k_0\omega_p + \kappa^2 = 0, \qquad (2.25)$$

expressing the rule of the conservation of energy; $\mathscr{E}_0$ is the ionization energy.

From the experimental point of view a more useful expression for the multiphoton ionization rate can be obtained by transforming expression (2.24) into the following form:

$$W_{f,g}^{(k_0)} = \beta^{(k_0)}(\omega_p)F^{k_0}, \qquad (2.26a)$$

where $\beta^{(k_0)}$ may be regarded as the generalized cross section of the multiphoton ionization and

$$\beta^{(k_0)}(\omega_p) = (1/4\pi^2)\omega_p^{k_0} |r_{f,g}^{(k_0)}|^2 \kappa. \qquad (2.26b)$$

Another form of the multiphoton ionization rate used especially in the experimental work can be given as

$$W_{f,g}^{(k_0)} = \beta_I^{(k_0)}I^{k_0}, \qquad (2.26c)$$

where $I$ is the light power density related to the photon flux by the relation

$$I = F\mathscr{E}_p.$$

$\mathscr{E}_p$ is the energy of the photon. The subscript $I$ of the generalized cross section $\beta_I^{(k_0)}$ refers to the using of the light power density in the expression of the multiphoton ionization rate.

While calculating the cross section using expression (2.23) or the rate of the $k_0$-photon ionization according to the expression (2.24), the main difficulty arises in the calculation of the transition matrix element $r_{f,g}^{(k_0)}$ [see expression (2.21)] because of the infinite summations over the energy eigenstates of the unperturbed atom. The situation is especially difficult in the case of the last summation because the energy of the $k$th virtual state

$\mathscr{E}_k = \mathscr{E}_{ag} + k\omega_p$, i.e. after $k < k_0$ photons are absorbed in the region of the spectrum where the atomic states are densely packed. Consequently the denominators of the terms of the last sum are small. Neglecting any terms in the sum, even if the one-photon matrix elements are small, can strongly influence the value of the resultant matrix element of $k_0$th order because of the great density of states. Furthermore, difficulty arises on taking into account the contribution of the continuous eigenstates of the atom of positive energy. Since the summations have to be performed over the complete system of the energy eigenstates of the atom, which also contains the states of positive energy, i.e. the continuous spectrum.

Because of this situation considerable effort has been made in theory to find a correct method of performing the summations needed. Four of these methods are presented, in common with that of multiphoton ionization, with specific advantages for different atoms.

## B. The Method of Bebb and Gold

The most simple approximative method for the calculation of the $k_0$-photon transition matrix elements is that of Bebb and Gold (1965, 1966; Gold and Bebb, 1965; Bebb, 1966, 1967).

The turning point of the calculation is the notion of the average frequency, the use of which can eliminate the summations. The average frequency $\bar{\omega}(v)$ is defined by the following equation

$$\sum_{a_v} \frac{\langle f \,|\, r^{k_0 - v} \,|\, v\rangle\langle v \,|\, r^v \,|\, g\rangle}{\omega_{v,\,g} - v\omega_p} = \frac{\langle f \,|\, r^{k_0} \,|\, g\rangle}{\bar{\omega}(v) - v\omega_p}. \tag{2.27}$$

Using expression (2.27) the transition matrix element of the $k_0$th order [expression (2.21)] can be simplified to the form

$$r_{f,\,g}^{(k_0)} = \langle f \,|\, r^{k_0} \,|\, g\rangle / \prod_{v=1}^{k_0-1} [\bar{\omega}(v) - v\omega_p]. \tag{2.28}$$

Continuing the procedure the general average frequency $(\bar{\omega})$ can be defined by the equation

$$\prod_{v=1}^{k_0-1} [\bar{\omega}(v) - v\omega_p] = \prod_{v=1}^{k_0-1} (\bar{\omega} - v\omega_p). \tag{2.29}$$

Substituting Eq. (2.29) into expression (2.28), the $k_0$th-order transition matrix element becomes

$$r_{f,\,g}^{(k_0)} = \langle f \,|\, r^{k_0} \,|\, g\rangle / \prod_{v=1}^{k_0-1} (\bar{\omega} - v\omega_p). \tag{2.30}$$

In the calculation of the average frequency only the dominant term was retained.

If nearly resonant terms occurred in the sum of the $k_0$th-order transition matrix element these resonant terms became separated. Suppose the resonances are in the $k$th order, the resultant expression for the $k_0$th-order transition matrix element is

$$r_{f,g}^{(k_0)} = \sum_k \langle f \mid r^{k_0-k} \mid k \rangle \langle k \mid r^k \mid g \rangle / (\omega_{k,g} - v\omega_p) \cdot 1 / \prod_{\substack{v=1 \\ v \neq k}}^{k_0-1} [\bar{\omega}(v) - v\omega_p], \quad (2.31)$$

where the new average frequencies are defined, if $v > k$, as

$$\sum_v \langle f \mid r^{k_0-k-v} \mid v \rangle \langle v \mid r^v \mid k \rangle / (\omega_{v,g} - v\omega_p)$$

$$= \langle f \mid r^{k_0-k} \mid k \rangle / \bar{\omega}_k(v) - v\omega_p, \quad (2.32)$$

and if $k > v$,

$$\sum_v \langle k \mid r^{k-v} \mid v \rangle \langle v \mid r^v \mid g \rangle / \omega_{v,g} - v\omega_p = \langle k \mid r^k \mid g \rangle / \bar{\omega}_k(v) - v\omega_p. \quad (2.33)$$

If more than one resonance or resonances of different order occur, the procedure is continued. The resonant terms are separated by analogy with expression (2.31) and new average frequencies are defined.

Near the resonances the finite level width ($\gamma_k$) of the atom also has to be taken into account by adding to every frequency difference in the denominators the factor $i\gamma_{j,g}$. The resulting expression for the transition matrix element of $k_0$ th order is

$$r_{f,g}^{(k_0)} = \sum_{a_{k_0-1}} \cdots \sum_{a_1} \langle f \mid r \mid k_0 - 1 \rangle \cdot \frac{\langle k_0 - 1 \mid r \mid k_0 - 2 \rangle \cdots \langle 1 \mid r \mid g \rangle}{[\omega_{k_0-1,g} - (k_0 - 1)\omega_p + i\gamma_{k_0-1,g}] \cdots [\ \ ]}.$$

$$(2.34)$$

Using the method of Bebb and Gold the level widths are taken into account only in the denominators of the separated dominant resonance terms.

The advantage of the method is that the number of infinite summations and the number of transition matrix elements to be calculated is greatly reduced using the notion of average frequency.

The result of the calculation of the average frequency ($\bar{\omega}$) showed that the numerical value is not affected by the dominant resonances occurring in the sum of the transition matrix element of $k_0$th order, $\bar{\omega}$ is approximately equal to the value of the first excitation energy of the atom.

This method was used for the calculation of the four- to twelve-photon ionization rate from the ground state of the hydrogen atom and the two-photon ionization rate of the metastable hydrogen (Bebb and Gold, 1966).

The ionization rate divided by $F^{k_0}$ is proportional to the transition matrix element of $k_0$th order [see expressions (2.23), (2.24), and (2.26a), (2.26)] and shows typical resonance-type dependence on the energy of the laser quanta. This function $[\beta^{(k_0)}(\mathscr{E}_p)]$, the generalized cross section of the multiphoton ionization as the result of the calculation, is plotted in Fig. 2 for

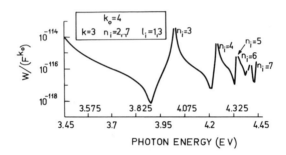

FIG. 2. The generalized cross section of four-photon ionization $\beta^{(k_0)}(\mathscr{E}_p) = W_{f,g}^{(k_0)}/F^{k_0}$ of the hydrogen atom in the ground state (Bebb and Gold, 1966). The photon energy $(\mathscr{E}_p)$ is plotted in electron volts.

the case of four-photon ionization of the hydrogen atom. The resonances are in the third $(k = 3)$ order at the principal quantum number $n_i = 2$ to 7. The contributions of the subsequent resonances to the sum have different signs which result in their cancellation and deep valleys between the resonances.

The multiphoton ionization rate was also calculated for the rare gas atoms (Bebb and Gold, 1966). The hydrogenic functions scaled to the atomic radius were used as the wavefunctions of the atoms. The final state wavefunction was regarded as a plane wave.

It is characteristic of the method that only a finite number of the leading resonant terms are separated at the different order $k$ of the summations. What is important is that the contribution of the continuous spectrum is never taken into account. The neglected terms have a great influence on the value of the $k_0$th order transition matrix element far from the resonances, in the valleys, resulting in several orders of magnitude discrepancy between the correct value and the result of the approximation (see Section II,D).

The deviation from the correct value is especially large at high-order processes. Attention should be drawn also to the arbitrariness in evaluating the average frequencies.

## C. The Method of Morton

It is felt that some consideration should be given here to the approximative method of computation made by Morton (1967).

Contrary to the method of Bebb and Gold, Morton averaged the numerators in the expression (2.21) for the transition matrix element, substituting only one effective atomic level instead of the set of levels in each infinite sum. So the one-photon transition matrix element in every sum is supposed to be

$$\langle \,|r|\, \rangle = (1/\Omega)^{1/2},$$

where

$$\Omega = N\omega_p .$$

The summation for the different denominators was performed by him for the leading, dominant, fairly resonant term, being determined after a thorough investigation of the spectrum of the atom.

Morton tabulated the result of the computation for the generalized cross section of the multiphoton ionization for many rare gases, alkalides, and for the hydrogen atom in both the ground and the metastable state.

## D. The Method of Schwartz and Thiemann

### 1. General Description

This method is the generalization of that used by Dalgarno and Lewis (1956) for the calculation of the energy correction term in the second approximation, and reformulated by Schwartz and Thiemann (Schwartz and Thiemann, 1959; Schwartz, 1959) in the calculation of the Lamb shift. In this section only the multiphoton ionization of the hydrogen atom is discussed.

The transition matrix element of $k_0$th order [expression (2.21)] can be transformed into the following form

$$r^{(k_0)}_{f,\,g} = \imath^{(k_0)}_{f,\,g} = \int \psi^{(k_0)}_g \psi_g \, dv, \tag{2.35}$$

where the integration is taken over the volume $(dv)$ and

$$\psi^{(k_0)}_g = \psi^*_f \sum_{a_{k_0-1}} \cdots \sum_{a_1} r|k_0 - 1\rangle\langle k_0 - 1|r \cdots r|1\rangle$$
$$\times \langle 1|r/[\mathscr{E}_{k_0-1,\,g} - (k_0 - 1)\mathscr{E}_p] \cdots [\mathscr{E}_{1,\,g} - \mathscr{E}_p]. \tag{2.36}$$

The wavefunctions of the states are

$$\psi^*_i = \langle a_i|\psi\rangle, \tag{2.37}$$

expressed in the coordinate representation.

The matrix element [expression (2.21)] can be calculated (instead of performing the multiple infinite summations) by solving the differential equation of the wavefunction $\psi_f^{(k_0)}$, which can be deduced from the Schrödinger equation, and by accomplishing the integration in the expression (2.35) afterwards. In the following the differential equation for $\psi_f^{(k_0)}$ is determined for the case of two-photon and later for that of $k_0$-photon ionization, and the ionization cross sections are also determined.

## 2. Two-Photon Case

The differential equation of the wavefunction $\psi_f^{(k_0)}$ was outlined first by Zernik (1964; Zernik and Klopfenstein, 1965) for the case of two-photon ionization of a metastable hydrogen atom. The total cross section of the two-photon ionization of the state characterized by the principal and orbital quantum numbers $n$ and $l$, respectively, can be expressed as

$$\sigma_{\kappa,nl}^{(2)} = (\pi^2\alpha/15)I \cdot \mathscr{E}_p\{C_1|\mathscr{R}_{\kappa l+2,nl}^{(2)}(l+1)|^2$$
$$+ C_2|\mathscr{R}_{\kappa l-2,nl}^{(2)}(l-1)|^2 + C_3|\mathscr{R}_{\kappa l,nl}^{(2)}(l+1)|^2 + C_4|\mathscr{R}_{\kappa l,nl}^{(2)}(l-1)|^2\} \quad (2.38)$$

after performing the integrations of the angular dependent part of the wavefunctions in the transition matrix element (2.21). $I$ is the power density of the light expressed in atomic units ($14.0 \times 10^{16}$ W/cm²). $\mathscr{R}_{f,g}^{(2)}(l_i)$ is the radial matrix element of second order with intermediate state having orbital quantum number $(l_i)$, and it is expressed in the form

$$\mathscr{R}_{\kappa l_f,nl}^{(2)}(l_i) = \lim_{\varepsilon \to 0} \sum_{k > l_i+1}^{\infty} \frac{\int R_{nl}(r)R_{kl_i}(r)r^3\,dr \int R_{kl_i}(r')R_{\kappa l_f}(r')r'^3\,dr'}{\mathscr{E}_{nl} - i\varepsilon - \mathscr{E}_{kl_i} + \mathscr{E}_p}. \quad (2.39)$$

$R_{nl}(r)$, $R_{kl_i}(r)$, and $R_{\kappa l_f}(\kappa, r)$ are the radial wavefunctions of the hydrogen in the initial, intermediary, and final continuum state. It is supposed that the bound state is expressed as

$$|i\rangle = |n_i, l_i\rangle = R_{n_i l_i}(r)Y_{l_i m_i}(\theta, \varphi), \quad (2.40a)$$

where $\theta$, $\varphi$ are the colatitude and the azimuth. $Y_{l_i m_i}(\theta, \varphi)$ are the spherical harmonics. The final state is the superposition of the outgoing plane wave with wave vector $\kappa$ and incoming spherical wave (Gordon, 1928)

$$\psi_f^* = \sum_{l_f} i^{l_f}\pi[2(2l_f + 1)]^{1/2}\kappa^{-1/2}e^{i\eta_{l_f}}R_{\kappa l_f}(\kappa, r)Y_{l_f 0}(\theta_\kappa) \quad (2.40b)$$

and

$$\eta_{l_f} = \arg \Gamma[l_f + 1 - (i/\kappa)]. \quad (2.40c)$$

$\theta_\kappa$ is the colatitude referred to the direction of the outgoing electron.

The constants $C_i$ depend on the orbital quantum number of the initial state and they are the result of the contribution from the angular dependent part of the wavefunctions $Y_{lm}(\theta, \varphi)$ of the atom.

$$C_1 = \frac{2(l+1)(l+2)}{(2l+3)(2l+1)}, \qquad C_2 = \frac{2l(l-1)}{(2l-1)(2l+2)},$$

$$C_3 = \frac{(l+1)(4l^2+8l+5)}{(2l+1)^2(2l+3)}, \qquad C_4 = \frac{l(4l^2+1)}{(2l+1)(2l-1)}.$$

The summation procedure outlined in expression (2.39) for the radial matrix element of second order can be avoided or, preferably, can be performed exactly by using the procedures presented in Eqs. (2.35), (2.36), and (2.37). New radial wavefunctions of second order will be defined by the following equality:

$$R_{nl}^{(2)}(l_i, r) =$$

$$\lim_{\varepsilon \to 0} \sum_{k > l_i+1}^{\infty} rR_{kl_i}(r) \int R_{kl_i}(r')R_{\kappa l_f}(\kappa, r')r'^3 \, dr'/(\mathscr{E}_{nl} - i\varepsilon - \mathscr{E}_{kl_i} + \mathscr{E}_p),$$

$$(2.41)$$

and the transition matrix element of second order can be expressed by analogy with expression (2.35) in the form

$$\mathscr{R}_{\kappa l_f, nl}^{(2)}(l_i) = \int R_{nl}^{(2)}(l_i, r)R_{nl}(r)r^2 \, dr. \qquad (2.42)$$

The radial wavefunction of second order given by expression (2.41) satisfies the differential equation

$$[\mathscr{E}_n - i\varepsilon + \mathscr{E}_p + \tfrac{1}{2}(d^2/dr^2) + (1/r) - \tfrac{1}{2}l_i(l_i+1)/2r^2]R_{nl}^{(2)}(l_i, r)$$

$$= R_{\kappa l_f}(\kappa, r)r^2. \qquad (2.43)$$

This is immediately apparent, while the function $rR_{kl_i}(r)$ is the solution of the Schrödinger equation

$$\frac{1}{2}\frac{d^2}{dr^2}[rR_{kl_i}(r)] + \left[\frac{1}{r} - \frac{l_i(l_i+1)}{2r^2}\right]rR_{kl_i}(r) = \mathscr{E}_k rR_{kl_i}(r). \qquad (2.44)$$

The differential equation (2.43) was numerically solved by Zernik and Klopfenstein performing a Laplace transformation and using a Taylor series expansion. The result of the calculation was displayed as the dependence of the generalized cross section of the ionization on the wavelength of the light (see Fig. 3).

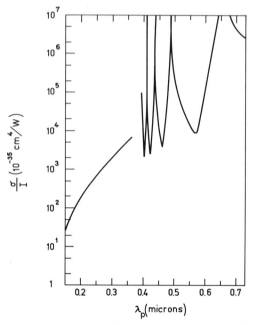

FIG. 3. The dependence of the generalized cross section $\sigma/I \sim \beta_I^{(k_0)}$ of the two-photon ionization of the hydrogen atom in the metastable state on the wavelength of the ionizing light $(\lambda_p)$, given by Zernik and Klopfenstein (1965).

## 3. Generalization for the Multiphoton Case

The method was generalized by Gontier and Trahin (1968a) for the case of more than two-photon ionization of atomic hydrogen. The total cross section of the ionization of the state with principal quantum number $n$ is as follows

$$\sigma_n^{(k_0)} = 4\pi^2 I^{k_0-1}\mathscr{E}_p \alpha(1/n^2) \sum_{l=0}^{n-1} 1/(2l+1)$$

$$\cdot \left| \sum_{m=-l}^{+l} \prod_{i=1}^{k_0-1} [A(l_i)\,\delta(l_i, l_{i-1}+1) + B(l_i)\,\delta(l_i, l_{i-1}-1)]\, \mathscr{R}_{\varkappa l_f, nl}^{(k_0)}(\{l_{ij}\}_1^{k_0-1}) \right|^2.$$

$$(2.45)$$

Here

$$A(l_i) = [(l_i - m + 1)(l_i + m + 1)/(2l_i + 1)(2l_i + 3)]^{1/2},$$

$$B(l_i) = [(l_i - m)(l_i + m)/(2l_i - 1)(2l_i + 1)]^{1/2},$$

$$l_0 = l, \quad l_f = l_{k_0}, \quad \text{and} \quad \{l_{ij}\}_j^{k_0} = l_j, l_{j+1}, \ldots l_f. \tag{2.46}$$

The radial transition matrix element of $k_0$th order $\mathscr{R}^{(k_0)}_{\varkappa l_f,\,nl}(\{l_i\}^{k_0-1}_1)$ is defined as the generalization of expression (2.42), i.e.

$$\mathscr{R}^{(k_0)}_{\varkappa l_f,\,nl}(\{l_i\}^{k_0-1}_1) = \int R^{(k_0)}_{nl}(\{l_i\}^{k_0-1}_1,\,r)R_{nl}(r)r^2\,dr. \tag{2.47}$$

where the radial wavefunction of $k_0$th order has the form

$$R^{(k_0)}_{nl}(\{l_i\}^{k_0-1}_1,\,r) = rR_{n_1 l_1}(r)\prod_{i=1}^{k_0-1}\sum_{\nu_i}\frac{\int R_{\nu_i l_i}(r_i)R_{\nu_{i+1}\,l_{i+1}}(r_i)r_i^3\,dr_i}{\mathscr{E}_g - \mathscr{E}_{\nu_i} - i\mathscr{E}_p}. \tag{2.48}$$

In the product on the right-hand side of expression (2.48) the supposition $R_{\nu_{k(0)}l_{k(0)}}(r) = R_{\varkappa l_f}(\kappa,\,r)$ was used. The generalized radial wavefunction of $k_0$th order, $R^{(k_0-j)}_{\nu_j l_j}(\{l_i\}^{k_0-1}_j,\,r)$, satisfies the differential equation

$$[\mathscr{E}_g + j\mathscr{E}_p + D_j]R^{(k_0-j)}_{\nu_j l_j}(\{l_i\}^{k_0-1}_j,\,r) = rR^{(k_0-j-1)}_{\nu_{j+1}l_{j+1}}(\{l_i\}^{k_0-1}_{j+1},\,r), \tag{2.49}$$

where

$$D_j = \frac{1}{2}\frac{d^2}{dr^2} + \frac{1}{r} - \frac{l_j(l_j+1)}{2r^2}. \tag{2.50}$$

The validity of Eq. (2.49) is immediately apparent on the basis of Eq. (2.44). Expression (2.49) can be regarded as the recursion formula for the generalized radial wavefunctions. Using the result of the computation of the wavefunction of second order made by Zernik (1964), expression (2.41), the differential equation (2.49) can be solved for the wavefunction of third order, etc.

The differential equation (2.49) was solved by Gontier and Trahin (1968a) performing a Laplace transformation and using a computer for the case of 2–8-photon ionization of the hydrogen atom in the ground state. The generalized cross sections of the multiphoton ionization were displayed by their dependence on the wavelength ($\lambda_p$) of the ionizing light.

## 4. Conclusions

The method gives fairly exact results for the generalized cross section of the multiphoton ionization but the calculation becomes difficult when resonance occurs at a level of high principal quantum number.

The result depends on the correct numerical calculation. This fact was demonstrated by the different values of the cross section of the two-photon ionization obtained on using the same method as that of Gontier and Trahin (1968a) and Zernik (1968) in the same range of wavelength of the ionizing light (see calculations of Chan and Tang, 1969; Klarsfeld, 1970; Rapoport et al., 1969; Gontier and Trahin, 1971).

The method of Schwartz and Thiemann makes it possible to check the accuracy of the approximative method of Bebb and Gold. The comparison (Gontier and Trahin, 1968a) shows the previously mentioned fact that there are considerable discrepancies between the exact and approximative values of the generalized cross sections, especially between the resonances in the valleys of the cross sections (Fig. 4).

(See comment on the method of Bebb and Gold near the end of Section II,B.)

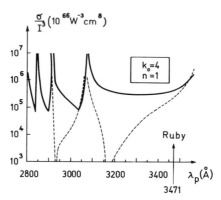

FIG. 4. The dependence of the generalized cross section $\beta_I^{(4)}(\lambda_p) \sim \sigma/I^3$ of the four-photon ionization of the hydrogen atom in the ground state on the wavelength of the ionizing light $(\lambda_p)$, given by Gontier and Trahin (1968a). The dashed line is the result of the calculation of Bebb and Gold (1966).

### E. Green's Function Method

An exact method of calculation of the transition matrix element of $k_0$th order is needed which can be applied equally to the hydrogen and other complex atoms. Such a method is the powerful method of computation of the multiphoton transition matrix element which involves the use of the Green's function of the electron (Baz et al., 1971).

By analogy with the transformation of the transition matrix element of $k_0$th order [expression (2.21)] performed in the preceding section [expressions (2.35), (2.42), and (2.47)], this transition matrix element of $k_0$th order given by the expression (2.21) can be transformed into another form (Zon et al., 1971a), i.e.

$$r_{f,g}^{(k_0)} = r_{f,g}^{(k_0)} = \int \psi_f^*(\mathbf{r}_{k_0}) r_{k_0} G_{\delta_g + (k_0-1)\delta_p}(\mathbf{r}_{k_0}, \mathbf{r}_{k_0-1}) r_{k_0-1}$$

$$\cdot G_{\delta_g + (k_0-2)\delta_p}(\mathbf{r}_{k_0-1}, \mathbf{r}_{k_0-2}) r_{k_0-2} \cdots$$

$$\cdot G_{\delta_g + \delta_p}(\mathbf{r}_2, \mathbf{r}_1) r_1 \psi_g(\mathbf{r}_1) \, dv_{k_0} \cdots dv_1 \qquad (2.51)$$

Here $G_{\mathscr{E}}(\mathbf{r}, \mathbf{r}')$ is the Green's function of the valence electron defined as

$$G_{\mathscr{E}}(\mathbf{r}, \mathbf{r}') = \sum_k \psi_k^*(\mathbf{r})\psi_k(\mathbf{r}')/(\mathscr{E}_k - \mathscr{E}). \qquad (2.52)$$

The evaluation of the matrix element is therefore reduced to the computation of the Green's function as the solution of a definite differential equation and to performing the integrations in the expression (2.51). The integration is an especially simple procedure if the Green's function can be expressed as the product of two functions which depend separately on only one variable, $\mathbf{r}$ or $\mathbf{r}'$. Then the transition matrix element has the form of the product of integrals.

In the general case of complex atoms the one-electron Green's function—as is well known—satisfies the differential equation

$$[-\tfrac{1}{2}(d^2/d\mathbf{r}^2) + {}^{(p)}H_s(\mathbf{r}) + (Z/r)]G_{\mathscr{E}}(\mathbf{r}, \mathbf{r}') = -\delta(\mathbf{r} - \mathbf{r}'). \qquad (2.53)$$

${}^{(p)}H_s(\mathbf{r})$ is the potential energy of the electron excess to the Coulomb potential energy $Z/r$ in the volume where $r < r_c$. $Z$ is the charge of the nucleus which the electron sees being far from the nucleus. Separating the angular dependent part of the Green's function,

$$G_{\mathscr{E}}(\mathbf{r}, \mathbf{r}') = \sum_{l, m} g_l(\mathscr{E}, r, r')Y_{lm}(\mathbf{r})Y_{lm}^*(\mathbf{r}'), \qquad (2.54)$$

the radial Green's function $g_l(\mathscr{E}, r, r')$ satisfies the following differential equation

$$\left\{\frac{1}{2}\left[\frac{1}{r^2}\frac{d}{dr}\left(r^2\frac{d}{dr}\right) - \frac{l(l+1)}{r^2}\right] + \frac{Z}{r} + \mathscr{E}\right\}g_l(\mathscr{E}, r, r') = -\frac{\delta(r - r')}{r \cdot r'}. \qquad (2.55)$$

In the case of hydrogen (${}^{(p)}H_s(r) = 0$; $Z = 1$) this last equation (2.55) can be solved exactly (Zon and Rapoport, 1968; Rapoport and Zon, 1968; Zon et al., 1968), i.e.

$$g_l(\mathscr{E}, r, r') = (v/r \cdot r')[\Gamma(1 + l - v)/\Gamma(2l + 2)]M_{v, l+1/2}(2r_</v)$$
$$\cdot W_{v, l+1/2}(2r_>/v). \qquad (2.56)$$

$M_{v, i}$ and $W_{v, i}$ are the Whittaker functions (Kertis, 1970). $v$ is the principal quantum number of the virtual state

$$v = (-2\mathscr{E})^{-1/2}. \qquad (2.57)$$

$r_<$, $r_>$ are, respectively, the smaller and the greater of the variables $r, r'$.

In the case of a complex atom the radial Green's function can be expressed by generalizing the quantum defect method of Bates and Damgaard (1950), resulting in the expression

$$g_l(\mathscr{E}, r, r') = \frac{v}{r \cdot r'} \left[ \frac{\Gamma(l + 1 - v)}{\Gamma(2l + 2)} M_{v,\, l+1/2}\left(\frac{2r_<}{v}\right) W_{v,\, l+1/2}\left(\frac{2r_>}{v}\right) \right.$$

$$\left. + \frac{\Gamma(l + 1 - v) \sin \pi[\mu_l(\mathscr{E}) + l]}{\Gamma(l + 1 + v) \sin \pi[\mu_l(\mathscr{E}) + v]} W_{v,\, l+1/2}\left(\frac{2r}{v}\right) W_{v,\, l+1/2}\left(\frac{2r'}{v}\right) \right].$$

(2.58)

if $\mathscr{E} < 0$ (Zon et al., 1969, 1971a); and

$$g_l(\mathscr{E}, r, r') = (v/r \cdot r')e^{i\pi v}$$

$$\times \left[ e^{2i[\Delta_l - \pi l/2]} W_{v,\, l+1/2}(2r/v) W_{v,\, l+1/2}(2r'/v) \right.$$

$$\left. - W_{-v,\, l+1/2}(-2r_</v) W_{v,\, l+1/2}(2r_>/v) \right],$$

(2.59)

if $\mathscr{E} > 0$. Here $\mu_l(\mathscr{E})$ is the quantum defect calculated by using the measured spectrum $\mathscr{E}_{nl}$ of the atom. The quantum defect is determined by the equation

$$\mathscr{E}_{nl} = -\tfrac{1}{2}[n - \mu_l(\mathscr{E})]^{-2}.$$

(2.60)

The phase of the partial wave $l$ is

$$\Delta_l = \delta_l + \eta_l.$$

(2.61)

$\delta_l$ is the distortion of the phase of the $l$th partial Coulombian wave by the non-Coulombian potential $^{(p)}H_s(r)$. The distortion is expressed by the extrapolation of the quantum defect into the continuous spectrum in the form

$$\delta_l = \pi\mu_l(\mathscr{E})$$

(2.62)

(Seaton, 1958; Burgess and Seaton, 1960).

In the calculation of the radial matrix element of $k_0$ th order it is the radial Green's function that takes the place of the function $G_{\mathscr{E}}(\mathbf{r}, \mathbf{r}')$ in expression (2.51). The wavefunctions $\psi_f^*$ and $\psi_g$ are substituted by the radial parts of them $R_{\kappa l_f}(\kappa, r)$ and $R_{nl}(r)$.

The integral over the angular dependent part of the transition matrix element can be performed separately using the usual algebra of the theory of angular momentum (Rose, 1957).

The calculation of the radial matrix element of second order was performed analytically for the two-photon ionization of the hydrogen atom by Rapoport et al. (1969) and Klarsfeld (1970; see the computation of Karule, 1971). The numerical value of the cross section of the ionization was compared with the results of the calculations made by Zernik (1968) and Chan and Tang (1969) and good agreement was found.

In the numerical calculation of the radial matrix element of $k_0$th order for the complex atom the Whittaker functions are expressed in the form of series expansions (Davidkin et al., 1971).

The generalized cross section was computed for the three-photon ionization of the hydrogen atom in the ground state, for the two-photon ionization of the metastable hydrogen atom, for the three-photon ionization of the helium metastables (Zon et al., 1971a), for the two-photon ionization of the potassium atom, for the three-photon ionization of cesium and sodium atoms, for the four-photon ionization of potassium, and for the five-photon ionization of the sodium atom (Manakov, 1971).

*Summarizing:* the method gives exact analytical results in the case of hydrogen and a good approximation in the case of the complex atom. The accuracy of the calculation is the same as the accuracy of the Bates and Damgaard quantum defect method for the one-photon transition. So the ambiguous procedure of the infinite summations is avoided.

## F. Conclusions

Comparing methods for the calculation of the transition matrix element of $k_0$th order the Green's function method seems to have the advantage of accuracy given by the method of Schwartz and Thiemann only in the case of hydrogen, and a wide spectrum of applicability with rather good accuracy. This accuracy is not achieved with the other approximative methods of Bebb and Gold or with that of Morton. This last method gives only informative values of the generalized cross sections of the multiphoton ionization.

However, it should not be forgotten that in the calculations the following important suppositions were made: (1) that the light field is strictly monochromatic and the interaction time is long enough for the interaction to be regarded as stationary; and (2) that the structure of the atom is not distorted by the light field so that the use of the approximation of the first nonvanishing order of the perturbation theory is justified for the description of multiphoton ionization.

## III. EXPERIMENTAL INVESTIGATIONS

### A. Multiphoton Ionization of Rare Gases

It is the multiphoton ionization of rare gases that has been investigated most thoroughly (Peressini, 1966; Voronov and Delone, 1965, 1966; Voronov et al., 1966a,b; Bistrova et al., 1967; Barhudarova et al., 1967; Chin et al., 1969; Agostini et al., 1968a,b, 1970a,b,c).

As we shall see later, in spite of these most careful experimental investigations concerned with the phenomena of the multiphoton ionization of atoms, most of the as yet unsolved problems are connected with the ionization of rare gases.

The laser-produced spark was observed in gases (Terhune, 1963; Maker *et al.*, 1964; Meyerand and Haught, 1963; Minck, 1964) and for the explanation of spark development, two mechanisms for the ionization of atoms were suggested (DeMichelis, 1969). The first electrons can be produced only by multiphoton ionization of atoms (Bunkin and Prokhorov, 1964; Tozer, 1965; Bebb and Gold, 1966). Then at a sufficiently high pressure of the gas, these electrons are accelerated by the light field during collisions with neutral atoms in the process of inverse bremsstrahlung (Askar'yan and Rabinovich, 1965; Wright, 1964; Zel'dovich and Raizer, 1965; Raizer, 1965). Therefore some part of the transverse oscillation energy of the electrons is transformed into longitudinal energy. Those electrons gaining enough energy in subsequent collisions for the ionization of the atoms will ionize them. This process leads to the doubling of the number of electrons and consequently to the avalanche ionization of the gas through repeated doubling (generation) of the number of electrons.

Naturally the need to investigate these two ionization processes arose separately—particularly the investigation of multiphoton ionization which had not been observed before. This process determines one of the important parameters of the laser-produced spark, namely, the threshold light power density of the creation of the spark.

The more important reason for the investigation of multiphoton ionization is that this process gives more information about some of the atoms in a high intensity laser field than the one-photon process. The aim of the investigations initiated by Voronov and Delone (1965) was to check the validity of the relation (2.26), to determine the value of the generalized cross section, and to compare this value with the result of the theoretical calculations in order to check on the validity of the method of calculation of the transition matrix element of $k_0$th order.

## 1. *General Description of the Measurement*

The typical measuring setup (Fig. 5) was constructed, taking into account the above-mentioned aims. The ionizing interaction of the high intensity light pulse of a $Q$-switched solid state laser with the atom of the gas was investigated in a chamber filled with very low pressure gas (Fig. 5). The pressure was low enough to avoid the collisions of electrons with neutral atoms and also their acceleration by inverse bremsstrahlung in the volume of the interaction during the laser pulse.

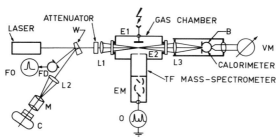

FIG 5. Experimental arrangement for the measurement of the multiphoton ionization of rare gases. This type of measuring setup was devised in the Lebedev Institute of Physics, Moscow (Voronov et al., 1966a,b; Voronov and Delone, 1966).

The chamber is evacuated thoroughly to a pressure of the order of $10^{-5}$–$10^{-6}$ mm Hg, in order to avoid impurity effects, before filling it with the gas to be investigated.

## 2. Detection of Ions

The laser radiation is focused by lens L1 into the middle of the gas chamber where the interaction takes place (Fig. 5).

The ions created in multiphoton ionization are subtracted by the field of the electrodes E1, E2 and pursued into the time-of-flight mass spectrometer. The mass spectrometer separates the impurity ions disturbing the measurement. For the sake of increasing the sensitivity of the ion detection an electron multiplier EM is used. The mass spectrum is displayed on the screen of the oscilloscope O.

In the interaction volume the ion density $\mathcal{N}_i$ created by the multiphoton ionization is determined by solving the rate equation

$$d\mathcal{N}_g/dt = -\mathcal{N}_g \beta_I^{(k_0)} I^{k_0}(\mathbf{r}) \tag{3.1}$$

and using the relation

$$\mathcal{N}_g + \mathcal{N}_i = \mathcal{N}_0. \tag{3.1a}$$

$\mathcal{N}_g$, $\mathcal{N}_i$, $\mathcal{N}_0$ are, respectively, the population of the initial state $|g\rangle$, the density of ions, and the density of atoms in the initial state at the time $t = 0$ when the perturbation is switched on.

Supposing that the rate of the $k_0$-photon ionization $W_{f,g}^{(k_0)}$ is small so that the effectiveness of the process

$$\Xi = \int_0^\infty W_{f,g}^{(k_0)}(t) \, dt \ll 1, \tag{3.2}$$

the density of ions can be expressed in the simple way

$$d\mathcal{N}_i(\mathbf{r}, t) = \mathcal{N}_0 \beta_I^{(k_0)} I^{k_0}(\mathbf{r}) \, dt \tag{3.3}$$

The distribution of the light power density is not uniform in the focal volume. Consequently the ion density also depends on the space coordinate (**r**).

All the ions created by multiphoton ionization are drawn out by the electric field of the electrodes E1 and E2 to the time-of-flight mass spectrometer. The mass spectrometer selects the ions of the atoms to be investigated, thus avoiding the disturbance of the possible impure atoms in the gas. The detection is performed by the electron multiplier. The sensitivity of the ion detection is about ten ions per light pulse.

The total number of ions reaching the first dynode of the electron multiplier (the collecting efficiency is regarded as unity) can be expressed by performing the integration on space and time coordinates in expression (3.3),

$$N_i = \mathcal{N}_0 \beta_I^{(k_0)} I^{k_0} v^{(k_0)} \tau^{(k_0)}. \tag{3.4}$$

On deducing expression (3.4) the light power density should have the form

$$I(\mathbf{r}, t) = I f(\mathbf{r}, t) = I f_1(\mathbf{r}) f_2(t). \tag{3.5a}$$

The time–spatial distribution function of the light power $[f(\mathbf{r}, t)]$ is normalized to unity at the maximum and it is supposed that it is separable; i.e. it can be expressed as the product of functions $f_1(\mathbf{r})$ and $f_2(t)$ depending only on the space and time variables separately. The spatial distribution $f_1(\mathbf{r})$ and the temporal distribution $f_2(t)$ are also normalized to unity.

$v^{(k_0)}$ and $\tau^{(k_0)}$ are the interaction volume and the interaction time of $k_0$ order, respectively,

$$v^{(k_0)} = \int f_1^{k_0}(\mathbf{r}) \, dv, \tag{3.5b}$$

$$\tau^{(k_0)} = \int_0^{\infty} f_2^{k_0}(t) \, dt; \tag{3.5c}$$

$I$ is the peak power density of the laser radiation.

The separability of spatial–temporal distribution is a very strong supposition in the case of the use of multimode pulsed lasers, especially that of a multimode ruby laser. Almost all the experimental investigations were performed using multimode lasers for which the first supposition of Section II,F about monochromaticity [resulting in expression (3.4)] is also questionable. The separability of the spatial–temporal distribution is the question to be investigated experimentally using, for instance, a streak camera in the concrete measuring setup (Agostini et al., 1970c).

The separability of the temporal and spatial distributions is equivalent to the stationarity of the spatial distribution and of the spectrum of the laser radiation, which is not always fulfilled (see Korobkin et al., 1967; Barjot, 1971).

As may be seen from expression (3.4) the dependence of the total number of ions on the light power density is the same as that of the rate of the ionization [see expression (2.26c)].

The generalized cross section of the ionization can also be determined. From expression (3.4) we get

$$\beta_I^{(k_0)} = (N_i/\mathcal{N}_0)1/I^{k_0}v^{(k_0)}\tau^{(k_0)}. \tag{3.6}$$

We have to measure the density of atoms in the ground state ($\mathcal{N}_0$) which is given by the gas pressure, the total number of ions ($N_i$) calibrating the sensitivity of the detecting system with Faraday cup. The peak light power density, the interaction volume, and the interaction time have to be calculated on the basis of Eqs. (3.5b) and (3.5c), measuring the spatial and temporal distributions and the energy of the laser pulse $Q$ [see expressions (3.7) and (3.8)].

If the power density of the light is high, i.e. the effectiveness

$$\Xi = W_{f,g}^{(k_0)}\tau^{(k_0)} \sim 1, \tag{3.2a}$$

the change of the population of the initial level also has to be taken into account. Then the density of ions after the light pulse has passed can be given by the expression

$$\mathcal{N}_i = \mathcal{N}_0(1 - \exp[-\beta_I^{(k_0)}I^{k_0}f_1^{k_0}(\mathbf{r})\tau^{(k_0)}]) \tag{3.3a}$$

instead of expression (3.3). This phenomenon is the saturation of the multiphoton ionization. The total number of ions can be expressed as

$$N_i = \mathcal{N}_0 \int (1 - \exp[-\beta_I^{(k_0)}I^{k_0}f_1^{k_0}(\mathbf{r})\tau^{(k_0)}])\, dv. \tag{3.4a}$$

Consequently the simple relation [expression (3.4)] between the number of ions and the ionization rate, and furthermore the light power density, ceases to exist in the case of saturation.

## 3. Measurement of the Parameters of the Laser Radiation

The method of measurement of the parameters of the laser radiation was worked out by Voronov and Delone (1966) and Barhudarova (1969, 1970). The energy of the laser pulse ($Q$) is measured by a calorimeter which detects, by the galvanometer VM, the rise in temperature of the little metallic ball B in which the light pulse is focused by the lens L3 and absorbed and which is in the vacuum chamber (Fig. 5).

The spatial distribution is measured by separating part of the light using the glass wedge W and by focusing this light with a lens similar to the L1 lens (L2). The light distribution in the focal point is magnified by microscope M and is photographed by the camera C. After developing, the pattern is scanned by microdensitometer and transformed to the light power density distribution using the separately measured characteristic curve of the film.

The temporal distribution of the light is measured using the second beam reflected by the glass wedge. The signal (of some gigahertz bandwidth) of the fast concentric photodiode FD, which is hit by this second beam of light, is supplied to a fast oscilloscope FO. The shape of the pulse is photographed on the oscilloscope screen.

The peak light power density can be calculated using the expression

$$I = Q/\mathscr{S}\tau^{(1)}, \tag{3.7}$$

where $\tau^{(1)}$ is given by expression (3.5c). The effective surface in the focal plane ($z = 0$) of the lens is expressed as

$$\mathscr{S} = \int f_1(x, y, z = 0) \, dx \, dy. \tag{3.8}$$

## 4. The Laser

The laser used in multiphoton ionization experiments is a $Q$-switched solid state, ruby and neodymium glass laser usually in multimode operation. In the case of the investigation of rare gases one- or two-stage light amplifiers are also used to reach the light power density of about $10^{12}$ W/cm² which is needed for ionization involving more than ten quanta absorption. (When ionizing the helium atom by neodymium glass laser radiation, the number of quanta absorbed in the process is twenty-one.) In order to widen the variety of wavelength of light available, nonlinear crystals such as KDP and ADP are used to produce the second harmonics of the ruby and neodymium glass laser radiations, respectively. Recently dye lasers have also been used, especially in investigations where the number of quanta absorbed is few.

## 5. Result of the Ionization Measurements

The primary aim of the measurements was the verification of that dependence of the ionization rate which is given by expression (2.26c), i.e., that of the number of ions which is proportional to the rate of the ionization [see expression (3.4)], on the light power density, i.e., on the peak light power density [see expression (3.4)]. The result of the measurement is plotted on a double logarithmic scale of the ionization rate, i.e. of the number of ions detected and of the peak power density of light. In this scale the theoretically expected curve is a straight line, the slope of which is the number of quanta absorbed in the ionization. From expression (3.4) we have

$$\log N_i = \log(\beta_I^{(k_0)} v^{(k_0)} \tau^{(k_0)}) + k_0 \log I. \tag{3.9}$$

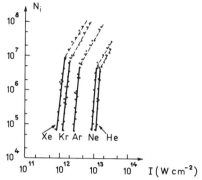

Fig. 6. The dependence of the number of ions of rare gases created in the multiphoton ionization on the peak light power density (Agostini *et al.*, 1970c). The wavelength of light is 1.06 $\mu$.

Figure 6 shows the experimentally observed variation of the number of ions of different rare gases with the light power density on a double logarithmic scale (Agostini *et al.*, 1970c). As is apparent, the linear relationship of expression (3.9) is fulfilled between log $N_i$ and log $I$. The discontinuity of the curves is caused by the effect of saturation [see expression (3.4a)]. The further increase in the number of ions detected is caused by the subsequent increase in the number of ions at the tail of the spatial distribution of the light power density where saturation has not yet taken place.

The characteristic feature of this result is that the slope of these lines ($k_{meas}$) does not agree with the theoretically expected value of the absorbed photons. This fact had already been observed in almost the first measurements (Voronov and Delone, 1966). Table I shows the measured values of the slopes of curves and the theoretically expected values for different rare gas atoms and different wavelengths of light used. It is apparent that all the measured slopes, except that for krypton atoms at the wavelength 0.53 $\mu$, are smaller than the theoretical value.

The three attempts to explain this effect follow:

*a. Quasi-continuous spectrum.* In the strong electromagnetic field of light the "dynamic" ionization potential of the atom is smaller than the "static" ionization potential used in the theoretical calculations. The cause of the "dynamic" ionization potential being smaller than the "static" one is the distortion of the atomic structure. The eigenenergies or level widths of the atom depend on the light power density. The result is the overlapping of levels in the energy region near the ionization limit. This quasi-continuous spectrum causes the decreasing ionization potential (Voronov *et al.*, 1966a).

*b. Multiphoton resonance.* In the case of resonance of $k < k_0$th order the dependence of the eigenenergies, figuring in the denominator (becoming

## TABLE I

The Measured Value ($k_{meas}$) of the Slope of the Linear Dependence of the Logarithm of the Rate of Multiphoton Ionization on the Logarithm of the Light Power Density

| Gas | Wavelength ($\mu$) | $k_0$ | $k_{meas}$ | $k = k_0 - k_{meas}$ | $F$ (cm$^{-2}$sec$^{-1}$) | References |
|---|---|---|---|---|---|---|
| He | 1.06 | 21 | 18 ±0.3 | 3 ±0.3 | | Agostini et al. (1970c) |
| | 0.53 | 11 | 9.2 ±0.3 | 1.8 ±0.3 | | Agostini et al. (1970c) |
| Ne | 1.06 | 19 | 13.7 ±0.3 | 5.3 ±0.3 | | Agostini et al. (1970c) |
| | 0.53 | 10 | 7.3 ±0.3 | 2.7 ±0.3 | | Agostini et al. (1970c) |
| Ar | 1.06 | 14 | 10.3 ±0.3 | 3.7 ±0.3 | | Agostini et al. (1970c) |
| | 0.53 | 7 | 5.7 ±0.3 | 1.3 ±0.3 | | Agostini et al. (1970c) |
| Kr | 1.06 | 12 | 9 ±0.3 | 3 ±0.3 | | Agostini et al. (1970c) |
| | | | 9 ±0.1 | 2.9 ±0.1 | $10^{31.4 \pm 0.3}$ | Bistrova et al. (1967) |
| | 0.53 | 6 | 5.5 ±0.6 | 0.5 ±0.6 | | Agostini et al. (1970a) |
| | | | 5.5 ±0.7 | 0.5 ±0.5 | $10^{32}$ | Delone and Delone (1968) |
| | 0.69 | 8 | 6.31 ±0.11 | 1.69 ±0.11 | $4.8 \times 10^{30}$ | Voronov et al. (1966b) |
| Xe | 1.06 | 11 | 8.7 ±0.3 | 2.3 ±0.3 | | Agostini et al. (1970c) |
| | | | 8.8 ±0.2 | 2.2 ±0.2 | $10^{31.4 \pm 0.3}$ | Bistrova et al. (1967) |
| | 0.53 | 6 | 4.1 ±0.3 | 1.9 ±0.3 | | Agostini et al. (1970a) |
| | | | 4.4 ±0.2 | 1.6 ±0.2 | $10^{32}$ | Delone and Delone (1968) |
| | 0.69 | 7 | 6.23 ±0.14 | 0.77 ±0.14 | $10^{30.25 \pm 0.25}$ | Voronov and Delone (1966) |
| | | | 7.44 ±0.77 | 0.44 ±0.77 | $10^{29.6}$ | Chin et al. (1969) |

zero) of expression (2.21) for the transition matrix element, on the light power density causes the generalized cross section of the multiphoton ionization to depend on the light power density. The rate of ionization turns out to be a complicated function of the light power density. The logarithm of this rate can be approximated by a linear function of the logarithm of the light power density only in its narrow range. While the measurements were performed in a small range of light power density this linear approximation was always allowed. But the slope of this line, i.e. that of the tangent of the log $W_{f,g}^{(k_0)}(\lambda_p, I)$ − log $I$ function depends equally on the wavelength of light and on the power density of the radiation (Voronov, 1966; Delone et al., 1972).

It should be noticed that on the basis of this explanation of the observed deviation of the slope from the theoretical value, the deviation in the direction of higher slopes is also to be expected. Experimentally, only the slope smaller in value than the theoretical slope was observed, where the multiphoton ionization of rare gas atoms were in the ground state. (For further details on this problem, see Section IV.)

c. Saturation. The third explanation of the measured effect that $k_{meas} < k_0$ is very simply the saturation of the process (Chin et al., 1969). But special care was taken to avoid the saturation in all the experiments listed. Therefore the saturation as the cause of the observed $k_{meas} < k_0$ can be ruled out.

The more probable explanation of the deviation of the measured $k_{meas}$ from the theoretical value is the multiphoton resonance. But it should not be forgotten that $k_{meas} > k_0$ is equally possible, and experimentally only $k_{meas} < k_0$ was measured.

Therefore the first explanation of the quasi-continuous spectrum should also not be excluded. The light field is very strong and the structure of the atom is heavily destroyed. For the description of the multiphoton ionization of rare gases the perturbative method discussed in Section II may not be applicable (for details concerning this problem, see Section V).

Finally we wish to draw attention to the interesting experimental fact that the energy of the number of photons which corresponds to the observed value of the slope of the log $N_i$–log $I$ linear function coincides with the energy of an atomic state with small energy discrepancy in every rare gas (Agostini et al., 1970c). Here the multiphoton ionization is explained as a two-step process—multiphoton excitation to the resonant level and subsequent ionization from this state.

The observed smaller slope of the log $N_i$–log $I$ linear function has been explained by the saturation of the second step. Namely, the ionization rate due to the smaller order of the last process is usually greater than the rate of the multiphoton excitation having greater order.

But this argument is not correct as it is only the rate of the multiphoton ionization that can be expressed as the product of the rate of excitation and of the rate of subsequent ionization in the case of resonance (Keldish, 1964; Kovarski, 1969b), i.e.

$$W_{f,g}^{(k_0)} = W_{f,r}^{(k_0 - k)} \cdot W_{r,g}^{(k)}.$$

The effectiveness to which the notion of saturation is connected [see Eq. (3.2a)] cannot be given as the product of effectiveness of the composing processes

$$\Xi \neq \Xi_{f,r} \cdot \Xi_{r,g}.$$

With certain rare gas atoms this resonant level cannot be excited by multiphoton excitation because of the selection rule.

Many authors determined the rate of the ionization and sometimes the generalized cross sections. But the fact that $k_{meas} < k_0$ shows that the theory of Section II is not adequate for the description of the process. So it is not surprising that these experimental values of the generalized cross sections do not agree with the theoretical predictions. It is worthwhile determining the generalized cross sections only in those cases where the value of $k_{meas}$ agrees with the theoretical value $(k_0)$.

There was hope that this would be the case if the value of $k_0$ was not so high. In this case the electric field strength of the light is also smaller than for the rare gases and there is less likelihood of distortion of the atomic structure. Therefore the multiphoton ionization of the alkaline and excited rare gas atoms was investigated.

## B. Alkaline Atoms

### 1. Experimental Arrangement

The experimental arrangement is shown in Fig. 7 (Delone et al., 1969). The difference in comparison with the experimental setup used in the investigation of rare gas atoms is in the vacuum chamber. In order to avoid covering the surfaces of different electrodes, walls, dynodes, and windows of the vacuum chamber with a layer of alkaline metals of low work function an atomic beam is used. The beam of atoms leaving the oven is formed by a multichannel collimator. The density of atoms attained by this source is of the order of $10^{12}$ atoms $cm^{-3}$. The atoms condense on the surface of the liquid nitrogen trap.

The ions produced in the multiphoton ionization are drawn out by the electric field of the electrodes E1 and E2. The direction of the electric field is perpendicular to the optical axis and to the direction of the beam. The ions

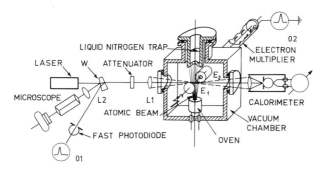

Fɪɢ. 7. Experimental setup for the investigation of multiphoton ionization of alkaline atoms.

are detected, as usual, after mass separation by a time-of-flight mass spectrometer, by an electron multiplier and oscilloscope.

The parameters of the laser beam, i.e. the energy of the laser pulse and the temporal and spatial distribution of the light power density are measured as described in Section III,A. The peak light power density can be calculated using expressions (3.7) and (3.8).

## 2. Generalized Cross Sections Measurement

The difference which occurs in the measurements of parameters of multiphoton ionization in the case of rare gases is in the determination of the generalized cross sections. In the measurement with alkaline atoms the first difficulty is concerned with the nonuniform distribution of the density of atoms and the measurement of this distribution. The second difficulty arises in the measurement of the intensity of the atomic beam.

These two difficulties can be overcome using, for example, the method of hot filament for the measurement of the spatial distribution of the atomic beam flux. The total flux of the beam can be determined by using an oscillating quartz crystal for the measurement of the mass of the atoms condensed on the surface of the crystal during a definite time. If there is a Maxwellian distribution of the atomic velocity corresponding to the temperature of the oven, the density distribution of the atoms in the beam can be calculated. Taking into account the complexity of this procedure, a new method for the determination of the generalized cross section is advised.

This new method uses the phenomenon of saturation (Delone *et al.,* 1971a). In the case of saturation, i.e. when the effectiveness $\Xi$ of the process approaches unity, it is not sufficient to take into account terms only up to the second one of the series expansion of the function under the integral of expression (3.4a). Calculating with only these two terms we get the familiar expression (3.4) for the total number of ions. The ratio of the total number of ions is obtained by using expression (3.4) ($N_i^{(2)}$) and also by taking into account the third term in the series expansion ($N_i^{(3)}$), i.e.

$$N_i^{(3)}/N_i^{(2)} = 1 + \tfrac{1}{2}\beta_I^{(k_0)} I^{k_0} \tau^{(k_0)} v^{(2k_0)}/v^{(k_0)} \tag{3.10}$$

(Chin and Isenor, 1970).

At the beginning of saturation it was supposed that the experimentally measured total number of ions as a function of the light power density can be approximated by the first three terms of the series expansion of the integrand of expression (3.4a), which is $N_i^{(3)}$. But at low light power density $N_i^{(3)} = N_i^{(2)}$. In the region of saturation $N_i^{(2)}$ can be calculated by extrapolation using the experimentally measured relationship (3.4) at small light power density.

The remaining quantities $I^{k_0}$, $\tau^{(k_0)}$, $v^{(2k_0)}$ in expression (3.10) can be calculated using the measured energy of the light pulse and spatial and temporal distribution of the light power density [see expressions (3.5b), (3.5c), (3.7), and (3.8)]. That is, the absolute value of the generalized cross section is given as

$$\beta_I^{(k_0)} = 2 \, \frac{v^{(k_0)}}{v^{(2k_0)}} \, \frac{N_i^{(3)} - N_i^{(2)}}{N_i^{(3)}} \, \frac{1}{I^{k_0} \cdot \tau^{(k_0)}}. \tag{3.6a}$$

## 3. Result of the Measurements

As expected the rate versus power density function of the multiphoton ionization of alkaline atoms was found to be a linear function on the double logarithmic scale and the slope of the straight line agrees with the theoretical value $k_0$ (Delone and Delone, 1969; Delone et al., 1969; Held et al., 1971, 1972). This means that the results of the perturbation theory of Section II are applicable for the description of the multiphoton ionization of alkaline atoms measured.

It is furthermore worthwhile comparing the experimentally measured values of the generalized cross sections with the theoretical predictions. Table II contains the collected experimental and theoretical values (Delone, 1973) for the comparison. As can be seen, the agreement between theory and practice is fairly good. [See further details in Delone et al. (1973).]

The only contradiction between the experimental results of two experimental groups occurs in the case of potassium (Held et al., 1972; Delone et al., 1971a). Taking into account that the other experimental values are in agreement with theory up to the order five of the process the result of Delone's group can be regarded as more probable. The cause of the discrepancy is not yet known.

TABLE II

COMPARISON OF THE EXPERIMENTALLY OBSERVED VALUES $\beta_{\text{meas}}^{(k_0)}$ OF THE GENERALIZED CROSS SECTIONS FOR DIFFERENT ATOMS WITH THE THEORETICALLY CALCULATED VALUES $\beta_{\text{th}}^{(k_0)}$ USING THE APPROXIMATION OF LOWEST ORDER OF THE PERTURBATION THEORY

| | | Atom: K | Cs | Na | K | Na | Hg | References |
|---|---|---|---|---|---|---|---|---|
| | | $\mathcal{E}_p$(eV): 2.36 | 1.78 | 2.36 | 1.18 | 1.18 | 1.78 | |
| | | $k_0$: 2 | 3 | 3 | 4 | 5 | 6 | |
| Experiment | | | | | | | $10^{-171.5\pm2.3}$ | Chin et al. (1969) |
| | | | | | $10^{-103.8\pm1}$ | | | Fox et al. (1971) |
| | | | $10^{-77\pm1.5}$ | | | | | Held et al. (1972) |
| | | $10^{-47\pm1}$ | | | | | | Evans and Thonemann (1972) |
| | | $10^{-48\pm0.8}$ | $10^{-77.3^{+0.3}_{-0.7}}$ | $10^{-79.6^{+1.1}_{-1.0}}$ | $10^{-107.3^{+1.7}_{-1.6}}$ | $10^{-138^{+1.7}_{-1.6}}$ | | Delone et al. (1971a) |
| Theory | | $10^{-48.6}$ | $10^{-77.2}$ | $10^{-79.2}$ | $10^{-108.5}$ | $10^{-140.3}$ | $10^{-174.9}$ | Bebb (1966, 1967) |
| | | $10^{-48.6}$ | $10^{-76}$ | $10^{-77.5}$ | $10^{-108}$ | $10^{-139.9}$ | | Morton (1967) |
| | | $10^{-48.8}$ | $10^{-75.8}$ | $10^{-79}$ | | | | Manakov (1971) |

In discussing the results it must be noted that the theoretical results are connected with the use of coherent linearly polarized light. The measurements were performed by the radiation of multimode lasers and the polarization properties were not clearly defined. The maximal deviation between the results for the use of coherent and incoherent light is $k_0$! which in the case of order up to five can reach a value of about two orders of magnitude. In the light of the previously mentioned circumstances the agreement of the theoretical predictions and the experimental results can be regarded as excellent.

The result for the generalized cross section of mercury is also included in Table II, in order to complement the list of generalized cross sections which has been measured and calculated until now.

We conclude that the theory of Section II gives both a quantitatively and numerically correct description of the multiphoton ionization of alkaline atoms up to the number five of the absorbed photons.

But it should not be forgotten that in the experiments special care was taken to avoid multiphoton resonance (Delone and Delone, 1969; Held *et al.*, 1972) or the distortion of the atomic structure and therefore a very small light power density was used (Evans and Thonemann, 1972) to attain qualitative agreement ($k_{\mathrm{meas}} = k_0$) with the theory. As is apparent from Table II, in the case of qualitative agreement with theory ($k_{\mathrm{meas}} = k_0$) the numerical agreement is also satisfactory.

## C. Excited Atoms

The other group of atoms is the excited atoms of rare gases, the ionization potential of which is relatively small. Therefore the order of multiphoton ionization ($k_0$) is also small using ruby or neodymium glass lasers. The ionization potential of these excited atoms is of just the same order of magnitude as that of the alkaline atoms. Consequently the hope is justified that the perturbation theory of Section II is applicable also to these atoms (Zon *et al.*, 1971a).

Only the metastable (singlet and triplet) helium atoms were investigated experimentally (Bakos *et al.*, 1970a,b, 1971).

### 1. Experimental Setup

The experimental setup is illustrated in Fig. 8.

The parameters of the laser radiation are measured using the glass wedge W, the fast photodiode and oscilloscope O1 for the determination of the temporal distribution, the L1 lens being similar to the L2 lens, and the microscope objective MO placed in front of camera C for the determination of the spatial distribution. The energy of the laser pulse is measured by the calorimeter.

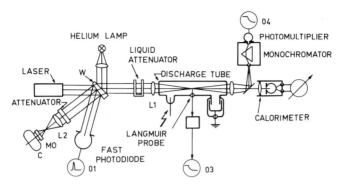

FIG. 8. Experimental setup for the investigation of the multiphoton ionization of metastable rare gases.

The helium atoms were excited to the metastable state in the glow discharge in the glass discharge tube. The laser radiation is focused into the middle of the tube where the interaction with the metastable atoms takes place. The discharge was extinguished some time before the occurrence of the laser pulse, and the interaction proceeds in the afterglow.

Further differences which occur in comparison with the earlier measuring setups are in the detection of the ions and in the determination of the population of the initial state of the ionization.

This population of the initial state of the process is determined by measuring the absorption of spectral lines of the helium lamp originating in the transition from the initial level. The suitable wavelength of the light of the helium lamp is selected by the monochromator. The dependence of the population on time in the afterglow, and furthermore the mean lifetimes of the metastable atoms, are determined by measuring the dependence of the absorption on time using the photomultiplier and oscilloscope O4.

## 2. Ion Detection

The ions are detected by a Langmuir probe placed in the middle of the discharge tube. The interaction of the laser radiation leads to the creation of the ion density distribution given by the time integral of the expression (3.3), in excess of the discharge ion density. This excess ion density is located far from the probe and can proceed to the probe by ambipolar diffusion in the afterglow plasma. On reaching the probe, this diffusing excess ion density causes a momentary increase in the ion density around the probe. This increase is added to the background ion density of the decaying afterglow. The result is a small pulse superposed on the exponentially decaying voltage on the load resistor ($R_L$) of the probe. The amplitude ($V_a$) of this pulse is regarded in the following discussions to be the multiphoton signal and can be calculated by solving the differential equation of the diffusion with the initial condition of ion density distribution given by expression (3.3). That is,

$$V_a = B\mathcal{N}_m^{(0)}e^{-\Delta t/\delta_m}\beta_I^{(k_0)}I^{k_0}\cdot\tau^{(k_0)},\tag{3.11}$$

where $\mathcal{N}_m^{(0)}$ is the density of metastable atoms at the time of switching off the discharge, $\Delta t$ is the time difference between switching off the discharge and the occurrence of the laser pulse, and $\delta_m$ is the mean lifetime of the metastable atoms the distribution of which is according to the normal mode of diffusion. The effectiveness of the ion detection $(B)$ is given by

$$B = \tfrac{1}{4}e\bar{v}S_p R_L \xi_{max}(\rho_p).\tag{3.12}$$

Here $e$ is the charge of the electron and $\bar{v}$ is the mean thermal velocity of the ions. This velocity is supposed to correspond to the room temperature. $S_p$ is the surface of the probe. $\xi_{max}(\rho_p)$ is the maximum value of the time function $\xi(\rho_p, t)$ given by the expression

$$\xi(\rho_p, t) = \sum_{ik} c_{ik} J_0(\eta_i\rho_p/R)\sin(k\pi/2)e^{-(t-t_L)/\delta_{ik}},\tag{3.13}$$

where

$$c_{ik} = \frac{\int_0^L dz \int_0^R \rho\,d\rho\, J_0(\eta_i\rho/R)\sin(k\pi z/L)\cdot J_0(2.4\rho/R)\sin(\pi z/L)f_1(\rho, z)}{L\int_0^R \rho\,d\rho\, J_0^2(\eta_i\rho/R)}.$$

$$\tag{3.14}$$

$\rho$ and $z$ are cylindrical coordinates; $\rho_p$ is the position of the probe in the middle of the tube $(z = L/2)$; $J_0(x)$ is the Bessel function with roots $\eta_i$; $L$ and $R$ are the length and the radius of the discharge tube; $\delta_{ik}$ is the mean lifetime of the ions the distribution of which is according to the $i, k$ mode of the diffusion; $t_L$ is the time of the laser pulse.

Expression (3.11) has the same form as expression (3.4) for the total number of ions. The differences between the two expressions are that the density of the atoms, being in the initial state of ionization, decays exponentially in time $(\mathcal{N}_m^{(0)}e^{-\Delta t/\delta_m})$ after switching off the discharge, and that the effectiveness of the ion detection which is equal to unity in expression (3.4) is determined by the diffusion process and $B \ll 1$.

In the measurement the multiphoton signal $(V_a)$ is measured as the function of the peak light power density and the time of the laser pulse relative to switching off the discharge.

The value of the effectiveness of the ion detection can be determined by measuring the usual laser radiation parameters, knowing the geometry of the discharge tube and using a computer.

## 3. Results

The expression (3.11) was verified experimentally. The rate of the multi-photon ionization versus light power density was measured using a ruby laser. The experimental points can be fitted by a straight line in the double logarithmic scale coordinate system. The value of the slope of the line agrees with the theoretical value $k_{meas} = k_0 = 3$.

The time dependence of the multiphoton signal $V_a$ was also investigated by changing the time interval between switching off the discharge and the occurrence of the laser pulse. The experimental points can be fitted by a straight line in the semilogarithmic scale coordinate system. The slope of this line is regarded as the "mean lifetime" of the multiphoton signal and agrees with the mean lifetime of the triplet metastable atoms (Fig. 9). This means that the three-photon ionization of the triplet metastable helium atom was observed (Bakos *et al.*, 1971).

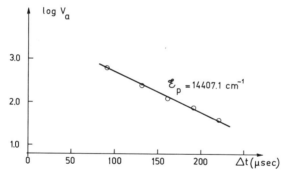

FIG. 9. Dependence of the multiphoton signal on the time difference between switching off the discharge and the time of the laser pulse in the afterglow. The circles are the experimental points of the multiphoton signal; the lines are the result of the lifetime measurements of the metastable atoms by the absorption of spectral lines originating in the transition from the metastable state (Bakos *et al.*, 1971).

The lifetime of the metastable atoms was measured by absorption of the helium line originating in the transition from the metastable state as described before. The observed lifetime also agrees with that measured by Phelps and Molnár (1953).

On changing the laser frequency by cooling the ruby rod, the generalized cross section of the ionization from the singlet metastable state overtakes that of the triplet metastable state (Zon *et al.*, 1971a). (For details concerning the resonances, see Section IV.) This could be observed so that the "mean lifetime" of the multiphoton signal abruptly changes and agrees with the lifetime of the singlet metastable state (Fig. 10).

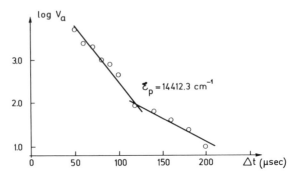

FIG. 10. The dependence of the multiphoton signal ($V_a$) of the metastable atoms of helium on time at the wavelength of light where the generalized cross sections of the triplet and singlet metastable atoms are equal. (See also Bakos *et al.*, 1973a,c.)

Comparing the theoretical values of the generalized cross sections, $\beta_{th}^{(3)} = 3.3 \times 10^{-78}$ cm$^6$ sec$^2$ for the singlet metastable state and $\beta_{th}^{(3)} = 3 \times 10^{-81}$ cm$^6$ sec$^2$ for the triplet metastable state, with the experimental ones we conclude that not such a big difference is observed between the two cross sections as predicted by theory (Zon *et al.*, 1971a); and the observed absolute values of the cross sections are several orders of magnitude larger ($\beta_{meas}^{(3)} = 10^{-72 \pm 2}$ cm$^6$ sec$^2$) than the theoretical ones.

The experimental generalized cross sections are calculated taking into account only the first term in the series expansion (3.13). So the experimental cross section can be regarded as a first approximation.

The discrepancy between the theoretical and experimental values needs to be investigated further in order to obtain an explanation.

### D. *Hydrogen Atoms*

The importance of the investigation of the hydrogen atom has no need of emphasis. It is the wavefunction of the hydrogen atom that is known exactly. The result of exact calculations makes possible the comparison with experimental data which may give interesting basically new results, and may reveal the influence of the light statistics on the multiphoton processes (see Section VI).

In spite of the importance of the investigation of the hydrogen atom only one experimental work has been concerned with this subject (Mainfray, 1973; Lu Van *et al.*, 1973). The reason for the absence of any investigation is probably due to the experimental difficulties resulting from hydrogen being, commonly, in molecular form.

The measurement of the multiphoton ionization of a hydrogen atom was performed with the usual setup (see Fig. 7). The beam of the hydrogen atom was produced, dissociating the hydrogen molecules in radio frequency discharge. The degree of the dissociation was about 80%. The vacuum chamber was continuously pumped by ionic pumps. (The liquid nitrogen trap of Fig. 7 is absent.)

The laser is a neodymium glass laser $Q$-switched by a rotating prism and frequency doubled by a nonlinear crystal ($\lambda_p = 0.53\ \mu$). Any resonances occur with the levels of the atoms in the process of six-photon ionization at this wavelength. Consequently the lowest order perturbation theory (see Section II) is applicable.

The measured log $W^{(6)}$ versus log $I$ function can be seen in Fig. 11. The results of the theoretical calculations of Gontier and Trahin (1971), Bebb and Gold (1966), and Morton (1967) are also plotted in the figure.

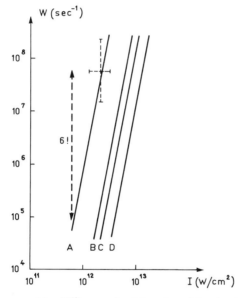

FIG. 11. The measured log $W^{(6)}$ versus log $I$ function of the six-photon ionization of the hydrogen atom at a wavelength $\lambda_p = 0.53\ \mu$ (curve A). Curve B is the result of the theoretical calculation of Gontier and Trahin (1971); curve C, that of Bebb and Gold (1966); and curve D, that of Morton (1967). (Lu Van et al., 1973.)

The slope of the linear function log $W^{(6)}$ versus log $I$ agrees with the theoretical value. The (absolute) value of the generalized cross section measured experimentally coincides with the theoretical one (Gontier and Trahin, 1971) in that case only if the factor $k_0!$ is taken into account as a consequence of the multimode operation of the laser. (See the experimental point with error bars and the value of $k_0! = 6!$ which is also indicated in the figure.)

The problem arising in connection with the determination of the generalized cross section was that the spatial distribution of the laser radiation
was not stationary during the laser pulse (Barjot, 1971). Consequently the
method of calculation of the generalized cross section as presented in
Section III,A,2 does not apply.

Further complications result from the dissociation and subsequent
ionization of the remaining (in the beam) $H_2$ molecules.

In conclusion, further experimental investigations of the multiphoton
ionization of hydrogen atoms are desirable.

### E. Multiphoton Detachment

In the very early days following the invention of the laser the investigation of multiphoton processes, namely the multiphoton photodetachment of
negative ions, was suggested by Hammerling at the Royal Society Conference on Optical Masers in 1962. Prompted by this suggestion, Geltman
(1963) calculated the two-photon photodetachment rate of $I^-$ ions. At the
same time as the publication of Voronov and Delone (1965) on the experimental investigation of multiphoton ionization of Xe atoms by a ruby laser,
Hall et al. (1965) published their work on the two-photon photodetachment
of negative ions.

The negative ions have the important feature that only one bound state
of the ions exists. Therefore the distortion of and the resonance with the
excited bound states (see Section IV) do not influence the multiphoton ionization (two-level model).

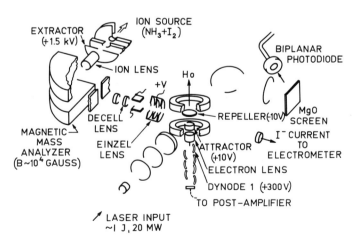

FIG. 12. Experimental arrangement for the measurement of the two-photon detachment of
$I^-$ ions (Hall et al., 1965).

The experimental arrangement can be seen in Fig. 12. The $I^-$ ions are extracted from the discharge of gas which is obtained from the mixture of $NH_3$ and $I_2$ gases. The beam is formed by an ion lens and mass-analyzed by a mass spectrometer. Passing some ion lenses the beam interacts with the radiation of a $Q$-switched ruby laser in the crossed beam apparatus. The detached electrons having been extracted by the field of the repeller electrode and formed by an electron lens are detected by an electron multiplier. The coincidence of the ion and laser beams is checked by a movable phosphor screen. The temporal distribution of the laser pulse is measured in diffusely reflected light from the MgO screen by a biplanar photodiode and fast oscilloscope.

The rate of the two-photon detachment of $I^-$ ions is measured to be the power function of the power density of the light. The power, i.e. the slope of the straight line of the function $W^{(2)}$ versus $\int I^2 \, dt$, equals unity as expected. This means that, using the usual display of results, the slope of the straight line of the function $\log W^{(2)}$ versus $\log I$ equals two. There is, accordingly, qualitative agreement with the perturbation theory of lowest nonvanishing order.

The value of the generalized cross section is also determined. After having performed some corrections for the nonuniform distribution of the ion beam in the interaction volume and for the coherence properties of the beam of the laser being in the multimode operation, the experimental value is found to be $\beta^{(2)}_{meas} = 1.8 \times 10^{-49}$ cm$^4$ sec. This experimental value deviates from the theoretical one, $\beta^{(2)}_{th} = 5 \times 10^{-51}$ cm$^4$ sec (Geltman, 1963). The refined theoretical calculation of Robinson and Geltman (1967) gives better agreement (see Fig. 13). As is apparent from the figure, the end of

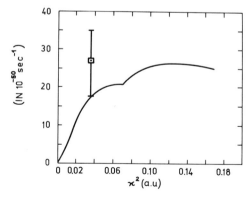

FIG. 13. The dependence of the generalized cross section of the two-photon detachment of $I^-$ ions on the square of the wavenumber of the detached electrons (in atomic units). The solid line is the result of the calculation of Robinson and Geltman (1967).

the error bar of the experimental point is just above the solid line, the result of the computation.

The difference between the two theoretical calculations (Geltman, 1963; Robinson and Geltman, 1967) is in the final state wavefunctions. The wavefunction distorted by the short-range potential of the atom is used in the latter calculation, applying the method of Schwartz and Thiemann (1959) for the calculation of the transition matrix element of second order. In the first calculation a simple plane wave is supposed to be the final state wavefunction.

The agreement between theory and experiment can be regarded as satisfactory.

## IV. RESONANCE MULTIPHOTON IONIZATION

### A. Rate versus Frequency Dependence

The transition matrix element of $k_0$th order [see expression (2.21)] and consequently the generalized cross section [see expression (2.26a)] is of typical resonance character, depending on the frequency of the light (see Fig. 2 for the case of the four-photon ionization of hydrogen atoms). In addition to the investigations of the multiphoton ionization rate on the light power density and the measurements of the absolute value of the generalized cross sections of the multiphoton ionization [see expression (2.26a)] which were performed in Section III it is worth investigating the resonance behavior of the generalized cross section. This investigation yields important information about the validity of the description of the multiphoton ionization process by the lowest order perturbation theory of Section II.

Such investigations of resonance multiphoton ionization were performed by Baravian et al. (1970, 1971; Benattar, 1971).

The thirteen-photon ionization of neon atoms was investigated by a ruby laser $Q$-switched by a rotating prism and frequency tuned by cooling the rod. The energy of the twelve ruby laser photons was in coincidence with the energy of the 11p $[3/2]_2$ state of the neon atoms (Racah notation is used). The interaction volume ($v^{(k_0)}$) and the effective area of the focus ($\mathscr{S}$) were calculated using the value of the beam divergency, the diameter of the beam on the focusing lens, and the focal length of the lens. The influence of the multimode operation of the laser on the measured generalized cross section was taken into account by calculating the instantaneous light power density knowing the mode structure of every laser pulse. The mode structure was photographed using a Fabry-Perot interferometer.

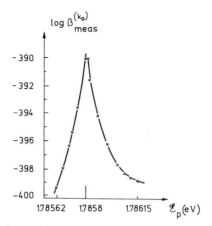

FIG. 14. The dependence of the generalized cross section of the 13-photon ionization of neon atoms on the energy of the laser quanta (in electron volts), given by Baravian *et al.* (1971).

The result of the measurement is seen in Fig. 14 where the measured generalized cross section of the multiphoton ionization is plotted as the function of the energy of the laser quanta $(\mathscr{E}_p)$.*

As is apparent, the resonance is very sharp. The increase in the rate of the ionization is more than ten orders of magnitude. The curve resembles similar curves calculated by Bebb and Gold (1966). The resonance energy of twelve photons, i.e. the energy of twelve photons at the peak of the resonance curve, almost coincides with the field free energy of 11p $[3/2]_2$ levels of the neon atoms. The difference in the energy of twelve photons at the resonance and the field free energy of the resonance level is some cm$^{-1}$. This difference is explained by uncertainties in the determination of the energy of the laser quanta (Baravian *et al.*, 1970) and by the shift of the level in the light field (Benattar, 1971).

In discussing the results of the measurement, it is especially important to understand the cause of the measured small value of the shift of the level very near to the ionization threshold of the atom ($\sim$ 1000 cm$^{-1}$); i.e. the smearing of levels (Voronov *et al.*, 1966a) is expected in that region of the spectrum (see Section III,A,5,a).

Therefore repeating the measurement or performing similar measurements of the dependence of the generalized cross sections on the frequency of light is highly desirable.

---

* Care must be taken in processing the experimental data in the case of resonance multiphoton ionization. The simple method of calculation of the generalized cross section of the multiphoton ionization using expression (3.6) cannot be applied because of the uncertainty in the interaction volume and interaction time ($v^{(k_0)}$, $\tau^{(k_0)}$).

### B. Shift and Broadening of Levels in the Light Field

In every discussion, until now it was supposed that the first nonvanishing $k_0$th-order approximation of the perturbation theory was also adequate for the description of the $k_0$-photon ionization of the atoms. The consequence of this supposition is that the product of the energy differences of the unperturbed atomic levels is in the denominator of the transition matrix element of $k_0$th order [see expression (2.21)]. Accordingly the transition matrix element $(r_{f,g}^{(k_0)})$ and the generalized cross section of the multiphoton ionization $\beta^{(k_0)}$ are constant and they are determined by the field free structure, the energy spectrum of the atom. The experimental investigations show that there is agreement between the theoretical formula for the multiphoton ionization rate [expression (2.26a)] in this approximation and the result of the measurement, only if the distortion of the atomic structure does not have to be taken into account; viz., if resonances of any order are carefully avoided, and if the order $(k_0)$ of the ionization and consequently the field strength of the light is small.

In this section the mechanism of the distortion of the atomic structure in the light field, i.e. the shift and broadening of atomic levels, is briefly discussed. A detailed description of the subject can be found in the review papers of Bonch-Bruevich and Khodovoi (1967), Bonch-Bruevich *et al.* (1973), and Papoular and Platz (1973).

The influence of this distortion on multiphoton ionization, which is especially big in the case of $k < k_0$-order resonance, is discussed in the following sections.

The motion of the atomic electron under the influence of the external time-dependent field is governed by the Schrödinger equation (semiclassical treatment)

$$i\dot{b}_k(t) = \sum_{s \neq k} V_{ks}(t)b_s(t)e^{i\omega_{ks}t}, \tag{4.1}$$

where $b_k(t)$ is the wavefunction of the electron in the energy representation of the unperturbed atom. The energy eigenfunction of the unperturbed Hamiltonian $H_0$ is

$$\psi_k^{(0)}(\mathbf{r}, t) = \varphi_k(\mathbf{r})e^{-i\mathcal{E}_k^{(0)}t} \tag{4.2}$$

in the coordinate representation with the energy $\mathcal{E}_k^{(0)}$. The wavefunction of the perturbed atom is expressed as

$$\psi(t) = \sum_s b_s(t)\varphi_s(\mathbf{r})e^{-i\mathcal{E}_s^{(0)}t}; \tag{4.3}$$

$V_{ks}(t)$ is the matrix element of the perturbation Hamiltonian. It is supposed that the atom is in the $n$ state at $t = 0$, i.e.

$$b_k(0) = \delta_{kn} . \tag{4.4}$$

Inserting the trial solution

$$b_n(t) = e^{-i\xi(t)} \tag{4.5}$$

into Eq. (4.1) results in the following expression (Heitler, 1954; Sobelman, 1963):

$$\dot{\xi}(t) = V_{nm} + \sum_{s \neq n} e^{i\xi(t)} V_{ns} b_s e^{i\omega_{ns} t},$$

$$\dot{b}_k(t) = e^{-i\xi(t)} V_{kn} e^{i\omega_{kn} t} + \sum_{s \neq n} V_{ks} b_s e^{i\omega_{ks} t}. \tag{4.6}$$

The phase $\xi(t)$ of the wavefunction $b_n(t)$ is usually a complex number. The system of equations (4.6) can be solved using the initial conditions (4.4) and the method of perturbation theory. The solution regarding the phase of the wavefunction $b_n(t)$ is given up to the second order by the formulas

$$\xi^{(1)}(t) = \int_0^t V_{nn}(t') \, dt' \tag{4.7}$$

and

$$\xi^{(2)}(t) = i \sum_{s \neq n} \int_0^t dt' \, V_{ns}(t') e^{i\omega_{ns} t'} \int_0^{t'} dt'' V_{ns}^*(t'') e^{-i\omega_{ns} t''}. \tag{4.8}$$

In the case of periodical perturbation and the long wavelength approximation the interaction Hamiltonian is given by expression (2.11). Suppose the light is linearly polarized and furthermore $\mathbf{r} \parallel \mathbf{E}$, then the interaction matrix element can be expressed as

$$V_{ns} = \tfrac{1}{2}E(e^{i\omega_p t} + e^{-i\omega_p t}) r_{n,s}^{(1)}, \tag{4.9}$$

where $E$ is the amplitude of the electric field strength of the light.

If the $n$ state is not degenerated the diagonal matrix element $r_{nn}^{(1)}$ is equal to zero because of the parity rule. The first nonzero phase term is given by the second-order approximation of the perturbation theory $[\xi^{(2)}(t)]$.

After performing the integration indicated in expression (4.8) over the period of the oscillation the averaged values of the real and imaginary parts of the phase of the wavefunction $b_n(t)$ are represented by

$$\text{Re } \xi(t) = \overline{\Delta \mathscr{E}_n(t)}$$

$$= \frac{1}{4} E^2 \sum_{s \neq n} \left\{ \frac{\langle n|r|s\rangle \langle s|r|n\rangle}{\omega_s - \omega_n - \omega_p} + \frac{\langle n|r|s\rangle \langle s|r|n\rangle}{\omega_s - \omega_n + \omega_p} \right\} \cdot t, \tag{4.10}$$

$$\text{Im } \xi(t) = -i \, \overline{\Delta \gamma_n(t)}$$

$$= \tfrac{1}{4} E^2 \sum_{s \neq n} |r_{ns}^{(1)}|^2 \, \delta(\omega_{ns} - \omega_p) t. \tag{4.11}$$

$\Delta\overline{\mathscr{E}}_n$, $\Delta\overline{\gamma}_n$ are the energy and level width correction terms of the quasi-stationary state $(\psi_n = \psi_n^{(0)} e^{-\gamma_n t})$ caused by the external perturbing field.

$$\Delta\overline{\mathscr{E}}_n(I) = c_n I, \tag{4.10a}$$

$$\Delta\overline{\gamma}_n(I) = d_n I, \tag{4.11a}$$

where

$$I = \tfrac{1}{2} c E^2$$

is the light power density and $c_n$ is the Stark constant of the level $n$.

The decay constant of the state $\gamma_n$ caused by the vacuum fluctuation of the electromagnetic field is introduced phenomenologically. The line function $f(\omega_{ns} - \omega_p, \gamma_n)$ has to be substituted in expression (4.11) instead of $\delta(\omega_{ns} - \omega_p)$ in the same way.

Consequently the amplitude of the initial state, the wavefunction $b_n(t)$, decreases exponentially in time with the time constant $1/\Delta\overline{\gamma}_n$ [see expression (4.5)]. Meanwhile the amplitude of the other $s \neq n$ states is slowly increasing. At the same time the energy of the initial $n$ state changes by an amount $\Delta\overline{\mathscr{E}}_n$ quadratically depending on the electric field strength of the light $(E^2)$.

The calculation of the shift requires the calculation of the second-order matrix element [see expression (2.21)] which can best be performed by using the Green's function method of Section II,D. That is,

$$\Delta\overline{\mathscr{E}}_n = \tfrac{1}{4} E^2 \int \varphi_n^*(\mathbf{r}_1) r_1 \{ G_{\mathscr{E}_n + \omega_p}(\mathbf{r}_1, \mathbf{r}_2) + G_{\mathscr{E}_n - \omega_p}(\mathbf{r}_1, \mathbf{r}_2) \}$$
$$\cdot r_2 \varphi_n(\mathbf{r}_2) \, d\mathbf{r}_1 \, d\mathbf{r}_2. \tag{4.12}$$

Using this method the Stark constant, i.e. the second-order matrix element, was calculated by Davidkin et al. (1971) and by others (e.g., Vetchinkin and Khristenko, 1968; Sestakov et al., 1972) for different states and for different atoms.

The shift of the levels in the light field can be interpreted as the self-energy change because of the following forward scattering processes (Chang and Stehle, 1971).

The $\omega_p$ photon is absorbed and the atom makes the transition to the virtual state $\mathscr{E}_n + \omega_p$. Afterwards the same photon is emitted and the atoms return to the original state $(n)$. This process is described by the first term in the expression (4.12) of the shift (Fig. 15a).

The $\omega_p$ photon is emitted first and afterwards absorbed. The atom makes the transition to the virtual state $\mathscr{E}_n - \omega_p$ and returns (Fig. 15b). This process is described by the second term of the expression (4.12).

This virtual process changes the energy of the initial state, the energy of which has to be "renormalized." The phenomenon was observed first by

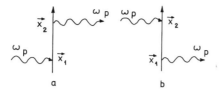

FIG. 15. Second-order processes leading to the energy shift of atomic levels.

Cohen-Tannoudji (1962) measuring the shift of the resonance frequency of the radio frequency transition of the $Hg^{199}$ atom caused by the field of the light wave.

The investigation of the shift of the level became possible in the visible part of the spectrum using high intensity solid state lasers. The shift of the center of the emitted line of the mercury atom was observed by Platz (1968, 1969, 1970) using the high power density radiation of a Nd glass laser.

The shift of the absorption line of a potassium atom was also measured by Bonch-Bruevich et al. (1969) using the radiation of dye laser as the probe beam. The shift was caused by the radiation of the ruby laser.

## C. Nonlinearity versus Frequency Dependence

The distortion of the atomic structure also occurs during the process of multiphoton ionization. The level of the atom in the $k$th-order resonance has a particularly strong influence on the multiphoton ionization rate because of its shift and broadening in high intensity laser radiation.

### 1. Level Distortion in Multiphoton Ionization

The level distortion can be taken into account substituting the field-dependent energies and level widths of the atomic states in the denominator of the transition matrix element [expression (2.21)]:

$$v_{f,g}^{(k_0)} = \sum \langle f \,|\, r \cdot \frac{|k_0 - 1\rangle\langle k_0 - 1\,|}{\mathscr{E}_{k_0-1}(I) - \mathscr{E}_g(I) - (k_0 - 1)\mathscr{E}_p + i\gamma_{k_0-1}(I)} \cdot r \,|$$

$$\vdots$$

$$\cdot \frac{|1\rangle\langle 1\,|}{\mathscr{E}_1(I) - \mathscr{E}_g(I) - \mathscr{E}_p + i\gamma_1(I)} \cdot r \,|\, g\rangle \tag{4.13}$$

(Keldish, 1964; Voronov, 1966). Expression (4.13) can be separated into two terms in the case of $k$th-order resonance with the $s$ level of the $k$ states, that is, if

$$\mathscr{E}_s^{(0)} - \mathscr{E}_g^{(0)} - k\mathscr{E}_p \sim 0. \tag{4.14}$$

Then

$$\imath^{(k_0)}_{f,g} = \imath^{(k_0-k)}_{f,s} \cdot \imath^{(k)}_{s,g} + \imath'^{(k_0)}_{f,g}; \qquad (4.15)$$

$\imath^{(k)}_{sg}$ is the transition matrix element of the $k$-photon excitation of the state $s$ and $\imath^{(k_0-k)}_{f,s}$ is the transition matrix element of the $(k_0 - k)$-photon ionization of the state $s$. The last term in expression (4.15) is the matrix element of the direct $k_0$-photon ionization of the initial level $g$, i.e.

$$\imath'^{(k_0)}_{f,g} = \sum_{v \neq s} \frac{\imath^{(k_0-k)}_{f,v}\imath^{(k)}_{v,g}}{\mathscr{E}_v(I) - \mathscr{E}_g(I) - k\mathscr{E}_p + i\gamma_v(I)}. \qquad (4.16)$$

It is supposed that the last expression (4.16) and its product with the first term on the right-hand side of expression (4.15) is negligible. Then the resonance $k_0$-photon ionization rate is found to be

$$W^{(k_0)}_{f(s)g}(I, \mathscr{E}_p) = \frac{1}{4\pi^2} \int d\Omega_\kappa \cdot \kappa$$

$$\cdot \frac{|\imath^{(k_0-k)}_{f,s}|^2 \cdot |\imath^{(k)}_{s,g}|^2}{(\mathscr{E}_s(I) - \mathscr{E}_g(I) - k\mathscr{E}_p)^2 + \gamma_s^2(I)} \cdot I^{k_0}. \qquad (4.17)$$

$\imath^{(k_0-k)}_{f,s}$ and $\imath^{(k)}_{s,g}$ do not depend on the intensity of light. Expression (4.17) can be transformed into the form

$$W^{(k_0)}_{f(s)g}(I, \mathscr{E}_p) = \tilde{\beta}^{(k_0)}_I \frac{I^{k_0}}{(\mathscr{E}_s^{(0)} - \mathscr{E}_g^{(0)} + c_s'I - k\mathscr{E}_p)^2 + (\gamma_s^{(0)} + d_s'I)^2} \qquad (4.18)$$

using expressions (4.10a) and (4.11a). $\tilde{\beta}^{(k_0)}_I$ is constant.

$$c_s' = c_s - c_g, \qquad (4.19a)$$

$$d_s' = d_s - d_g. \qquad (4.19b)$$

Expression (4.18) is obviously higher than the $k_0$th-order approximation of the perturbation theory for the description of the $k_0$-photon ionization. The shift of the levels is taken into account in the second-order approximation. Consequently the number of quanta which are absorbed or emitted in the process of $k_0$-photon ionization is $k_0 + 2$. So the order of approximation can be regarded to be $k_0 + 2$ in expression (4.18).

As is apparent, the simple power function relationship of the $k_0$-order approximation [see expression (2.26c)] is not valid between the ionization rate and the light power density. The ionization rate is a complicated function of the light power density. This resonance ionization rate can be expressed as the product of the power function $I^{k_0}$ and a resonance function of the light power density in the case of second-order approximation of the shift of the resonance level.

## 2. Nonlinearity of the Multiphoton Ionization

It is the ionization yield $[N_i(I)]$ versus light power density function which is proportional to the rate of the ionization that is measured in the experiment (see Section III,A,1). For the sake of simplicity it is supposed that the light power density is independent of the space–time coordinates and is monochromatic [see the supposition in the paragraph prior to expression (2.20)]. The result of the measurement is usually displayed on a double logarithmic scale (see Section III,A,5).

The range of the light power density in which the measurements are performed is small, i.e. the power function $I^{k_0}$ is a very steep function and a small variation in the light power density causes a change in the ionization rate of many orders of magnitude. But the range of the measurement of the ionization yield is limited to two or three orders of magnitude. If the light power density is high the process is saturated. On the other hand, the lower limit of the measurement of the yield is determined by the sensitivity of the ion detection.

The logarithm of the ionization rate can be approximated by a Taylor series expansion, retaining only the first two terms of the series in the small range of the light power density of the measurement. This means that the complicated function of the logarithm of the ionization rate is approximated by its tangent, i.e. by a linear function of the logarithm of the light power density ($\log I$). It is the slope of this line which is regarded to be the nonlinearity of the process in the following. This nonlinearity is measured in the experiment and is found to be the $k_{\text{meas}}$.

$$k_{\text{meas}}(I_t, \mathcal{E}_p) = (\partial \log W(I, \mathcal{E}_p)/\partial \log I)_{\substack{\mathcal{E}_p=\text{const.} \\ I=I_t}}, \qquad (4.20)$$

or, using expression (4.18),

$$k_{\text{meas}} = k_0 - 2I_t \frac{c_s'(\mathcal{E}_s^{(0)} - \mathcal{E}_g^{(0)} - k\mathcal{E}_p + c_s'I_t) + d_s'(\gamma_s^{(0)} + d_s'I_t)}{(\mathcal{E}_s^{(0)} - \mathcal{E}_g^{(0)} - k\mathcal{E}_p + c_s'I_t)^2 + (\gamma_s^{(0)} + d_s'I_t)^2}.$$

$$(4.20a)$$

$I_t$ is the value of the light power density at which the measurement is performed and around which value the Taylor series expansion is made. The plot of this function has been published by Delone et al. (1972) taking a different ratio of the constants $d_s'$ to $c_s'$. This function can be seen in Fig. 16.

The shape of the function $k_0 - k_{\text{meas}}(I_t; \mathcal{E}_p)$ depending on the energy of the light quanta ($\mathcal{E}_p$) is absorption-like if $d_s' \gg c_s'$. The rate of the ionization increases more slowly than the function $I^{k_0}$ by increasing the light power density because of spreading and therefore diminishing the influence of the resonance level on the multiphoton ionization. The broadening of the level exerts the greatest influence on the nonlinearity of the process at zero field-free detuning of the resonance, i.e.

$$\Delta_0 = \mathscr{E}_s^{(0)} - \mathscr{E}_g^{(0)} - k\mathscr{E}_p \,, \tag{4.21}$$

lowering its value considerably. The result of the broadening of the resonance level is always a decrease in the nonlinearity of the process. The most significant decrease in the nonlinearity is 2 which is caused by the smearing off of the resonance level.

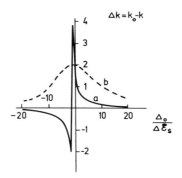

FIG. 16. The deviation ($\Delta k$) of the nonlinearity ($k_{meas}$) of the multiphoton ionization from $k_0$ depends on the field-free resonance detuning $\Delta_0$, expressed in the units of the shift of the resonance level in the light field $\Delta\mathscr{E}_s$. Curve a: the shift of the level is bigger than its broadening; curve b: the broadening is larger than the shift (Delone et al., 1972).

The shape of the curve, $k_0 - k_{meas}$ $(I_t = \text{const}; \mathscr{E}_p)$ is dispersion-like if $c_s' \gg d_s'$, i.e. if the shift of the resonance level is greater than its broadening. If the resonance level is shifted into resonance by the light field the " dynamic " detuning of the resonance

$$\Delta = \Delta_0 + c_s' I \tag{4.22}$$

approaches zero in the case of increasing the light intensity. So the rate of the ionization increases faster than $I^{k_0}$ and consequently the nonlinearity of the process is greater than $k_0$. On changing the sign of the field-free detuning, i.e. if the energy of resonance number of quanta is on the opposite side of the level, the resonance state moves away from the resonance in the case of increasing the light power density. The ionization rate takes place more slowly than $I^{k_0}$, and therefore the nonlinearity of the process is smaller than $k_0$.

## 3. Experimental Investigations

It was suspected by Delone and Delone (1968) that this resonance process with field-dependent shift and broadening of the resonance level plays a role in the six-photon ionization of xenon atoms, lowering the nonlinearity of the process by a factor of about 2 (see Table I). No lowering of the nonlinearity of the six-photon ionization could be observed in the case of Kr atoms at the same time. There are no resonances in the last process.

*a. Potassium atoms.* The behavior of resonance multiphoton ionization was investigated in the case of four-photon ionization of potassium atoms by Delone and Delone (1969) and Delone et al. (1972). There is three-photon resonance with the 4f states of the atom.

The radiation of the Nd glass laser is used to ionize the atom. The laser is Q-switched by a rotating prism. The frequency of the radiation is tuned in the range of the fluorescent bandwidth of the Nd atoms in the glass whose bandwidth is about $100 \text{ cm}^{-1}$. The frequency tuning is performed by two Fabry-Perot interferometers within the laser cavity. The interferometer having the smaller free spectral range serves simultaneously as the output mirror of the laser. Rotation of the second interferometer about the axis perpendicular to the optical axis of the laser resonator tunes the laser frequency. The bandwidth of the radiation is about $4 \text{ cm}^{-1}$.

The experimental setup, part of which is similar to that described in Section III,B,1, is used also for the resonance ionization measurement (see Fig. 7).

The ion yield (i.e. the ionization rate) versus light power density dependence is measured at different parameter values of the light frequency. The slope of the straight line of the ion yield versus light power density function is determined in the logarithmic scale from the experimental curves. Then the values of this slope, i.e. the nonlinearity of the process, are plotted as a function of the field-free resonance detuning of the three-photon resonance ($\Delta_0$); see expression (4.21). The result of the measurement can be seen in Fig. 17.

As is apparent, the shape of the measured nonlinearity agrees with the theoretical predictions if the shift and broadening of the resonance level is supposed to be of the same value or the broadening is a little more than the shift. The asymmetric shape of the curve is unexplained as yet.

The result of the measurement shows the important role which multiphoton resonance plays in the multiphoton ionization process. It can be concluded that multiphoton resonance may be one of the causes of the decrease in the nonlinearity of the multiphoton ionization of the rare gases. However, the problem remains that the increase in the nonlinearity also follows from the feature of the resonance multiphoton ionization process (see Fig. 16).

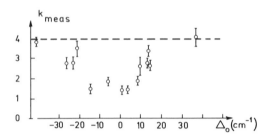

Fɪɢ. 17. Experimentally measured dependence of the nonlinearity $k_{meas}$ of the four-photon ionization of potassium atoms on the field-free resonance detuning $(\Delta_0)$ of the three-photon resonance excitation of the 4f levels (Delone *et al.*, 1972).

*b. Metastable helium atoms.* The resonance five-photon ionization of a triplet metastable helium atom was investigated (Bakos *et al.*, 1972a) using a Nd glass laser $Q$-switched by a rotating prism.

The frequency tuning of the laser is accomplished essentially by the same means as in the work of Delone *et al.* (1972). The bandwidth of the laser radiation is about $0.2$ cm$^{-1}$. The metastable atoms are created in a mild gas discharge. The interaction of the metastable atoms with the high intensity laser beam also occurred in the discharge and not in the afterglow, contrary to the earlier experimental arrangement of Fig. 8. The ions are detected in the usual way, by a Langmuir probe. The other part of the experimental setup is similar to that described in Section III,C,1.

Because of the permanent population of different excited levels of the atom by the continuous discharge, multiphoton ionization processes of different order can be observed to take place simultaneously from the different excited levels (Bakos *et al.*, 1972b). These processes can be separated from each other because the processes of different order have different generalized cross sections and therefore give a dominant contribution to the total yield of ions at different ranges of the light power density.

Four-photon resonance can be attained consecutively with the different $n^3$ S, D, G states of principal quantum number $11 < n < 16$. The investigation of these resonance states is especially important because these levels are in the region of the energy spectrum of the atom where the states are densely packed. The ionization limit of the atom is also over a small energy distance $(\sim 600$ cm$^{-1})$. Therefore the overlapping of levels due to the broadening and shifting, i.e. the formation of the quasi-continuous spectrum and the lowering of the ionization potential, is most likely (Voronov *et al.*, 1966a).

The rate of the ionization or, preferably, in this case, the multiphoton signal versus light power density function, is measured (see Section III,C,2) for different values of the frequency of the light, as in the earlier described experiment of Delone *et al.* (1972). The experimental points are fitted by a straight line on the double logarithmic scale. The slope of this line is the measured nonlinearity of the process.

The result of the measurement may be seen in Fig. 18 and it is apparent from the figure that the nonlinearity ($k_{meas}$) of the four-photon ionization of the triplet metastable helium atom depends significantly on the energy of four quanta. This fast function also gives evidence of existing sharp levels in that region of the spectrum. This means that the levels are not smeared off even near the ionization threshold at the light power density $\sim 10^9$ W/cm$^2$.

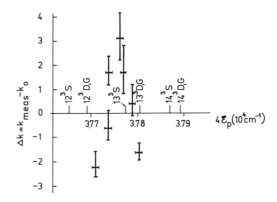

FIG. 18. The dependence of the deviation ($\Delta k'$) of the measured nonlinearity ($k_{meas}$) of the five-photon ionization of the triplet metastable He atom from $k_0$ on the energy of four light quanta ($4\mathscr{E}_p$) (Bakos et al., 1972a).

The value of the nonlinearity is greater than $k_0$ in one region of the energy of four photons, justifying the validity of the description of the rate of the resonance multiphoton ionization by the formula (4.18) or (4.20a), i.e. the values $k_{meas} \gtrless k_0$ are equally possible on the basis of expression (4.20a). Experimentally the greater value is observed the first time in the case of metastable He atoms.

c. *Cesium Atoms.* Further investigation of the resonance four-photon ionization of cesium atoms confirmed the existence of a higher than $k_0$ value of the nonlinearity of the process (Held et al., 1973).

The experimental setup is the slightly modified variant of the measuring apparatus described in Section III,B,1. The laser is Nd glass Q-switched by a rotating prism. The frequency of the radiation is tuned by a birefringent filter placed in the laser cavity. The bandwidth of the radiation is 0.4 cm$^{-1}$. The three-photon resonance is accomplished with the 6f levels.

The nonlinearity of the process is determined by measuring the ionization rate versus light power density function. The result of the measurement of the nonlinearity can be seen in Fig. 19. The curve is dispersion-like, as is

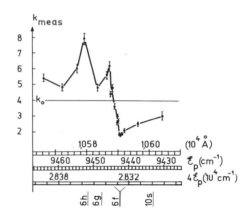

FIG. 19. The nonlinearity versus photon energy function of the four-photon ionization of Cs atoms. The ionization is caused by Nd glass laser radiation. The photon energy, the energy of three quanta, and the position of the atomic levels are simultaneously plotted on the horizontal axis in addition to the wavelength of the light quanta (Held *et al.*, 1973).

expected theoretically for the case of a small broadening to shift ratio. The resonance with the other levels near the 6f state is not possible because of the electric dipole selection rule. Therefore the resonance can be regarded to occur only with one level.*

The ionization rate versus light frequency function is also measured†‡ (Fig.20). The position of the maximum of the curve, according to the present author, fully coincides with the position of the unperturbed atomic state. There is consequently no shift of the resonance level. The rate versus light frequency function also has a secondary maximum which can be ascribed to resonance with any levels of the atom.

It can be seen that there are some difficulties with regard to the interpretation of the result. Namely, if there is no shift of the resonance level the existence of the greater than $k_0$ value of the measured nonlinearity is difficult to understand.

The explanation of the complicated structure of the rate versus frequency function needs a new physical concept which is not contained in the theoretical description which has been outlined in this and the preceding sections.

*d. Rare Gas Atoms.* The eleven-photon ionization of Xe and the twelve-photon ionization of the Kr atom was investigated (Alimov *et al.*, 1971; Delone, 1973) in order to reveal the real cause of the smaller than $k_0$ value of

---

* The doublet level 6f can be regarded as singlet if the laser line width is larger than the spin-orbit coupling.

† See the footnote on page 99.

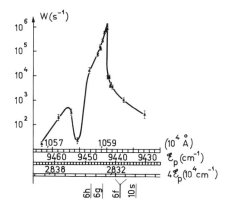

FIG. 20. The rate of the four-photon ionization of Cs atoms versus energy of the light quanta function. The wavelength of the radiation, the energy of three quanta, and the field-free position of the atomic levels are also plotted on the horizontal axis in addition to the energy of the quanta (Held *et al.*, 1973).

the nonlinearity of the rare gas atoms observed in several experiments and discussed in detail in Section III,A. (See further details Alimov *et al.*, 1973.)

The experimental apparatus used is essentially the same as described in Section III. The difference in this setup is in the Nd glass laser which is similar to that used in the resonance multiphoton investigation of the potassium atoms (see Section IV,C,3,a). A further difference, compared with the laser used in the last resonance investigations, is that a multiple stage light amplifier was used in order to attain the light power density needed for the ionization with the absorption of such a large number of quanta. The single light pulse output energy was about one joule. The duration of the pulse is ~ 30 nsec. The range of the frequency tuning is 24 cm$^{-1}$ and the bandwidth of the radiation is ~ 3 cm$^{-1}$.

The ionization rate versus light power density function is measured at different frequencies of the light. The nonlinearity of the process is determined as the logarithmic derivative of the measured curve as before. The result of the measurement of the nonlinearity is plotted in Fig. 21. The nonlinearity of these atoms shows sharp resonance character. This character verifies the existence of sharp atomic levels even at such a high light power density as $3 \times 10^{12}$ W/cm$^2$, i.e. at the electric field strength of the light of $5 \times 10^7$ V/cm.

The results of the earlier measurement listed in Table I are also plotted in the figure and these results are in good agreement with the curve measured in the resonance investigations. It is concluded from the large variation of the nonlinearity with frequency, and the narrow region of frequency where the value of the nonlinearity agrees with $k_0$, that the measured smaller than

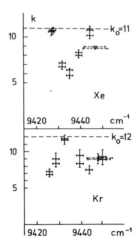

FIG. 21. The nonlinearity versus frequency dependence of the multiphoton ionization of Xe and Kr atoms (Delone, 1973).

$k_0$ value of the nonlinearity in the earlier measurement has a random nature which is not characteristic for the atoms of rare gases but depends rather on the experimental circumstances.

### D. General Light Power Density Function of Ionization Rate

The first two terms in the Taylor series expansion of the logarithm of the multiphoton ionization rate are an inadequate approximation if the Stark constant is large. The slope of the straight line of the log $W$–log $I$ function, the nonlinearity of the process, then loses its meaning (Bakos et al., 1972c). The expression (4.18) has to be used in its general form to describe the ionization rate versus light power density dependence. As already noted, the rate is the product of a power function $I^{k_0}$ and a resonant function. This last resonant function contains information about the dependence of the energy of the resonant level on the light power density. Namely, if the energy of the resonant state approaches the resonance in the case of increasing light power density the energy of the level can be determined. (For further details see Bakos et al., 1973b.) The "dynamic" detuning [expression (4.22)] is zero at exact "dynamic" resonance and the energy of the resonance level

$$\mathscr{E}_s(I_R) = k\mathscr{E}_p \tag{4.23}$$

is equal to the energy of the $k$ photon. To deduce expression (4.23) the expression (4.21) was used and it was supposed that the energy of the initial level $\mathscr{E}_g(I)$ was zero. The intensity at which the dynamic resonance takes place, the resonance intensity $(I_R)$, can be determined as follows. The resonance peak occurs on the ionization rate versus light power density function as the consequence of the power function multiplying the resonance function (Fig.22). The resonance intensity $(I_R)$ is determined by the position of this resonance peak on the light power density axis.

If the resonance peak is not clearly expressed because of the large value of the resonance breadth, $\gamma_s(I)$, the cross-point of the tangents to the rate function before and after the resonance point determines the resonance intensity. This is a quick procedure for orientation. If only a single level is in resonance, the general cross section of the ionization $\beta_I^{(k_0)} = W(I)/I^{k_0}$ is the resonance function multiplying the power function, and the peak of this resonance function clearly determines the value of the resonance intensity (Bakos *et al.*, 1973d).

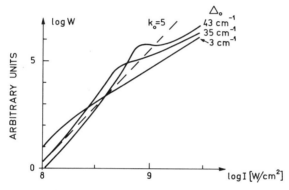

FIG. 22. Theoretical five-photon ionization rate versus light power density dependence of the triplet metastable helium atom in the case of fourth-order resonance. The parameter is the field-free resonance detuning (Bakos *et al.*, 1972c).

Changing the frequency of the light causes the resonance to occur at another light power density. The inverse of the function $I_R(k\mathscr{E}_p)$ is the energy dependence of the resonance level on the light power density [see expression (4.23)].

## 1. Triplet Metastable Helium Atom

The five-photon ionization of the metastable He atom is investigated over a wide range of the light power density using the modified experimental setup described in Section III,C. The modification is given in Section IV,C,3,b.

The measured ionization rate versus light power density can be seen in Fig.23. The points are experimental points and the solid line is the result of

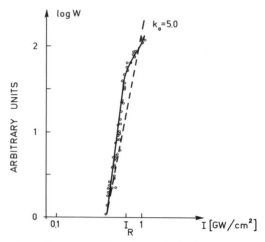

FIG. 23. Experimentally measured five-photon ionization rate versus light power density function of the metastable helium atom in the case of four-photon resonance with the $13^3S$ state. The points are experimental points; the curve is the theoretical fit on the basis of expression (4.18) (Bakos et al., 1972c).

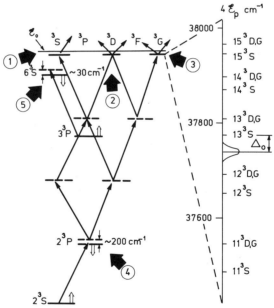

FIG. 24. The energy scheme of the triplet system of the helium atom and the diagram of the five-photon ionization of the triplet metastable level. Black arrows 1, 2, 3 show the fourth-order resonances. The corresponding range of the spectrum is plotted in an enlarged scale on the right-hand side of the figure where also the position of the "four-photon laser line" is indicated. Black arrows 4 and 5 show the levels in first-order resonance. The solid line arrows indicate the transitions followed by the absorption of one photon. The hollow arrows show the direction of the shift of the levels.

theoretical calculation using the expression (4.18). The values $c'_s$ and $d'_s$ are chosen to fit the measured points. It can be concluded that the theoretical model of the resonance ionization [expression (4.18)] is adequate.

Investigating the energy spectrum of the helium atom it is possible to recognize that the resonance levels $n^3$S, D, G are not shifted appreciably but the initial state $2^3$S is as a result of the excess first-order resonance with the $2^3$P state (see Fig. 24).

The resonance energy difference of different resonance levels and the initial one versus the light power density function were measured in the five-photon ionization of triplet metastable He atoms. The light field dependent energy of the levels was determined by the method outlined above. The dependence of the energy of the helium states on the light power density is seen in Fig. 25. It is customarily supposed in the plot of the figure that the initial level is not shifted. (As mentioned earlier, the contrary is true because of the first-order resonance with the $2^3$P state.) The observed shifts are big, amounting to some tenths of cm$^{-1}$.

The result cannot be explained simply by the fourth-order resonances with the $n^3$S, D, G states and by the first-order resonance with the $2^3$P state. The measured Stark constants $c'(n^3$S, D, G) behave singularly at the frequency indicated by the arrow in the figure. This frequency coincides with

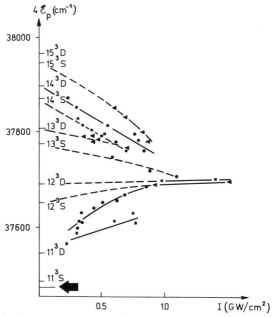

FIG. 25. The level energy versus light power density dependence of the helium atom. The energy is counted from the triplet metastable state.

the resonance frequency of the transition $3^3D$–$6^3S$, having nothing to do with the five-photon ionization in the first approximation.

It can be concluded that the structure of the atom is heavily distorted by the high intensity laser light of the multiphoton ionization in the case of resonances of different order. Even resonances having nothing to do with the multiphoton ionization process in the first approximation strongly influence the multiphoton ionization process.

## 2. Cesium Atoms

The resonance three-photon ionization of cesium atoms was similarly investigated by Evans and Thonemann (1972, 1973) over a wide range of light flux using a single pulse ruby laser. The two-photon resonance takes place with the $9^2D$ state of the atom. The result of the ion yield versus light flux measurement can be seen in Fig. 26. The resonance peak is clearly seen

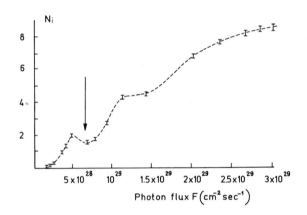

FIG. 26. The three-photon ionization rate versus light flux function of cesium atoms in the case of two-photon resonance with the 9D state using a ruby laser. The arrow indicates the level shift caused by a decrease of the rate (Evans and Thonemann, 1972).

on the curve. The slope of the tangent of the curve before the resonance peak equals $k_0$. Therefore the decreasing yield in the case of increasing light flux is explained by shifting the resonance level out of resonance. The influence of the decreasing resonance function on the ion yield is greater than that of the increasing power function $I^{k_0}$. The yield consequently decreases. The further oscillation of the curve is explained by the structure of the focal plane distribution of the light flux and by the saturation of the process.

The result of the measurement confirmed qualitatively the theoretical model given by expression (4.18).

It should be noticed that the shifting of levels out of resonance leads to the bleaching of the material which absorbs the light in multiphoton transition. The phenomenon is treated theoretically by Manakov *et al.* (1971a) and Zon *et al.* (1971b) and was earlier predicted by Keldish (1964).

## E. Strong Field Resonance Ionization

All the results of the experimental investigations of the resonance multiphoton ionization are explained using the method of perturbation theory. The multiphoton ionization is more simply described in the $k_0$th-order approximation. As the next step in the hierarchy of the subsequent approximations, the distortion of the energy spectrum of the atom was taken into account in the second-order approximation. The $(k_0 + 2)$th-order approximation of the $k_0$-photon ionization rate is obtained after substituting the result achieved for the atomic distortion in the formula describing the $k_0$-photon ionization rate. The experimental results generally harmonize with the $(k_0 + 2)$th-order approximation but some of them cannot be explained adequately. These experimental results are: (1) the unexplained higher than $k_0$ value of the nonlinearity of the four-photon ionization of the cesium atom while the resonance level is not shifted (Held *et al.*, 1973); (2) the secondary peak in the rate versus frequency function of the cesium atom which cannot be ascribed to any level of the atom (Held *et al.*, 1973); (3) the big shifts and singularity of the shift of the resonance levels of the He atoms at the resonance frequency of the transition which has nothing to do with the five-photon ionization of the metastable He atoms (Bakos *et al.*, 1972c); (4) the as yet unexplained smaller than $k_0$ value of the nonlinearity of the multiphoton ionization of rare gases (Section III,A); and (5) the sharp resonances of the nonlinearity of the rare gases (Alimov *et al.*, 1971) which are not identified as resonances with definite atomic levels and observed at very high light power density.

It seems that the interpretation of these experimental results cannot be regarded as satisfactory on the basis of the $(k_0 + 2)$th-order approximation. Some theoretical attempts have been made to treat the multiphoton ionization in a higher order approximation but the theoretical and experimental results have not yet been compared.

There are different methods of obtaining more accurate theoretical results, namely: (a) to correctly take into account higher order energy correction terms of the perturbation theory (Chang and Stehle, 1971; Gontier and Trahin, 1968b, 1973b); (b) to accurately solve the equation of motion of atomic levels, especially that of the degenerated atomic levels (Kovarski and Perelman, 1971b; Delone *et al.*, 1971b), and to use the resulting new levels in the modified perturbation theory (Pert, 1972; Kotova and Terentev, 1967;

Kovarski, 1969a; Lebedev, 1971); and (c) to accurately solve the Schrödinger equation leading to the notion of quasi-energy states (Zel'dovich, 1966; Kovarski, 1973).

## 1. *Higher Order Energy Corrections*

As described in Section IV,B, the Stark shift and level broadening can be considered as the consequence of the absorption and subsequent emission or first the emission and afterward the absorption of the same quantum of energy. This process is said to be the forward scattering of the photon (Chang and Stehle, 1971). The process can be expressed as the application of the operator

$$
\begin{aligned}
M(x_1, x_2) = {}& \gamma_\mu A_\mu^{(-)}(x_1) \cdot G(x_1, x_2) \cdot \gamma_\mu A_\mu^{(+)}(x_2) \\
& + \gamma_\mu A_\mu^{(+)}(x_1) \cdot G(x_1, x_2) \gamma_\mu A_\mu^{(-)}(x_2)
\end{aligned}
\tag{4.24}
$$

to the atomic electron (Gontier and Trahin, 1973b). $\gamma_\mu$ are the Dirac's matrices. $x$ is the space-time point. $A_\mu^{(+)}(x)$, $A_\mu^{(-)}(x)$ are the positive and the negative parts of the vector potential, i.e.

$$
\begin{aligned}
A_\mu^{(+)}(x) &= (A_\mu^{(-)}(x))^+ \\
&= \sum_\lambda \int [q_\lambda(K_\lambda)/(2K_\lambda)^{1/2}] e_{\lambda\mu} e^{iKx} \, dK_\lambda
\end{aligned}
\tag{4.25}
$$

[see expression (2.12)]. The operator $M(x_1, x_2)$ can be applied at any stage and time of the multiphoton process (see Fig. 27). This appears as the distortion of the Green's function under the influence of the light. This new Green's function

$$
\begin{aligned}
G'(x_1, x_2) = {}& G(x_1, x_2) \\
& + \int G(x_1, x_3) M(x_3, x_4) G'(x_4, x_2) \, dx_3 \, dx_4
\end{aligned}
\tag{4.26}
$$

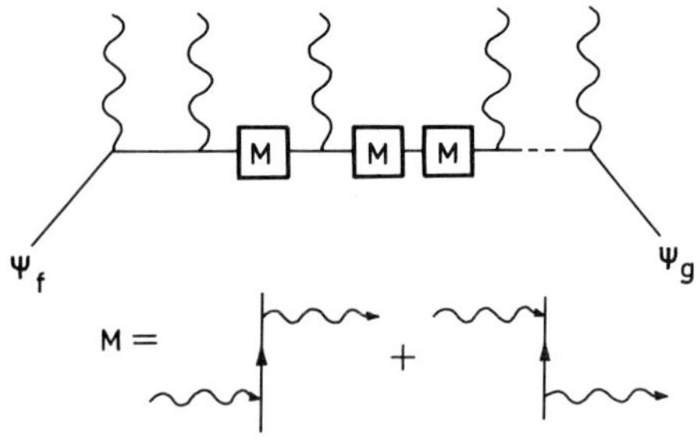

FIG. 27. The distortion of the atomic structure in the multiphoton ionization process can be taken into account in any accuracy by repeated application of the "mass operator $(M)$" to the virtual states.

(Feyman, 1949; Low, 1952) can be used for the evaluation of the transition $S$ matrix element (Roman, 1965).

$$S_{f,g}^{(k_0)} = \int dx_{k_0} \int dx_{k_0-1} \cdots \int dx_1 \langle n_p - k_0; a_{k_0} | \gamma_\mu A_\mu^{(+)}(x_{k_0})$$

$$\cdot G'(x_{k_0}, x_{k_0-1}) \gamma_\mu A_\mu^{(+)}(x_{k_0-1}) \cdots G'(x_2, x_1) \gamma_\mu A_\mu^{(+)}(x_1) | a_g, n_p \rangle$$

$$(4.27)$$

[see the corresponding nonrelativistic equation (2.51)].

Using this method the expression (4.17) for the resonance multiphoton ionization rate is deduced as the approximation of the application of the mass operator to the intermediary resonance state only once (Gontier and Trahin, 1968b).

The transition matrix elements between the levels of a two-level system, the shift and the broadening of the levels, were also calculated by Chang and Stehle (1971) using this method. The peculiar behavior of the "saturation" of the shift and broadening is predicted (Fig. 28a,b). This first phenomenon was also treated by Kovarski (1973).

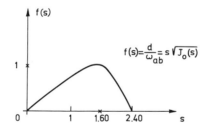

FIG. 28a. The width $(d)$ of the single quantum transition of the two-level atom and dependence on the "field strength parameter" of the light $s = (1/\omega_{ba})(n/\omega_p)^{1/2} \langle a | \mathbf{e}_\lambda e^{\mathbf{Kr}} | b \rangle$ (Chang and Stehle, 1971).

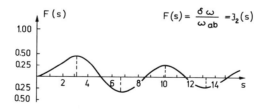

FIG. 28b. The shift $\delta\omega$ of the transition frequency $\omega_{ab}$ of the two-level atom and dependence on the field strength parameter of the light wave. (The definition can be found in the legend of Fig. 28a.) (Chang and Stehle, 1971).

A consequent application of this theory to the multiphoton ionization of hydrogen atoms (Gontier and Trahin, 1973a) results in a shift of the virtual state, and not only the shift of the near resonant level has to be taken into account. This requirement is immediately apparent from Fig. 27. The notion of the shift of the virtual state includes the resulting consequences of the shift of all the levels of the atom simultaneously.

Naturally if the energy of the virtual state $(\mathscr{E}_0 + k\mathscr{E}_p)$ is very near to that of one level of the atom the shift of this level plays a decisive role in the process. Then the treatment of the resonance multiphoton ionization made in the preceeding sections is regarded as valid. But in other cases the shifts of all the levels appear simultaneously with the shift of the virtual state. Then large shifts are observable. The multiphoton resonance induces resonance in some shifts of the virtual state (Gontier and Trahin, 1973b). As a consequence of this induced resonance in the shift a secondary maximum appears in the rate versus light frequency function. This secondary maximum cannot be attributed to resonance with any level of the atom (see Fig.29a). This phenomenon could be the explanation of the already observed secondary maximum on the ionization rate versus frequency function of the Cs atom (Held et al., 1973).

Attention must be drawn to the curves of nonlinearity in Fig. 29a. These curves resemble the curves of Fig. 16 of the paper of Delone et al. (1972) and confirm them.

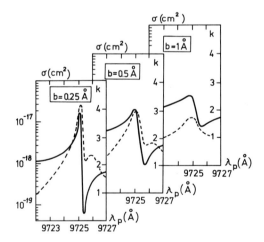

FIG. 29a. The dependence of the nonlinearity and the cross section of the three-photon ionization of the atomic hydrogen on the wavelength of the ionizing light at different values of the bandwidth of the light ($b$). The model of the amplitude stabilized laser light was supposed (Gontier and Trahin, 1973b).

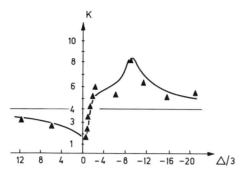

FIG. 29b. The nonlinearity versus light frequency function of the four-photon ionization of a Cs atom. The solid triangular points are experimental ones measured by Held *et al.* (1973). The solid line curve is the result of the theoretical calculation of Chang and Stehle (1973). Three-photon resonance with the 6f level takes place at $\triangle = 0$. $\triangle$ is the "dynamical" detuning of the resonance. (Chang and Stehle, 1973.)

The application of this theory to the five-photon ionization of the triplet metastable He atom can probably explain the observed shift of the levels.

Using the above described theory Chang and Stehle (1973) calculated the rate and nonlinearity of the multiphoton ionization of the Cs atom as a function of the light frequency (see Fig. 29b). A small shift of the three-photon resonance level (6f) was found which can already explain the observed higher than theoretical value $(k_0)$ of the nonlinearity (Held *et al.*, 1973).

It is surprising that the secondary maximum in the ionization rate versus light frequency function is not shown by this calculation, contrary to the prediction given by Gontier and Trahin (1973b) for the case of the hydrogen atom.

## 2. Modified Perturbative Methods

The perturbative method of the preceding section cannot be applied if the interaction with the light is so strong that the series expansion of the perturbation theory is not convergent or the interaction energy is bigger than the separation between the levels. The interaction is especially strong in the case of resonances. The levels connected with multiphoton transition can also be strongly influenced—the transition is "saturated" by the intense light of the laser leading, eventually, to multiphoton ionization of the atom.

The original degenerate levels are split, and the degeneration is resolved. New states are formed from the field-free states which do not have further harmonic time dependence. These new states, with the older uninfluenced states, can be used as the basic states of the new perturbation theory.

Already the first term of this new perturbation expansion, which corresponds to the one-photon transition amplitude in the older perturbation theory, gives the summed-up contribution to the transition amplitude of the multiphoton transitions of different order. In these multiphoton transitions the first photon absorbed by the atom is supplemented by the remaining number of photons needed for the transition, which have already been spent for the distortion of the structure of the atom, to form new atomic states.

    a. *Degenerate States.* The multiphoton ionization of the hydrogen atom is discussed as an example (Kovarski and Perelman, 1971a; Delone *et al.*, 1971b). The influence of the resolution of the degeneracy of the states is investigated in the multiphoton ionization. The system of equations of motion of these states is solved unequivocally. The new states are

$$\psi_{v_1(v_2)}(\mathbf{r}, t) = (1/\sqrt{2})[\psi_\alpha(\mathbf{r}) \pm \psi_\beta(\mathbf{r})]$$
$$\cdot \exp(-i\mathscr{E}_{\alpha(\beta)}t - i\rho_{v_1(v_2)} \sin \omega_p t); \qquad (4.28)$$

in the case of principal quantum number $n = 2$, for instance. The $\psi_\alpha(\mathbf{r})$, $\psi_\beta(\mathbf{r})$ are the wavefunctions of the states $2S_{1/2}$ and $2P_{1/2}$ in the absence of the light wave. The expression

$$\rho_{v_1(v_2)} = (\pm)V_{\alpha\beta}/\omega_1 \qquad (4.29)$$

is the solution of the secular equation

$$\mathrm{Det}\,|\omega_p\rho_{v_i}\,\delta_{\alpha\beta} + V_{\alpha\beta}| = 0, \qquad (4.30)$$

where

$$V_{\alpha\beta} = Ez_{\alpha\beta} \qquad (4.31)$$

is the interaction Hamiltonian in the case of $z$ polarization and the electric dipole approximation. The Green's function of the atomic electron

$$G = G_1 + G_2. \qquad (4.32)$$

The contribution of the bound states to the Green's function is

$$G_1(\mathbf{r}_1, t_1; \mathbf{r}_2 t_2) = -i\theta(t_1 - t_2) \sum_{n,\,\alpha,\,\beta} \psi_{n\alpha}^{(v)}(\mathbf{r}_1)\psi_{n\beta}^{*(v)}(\mathbf{r}_2) \cdot c_{n\alpha}^{(v)} \cdot c_{n\beta}^{(v)}$$
$$\cdot \exp[-\mathscr{E}_n(t_1 - t_2) - i\rho_{nv}(\sin \omega_p t_1 - \sin \omega_p t_2)]. \qquad (4.33)$$

$G_2(\mathbf{r}_1, t_1; \mathbf{r}_2 t_2)$ is the part of the Green's function connected to the continuum state. It is supposed that the wavefunction of the continuum state is Volkov's solution of the wave equation (Volkov, 1935). The first part of the Green's function $G_1(\mathbf{r}_1, t_1; \mathbf{r}_2, t_2)$ is not harmonic in time and can be expressed as a double harmonic series if the function

$$\exp(i\rho \sin \omega_p t) = \sum_n J_n(\rho)\exp(in\omega_p t) \qquad (4.34)$$

is expanded into a harmonic series with Bessel functions $[J_n(\rho)]$ as expansion coefficients.

The multiphoton ionization rate can be obtained using the Green's function given by expressions (4.32), (4.34), and the $k_0$ th-order expansion of the $S$ matrix

$$W_{\kappa,g}^{(k_0)} = 2\pi \sum_{\kappa} \sum_{N_1 \cdots N_{k-1} = -\infty}^{\infty} |A_{g\kappa}(N_1, N_2, \ldots, N_{k-1})|^2$$
$$\cdot \delta[\mathscr{E}_\kappa - \mathscr{E}_g - k\omega_p - 2(N_1 + N_2 + \cdots + N_{k-1})\omega_p], \quad (4.35)$$

where

$$A_{g\kappa}(N_1, N_2, \ldots, N_{k-1}) = \kappa \sum_{j_1, \ldots, j_{k-1}} \sum_{l_1, \ldots, l_{k-1} = -\infty}^{\infty} V_{gj_1} R(j_1, l_1, N_1)$$
$$V_{j_1 j_2} R(j_2 l_2 N_2) \cdots V_{j_{k-1}\kappa} \quad (4.36)$$

and

$$R(j_s, l_s, N_s) = c_{js} J_{2N_s + l_s}(\rho_{js}) J_{l_s}(\rho_{js})/[\mathscr{E}_{js} + \mathscr{E}_g + (l_s - s)\omega_p]. \quad (4.37)$$

The expression (4.36) is the new matrix element of $k_0$ th order of the interaction Hamiltonion $V$. The modification in comparison with expression (2.19) is the multiplication by the $R(j_s, l_s, N_s)$ expressions which shows the distortion of the atomic levels. The $\delta(x)$ function in expression (4.35) states the rule of conservation of energy. The energy difference between the final $(\mathscr{E}_\kappa)$ and the initial state $(\mathscr{E}_g)$ is equal to the energy of the absorbed quanta $(k_0)$ if there is no distortion of the atomic states $(N_1, N_2, \ldots, N_{k-1} = 0)$. If there is distortion, i.e. some $N_i \neq 0$, apparently a "smaller number of quanta $k < k_0$" is sufficient for the transition. Expressing this more exactly, a smaller than $k_0$ th-order expansion term of the $S$ matrix element gives a contribution to the transition rate in the perturbation theory using the distorted level wavefunctions as the basic set. Naturally the total number of quanta absorbed in the process satisfies the energy conservation as the $\delta(x)$ function clearly shows in expression (4.35).

b. *Strongly Coupled States.* If two atomic states are coupled by any photon transition, the system of equations of motion of these states is solved separately from that of the other levels. The resulting new states are used in the "modified" perturbation theory. The equation of motion of two states was solved for a one-photon transition some time ago (Landau and Lifshitz, 1958). The equation of motion of states connected by multiphoton transition was set, for instance, by Mollow (1972), Lebedev (1971), and Kotova and Terentev (1967).

Lebedev (1971) treated the resonance multiphoton ionization of potassium atoms and the result is compared with experiment.

## 3. Quasi-Energetic States

The distorted wavefunction of the degenerate states of Section IV,E,2,a is the special case of the quasi-energetic states, the solution of the wave equation in a periodic electric field of light. Generally this solution has the form (Zel'dovich, 1966, 1973):

$$\psi_k(r, t + T) = \exp(-i\mathcal{E}'_k T)\psi_k(r, t). \tag{4.38}$$

This expression can be transformed by the substitution

$$\psi_k(r, t) = \exp(-i\mathcal{E}'_k t)\varphi(r, t), \tag{4.39}$$

and this transformation results in the expression

$$\varphi(r, t + T) = \varphi(r, t). \tag{4.40}$$

Consequently $\varphi_k$ is exactly periodic but not a harmonic function of time and can be expanded into a Fourier series. After performing this expansion

$$\psi_k(r, t) = \sum_{n=-\infty}^{\infty} \sum_{m=-\infty}^{\infty} c_k^{nm} j_m(r)\exp[i(\mathcal{E}'_k + n\omega_p)t], \tag{4.41}$$

where $j_m(r)$ is the eigenfunction of the atom without the time-dependent part of the Hamiltonian. $\mathcal{E}'_k$ is the quasi-energy and $\psi_k(r, t)$ is the wavefunction of the quasi-energetic state. The quasi-energy is defined up to a multiple of the light quantum as can be seen from expression (4.41). $\mathcal{E}'_k + m\omega_p$ is the energy of just the same quasi-energetic state where $m$ is a whole number. The transition frequency between two quasi-energetic states is

$$\omega_{kl} = \mathcal{E}'_k - \mathcal{E}'_l + (m_k - m_l)\omega_p. \tag{4.42}$$

The absorption of the $\omega_{kl}$ quantum is supplemented by the "virtual" absorption of $(m_l - m_k)\omega_p$ quanta of the light, causing the distortion of the atom. So the conservation of energy is fulfilled (Kovarski and Perelman, 1971b). The interpretation of expression (4.35) by the notion of quasi-energy gives a clear physical picture of the process of multiphoton ionization in the case of strong distortion of the atomic structure.

## V. STRONG FIELD PHOTOEMISSION

The influence of the distortion of the atomic structure on multiphoton ionization has been investigated in the preceding section. The distortion is caused by the field of light which ionizes the atom. The final and the initial state wavefunctions are generally supposed to agree with the field-free wavefunctions and only the intermediary state wavefunctions are regarded as being distorted.

It was supposed that the final state wavefunction was a plane wave (Bebb and Gold, 1966), a Coulomb wavefunction of the continuous spectrum, i.e. an incoming spherical wave and outgoing plane wave (Zernik, 1964), or a quantum defect wavefunction (Bebb, 1966, 1967; Zon et al., 1971a) for the calculation of the resonance and more simple than the resonance multiphoton ionization probability. These wavefunctions do not contain the interaction with the intense light wave.

The strong field resonance multiphoton ionization has already been calculated by Kovarski (1973) who not only calculated the distorted bound state wavefunctions but also used the distorted by the light field continuum state wavefunction, i.e. Volkov's (1935) solution of the Dirac equation, as the final state wavefunction. But the influence of the Coulomb field of the residual atom on the continuum state wavefunction was not taken into account.

Naturally the strongest distortion is theoretically expected for the case of the continuum state wavefunction. Therefore it is curious that almost all the experimental results are explained by a theory which does not use distorted continuum state wavefunctions, and the experimental results show only the influence of the distortion of the intermediary state wavefunctions.

The dominant role of the distortion of the final state wavefunction in the strong light field photoemission had already been realized by Keldish (1964) and, in subsequent work, by Perelomov et al. (1966a,b, 1968), Perelomov and Popov (1967). Popov et al. (1967), and Nikishev and Ritus (1966, 1967), revealing the intrinsic relationship between multiphoton ionization and the field emission (tunnel effect). Unfortunately interesting features of the strong field photoemission predicted theoretically have not yet been investigated experimentally.*

Further, the claim to treat theoretically strong and weak field photoemission and transitions uniformly arose, and the need to obtain an expression for the ionization probability which is also valid for a very intense light field. The new theoretical method of "momentum translation" found by Reiss (1970a,b) is adequate to give results in the case of a very intense light field. Interesting new phenomena are predicted in many works (Reiss, 1971, 1972a,b; Rahman and Reiss, 1972) using this method. Unfortunately these new theoretical results have not yet been confirmed by experiment.

A theoretical effort similar to that of Reiss was also made by Henneberger (1968).

---

* The transition of the multiphoton photoeffect taking place from the surface of metals into the tunnel effect was investigated by Farkas et al. (1972), continuing the study of the photoeffect commenced earlier by Farkas et al. (1967).

## A. Quasi-Classical Theory

### 1. Short Range Binding Forces

The solution of Dirac's equation is known in a temporarilly periodic electric field (Volkov, 1935). The outgoing electron from the atom moves not only under the influence of the external electric field of the light wave but also under the influence of the field of the residual atom. The solution of the wave equation, including also the Hamiltonian of the last interaction, is not yet known. The influence of this last interaction can be calculated by using the perturbation theory. The most simple procedure is if the field of the residual atom is that of a short range force. Then the final state wavefunction of the photoemission process can be expressed (Perelomov et al., 1966a,b) as follows:

$$\psi(\mathbf{r}, t) = i \int_{-\infty}^{t} dt' \int d\mathbf{r} G(\mathbf{r}, t; \mathbf{r}'t')^{(\mathrm{P})} H_s(\mathbf{r}')\psi(\mathbf{r}', t'); \tag{5.1}$$

$G(\mathbf{r}, t; \mathbf{r}'t')$ is the Green's function of the electron moving under the influence of the light field and the atom is described by the short range potential $^{(\mathrm{P})}H_s(\mathbf{r})$. Due to this short range potential the $\psi$ function under the integral sign can be assumed to agree with the unperturbed wavefunction of the atom.

The Green's function of the electron of momentum $p$ is

$$G(\mathbf{r}, t; \mathbf{r}'t') = [\theta(t - t')/2\pi] \int_{-\infty}^{\infty} d\mathbf{p}$$

$$\cdot \exp\left\{i\boldsymbol{\pi}(t)\mathbf{r} - i\boldsymbol{\pi}(t')\mathbf{r}' - \frac{1}{2}\int_{t_0}^{t} \boldsymbol{\pi}^2(\tau)\, d\tau\right\} \tag{5.2}$$

in the momentum $(p)$ representation and

$$\boldsymbol{\pi}(t) = \mathbf{p} - \boldsymbol{\vartheta}(t), \qquad \boldsymbol{\vartheta}(t) = \int_{t_0}^{t} \mathbf{f}(t')\, dt', \tag{5.3}$$

where $\mathbf{f}(t)$ is the force of the external field of light acting on the electron.

a. Photoemission Probability. Suppose there is a periodical time dependence of the force

$$\mathbf{f}(t) = \mathbf{E} \cos \omega_p t, \tag{5.4}$$

and linear polarization

$$\mathbf{E} = (E, 0, 0), \tag{5.5}$$

the current of the photoelectrons can be given as usual by

$$j(x, t) = \tfrac{1}{2}i[\psi\, \partial\psi^*/\partial x - \psi^*\, \partial\psi/\partial x].$$ (5.6)

The probability of the photoemission is the value of the current at infinity, $x \to \infty$, and averaged over the period of the light. Substituting expressions (5.1) and (5.2) into (5.6) the resulting expression is not a harmonic function of time. After Fourier transformation and performing the integrations needed the probability of the photoemission can be expressed as the sum of multiphoton photoemission processes of different order

$$\overline{P(E, \omega_p)} = \sum_{k > k_0'} \overline{P_k(E, \omega_p)}.$$ (5.7)

The probability of $k$-photon absorption causing the $k$-photon photoemission from the atom is

$$\overline{P_k(E, \omega_p)} = L(l, m, \omega_p, \xi, E, \omega_0)M_k(\omega_0, \xi)\exp[-\tfrac{2}{3}(E_0/E)g(\xi)],$$ (5.8)

where

$$L(l, m, \omega_p, \xi, E, \omega_0) = [\omega_0 \,|\, c_{kl} \,|^2 4\sqrt{2}/\pi]\left[\frac{(2l + 1)(l + |m|)!}{2^{|m|}(|m|!)^2(l - |m|)!}\right]$$

$$\cdot \left[\frac{\xi^2}{1 + \xi^2}\left(\frac{E(1 + \xi^2)^{1/2}}{2E_0}\right)^{|m| + (3/2)}\right];$$ (5.9a)

$$M_k(\omega_0, \xi) = \exp[-a(k - k_0')]\Lambda(b(k - k_0')^{1/2};$$ (5.9b)

and here

$$\Lambda(x) = \int_0^x \exp(y^2 - x^2)(x^2 - y^2)^{|m|}\, dy.$$ (5.9c)

The other symbols have the following meaning

$$g(\xi) = (3/2\xi)\{[1 + (1/2\xi^2)]\operatorname{arsh} \xi - [(1 - \xi^2)^{1/2}/2\xi]\},$$ (5.9d)

$$a = 2[\operatorname{arsh} \xi - \xi/(1 - \xi^2)^{1/2}],$$ (5.9e)

$$b = 2\xi/(1 + \xi^2)^{1/2},$$ (5.9f)

$$k_0' = (\omega_0/\omega_p)[1 + (1/2\xi^2)],$$ (5.9g)

$$\xi = \omega_p/\omega_t = (\omega_p/\omega_0)(E_0/2E),$$ (5.9h)

$$E_0 = (2\omega_0)^{3/2}.$$ (5.9i)

$\omega_0$ is the ionization energy of the initial level of the orbital and magnetic quantum numbers $l$ and $m$. $E_0$ is the intra-atomic electric field of the nucleus defined by the expression (5.9i). $[k_0' + 1]$ is the minimal number of photons needed for the ionization. The symbol $[\![x]\!]$ means whole number of $x$. The energy of the $[\![k_0' + 1]\!]$ photons is spent in supplying the ionization energy $(\omega_0)$ and the transversal oscillation energy of the electron

$$\omega_0/2\xi^2 = (E^2/4\omega_p^2)\omega_0 = \tfrac{1}{2}\overline{\vartheta^2}(t). \qquad (5.10)$$

$c_{kl}$ is a dimensionless multiplication factor of the exact final state wavefunction. This factor can be determined exactly only by solving the Schrödinger equation.

The photoemission probability is a maximum at $m = 0$ [see expressions (5.9a), (5.9b), and (5.9c)]. The axis of the quantization coincides with the direction of the electric field strength of the light.

$\xi$ is the stationarity parameter and $\omega_t$ is the critical frequency of ·the tunnel effect.

If the frequency of the light $\omega_p < \omega_t$ the stationary wavefunction describing the tunnel effect in the case of the static electric field (Landau and Lifshitz, 1958) can be regarded as adequate for the description of the motion of the electron also in the alternating field with the instantaneous electric field strength $E(t) = E \cos \omega_p t$ (Perelomov et al., 1966a). The probability of the tunnel transition is the average value of the transition probability in the instantaneous "static" field over the period of the light.

If $\omega_p > \omega_t$, i.e. $\xi \gg 1$, the change of the wavefunctions of the electron before, under, and over the potential wall cannot be neglected during the time of tunneling. Therefore the corresponding time-dependent Schrödinger equation has to be solved so as to find the transition probability. Then the photoemission occurs through the multiphoton ionization process.

(i) *Field Emission.* Actually if $\xi \ll 1$, i.e. in the range of the tunnel effect, the contribution of $k$-photon emission where $k \gg [\![k_0' + 1]\!]$ is significant to the total photoemission probability [see expressions (5.7) and (5.8)]. The photoemitted electron has relatively large transversal oscillation energy. The corresponding expression for the photoemission probability is

$$\overline{P^{(t)}(E, \omega_p)} = \omega_0 \, |c_{kl}|^2 (6/\pi)^{1/2}(2l + 1)(l + |m|)!/2^{|m|}(|m|!)(l - |m|)!$$
$$\cdot (E/2E_0)^{|m| + (3/2)} \exp[-(2E_0/E)(1 - \tfrac{1}{10}\xi^2)].$$

$$(5.11)$$

(ii) *Multiphoton ionization.* If the frequency of the light increases so that $\xi \gg 1$ the $g(\xi)$ function in the exponent of expression (5.8) decreases steeply. Consequently the probability of the photoemission increases. The multiphoton ionization probability is orders of magnitude higher than the corresponding probability of the tunnel effect at the same electric field strength of the light. The dominant contribution to the photoemission probability arises from the $[\![k_0' + 1]\!]$-photon absorption process [see expression (5.8)], i.e. from the first term only in the sum of the expression (5.7). The factor preceding the exponential term has threshold behavior at

$$k\omega_p = \omega_0 + (\omega_0/2\xi^2) \tag{5.12}$$

(see Fig. 30).

The probability of multiphoton ionization of the initial state of magnetic quantum number $m = 0$ can be expressed as

$$\overline{P^{[k_{0'}+1]}}(E, \omega_p) = |c_{kl}|^2 (4\sqrt{2}/\pi)(2l+1)\omega_0(\omega_p/\omega_0)^{3/2}$$

$$\exp\left(2\left[\frac{\tilde{\omega}_0}{\omega_p}+1\right] - \frac{\tilde{\omega}_0}{\omega_p}\right)\Lambda\left(2\left[\frac{\tilde{\omega}_0}{\omega_p}+1\right] - \frac{\tilde{\omega}_0}{\omega_p}\right)^{1/2}$$

$$\cdot \left(\frac{\omega_0}{\omega_p}\frac{E}{E_0}\right)^{2[(\tilde{\omega}_0/\omega_p)+1]}, \tag{5.13}$$

where

$$\tilde{\omega}_0 = \omega_0[1 + (1/2\xi^2)] \tag{5.14}$$

is the effective ionization energy which is increased by the transversal kinetic energy of the electron in the field. The dominant term in the expression

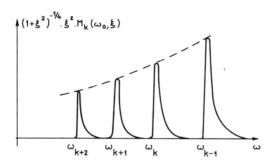

FIG. 30. Threshold behavior of the factor preceding the exponential term of expression (5.8) (Perelomov et al., 1966a).

(5.13) is the last one, the power function of the field strength $E$. The power is double the minimum number of photons needed for the ionization. Taking into account that $cE^2/\omega_p = F$, expression (5.13) agrees in character with expression (2.26a) deduced using the perturbation theory.

The common character of multiphoton ionization and the tunnel effect is clearly seen from the discussion above. The only difference which occurs in the tunnel effect in comparison with multiphoton ionization is that the energy spent for the transversal oscillation energy of the electron has to be calculated in excess of the ionization energy of the atom.

*b. Polarization Dependence.* Linear polarization of the light is supposed in all the preceding discussions. If the light is elliptically polarized the electron absorbs light quanta not only to gain transversal oscillation energy over the binding ionization energy but the electron obtains longitudinal momentum too (Perelomov *et al.*, 1966b). The most probable value of the momentum of the electron is

$$\mathbf{p}_0 = (0, p_0, 0),\qquad\qquad(5.15)$$

where

$$
p_0 = \begin{cases}
(2\omega_0)^{1/2}\varepsilon/\xi & \text{if } \xi \ll 1 \\
& \text{(tunnel effect),}\qquad(5.16) \\[2ex]
(2\omega_0)^{1/2}\dfrac{\varepsilon}{(1+\varepsilon^2)^{1/2}}\left(\ln\dfrac{2\xi}{(1-\varepsilon^2)^{1/2}}\right)^{-1} & \text{if } \xi \gg 1 \\
& \text{(multiphoton ionization).}
\end{cases}
$$

It is supposed that the electric field strength of the light has the form

$$\mathbf{E} = (E\cos\omega t, \varepsilon E\sin\omega t, 0).\qquad\qquad(5.17)$$

$-1 < \varepsilon < 1$ is the degree of ellipticity of the polarization. It is remarkable that the electron is emitted perpendicular to the direction of the maximal value of the electric field strength.

In the case of circular polarization the electron is emitted isotropically in the plane of the polarization.

The probability of photoemission is always less in the case of elliptically or circularly polarized light than in the case of linearly polarized light (see also Nikishev and Ritus, 1967).

## 2. Coulomb Field Contribution

The short range potential $^{(P)}H_s(\mathbf{r})$ does not describe the real atom adequately. The Coulomb potential of the positively charged residual atom also has an effect on the outgoing electron. This potential has a long tail—its value is significant also at great distances. The volume of the integration in expression (5.1) which gives a significant contribution is enlarged. Consequently the photoemission probability increases.

Perelomov *et al.* (1966b) and Popov *et al.* (1967) elaborated the method of quasi-classical approximation of the wavefunction (WKB method) for the case of an alternating electric field (see also Baz *et al.*, 1971). The Green's function of the electron is expressed as

$$G(\mathbf{r}, t; \mathbf{r}'t') = \{\theta(t-t')/[2\pi i(t-t')]^{3/2}\}\exp[iS(\mathbf{r}, t; \mathbf{r}', t')],\qquad(5.18)$$

where $S(r, t; r', t')$ is the classical action. The wavefunction is obtained from (5.1) in the form

$$\psi(\mathbf{r}, t) \sim \exp(i\mathscr{E}^{(0)}t) \int_{-\infty}^{t} dt \int d\mathbf{r} \exp[i\tilde{S}(\mathbf{r}, t; \mathbf{r}', t')^{(p)}H_s(\mathbf{r}')\varphi^{(0)}(\mathbf{r}'), \quad (5.19)$$

where $\tilde{S}(\mathbf{r}, t; \mathbf{r}'t')$ is the classical shortened action

$$\tilde{S}(\mathbf{r}, t; \mathbf{r}'t') = S(\mathbf{r}, t; \mathbf{r}', t') + \mathscr{E}^{(0)}(t - t'). \quad (5.20)$$

The Coulomb interaction is calculated as the perturbation term in the shortened action (Perelomov and Popov, 1967; see also Perelomov et al., 1968), i.e.

$$\tilde{S} = \tilde{S}^{(0)} + \delta\tilde{S}. \quad (5.21)$$

The contribution of the last correction term is

$$\exp(-2 \operatorname{Im} \delta\tilde{S}) = (2E_0/E)^{2\lambda} \quad (5.22)$$

to the probability of the photoemission

$$P \sim \exp(-\operatorname{Im} \tilde{S}^{(0)}) \quad (5.23)$$

[see also expression (5.11)]. Here $\lambda$ is the Coulomb parameter

$$\lambda = (\omega_{0c}/\omega_0)^{1/2}; \quad (5.24)$$

$\omega_{0c}$ is the ionization energy of the atom in the case of the Coulomb interaction only.

Introducing the correction term (5.21) into the probability of the photoemission, the asymptotic expression for the static field contains the correct factor before the exponential term of

$$(2E_0/E)^{2\lambda - 1}, \quad (5.25)$$

contrary to expression (5.11).

The Coulomb interaction enlarges the probability of the photoemission. This increase is about six orders of magnitude in usual experimental circumstances.

The Coulomb contribution was also calculated by Nikishev and Ritus (1967). The expression for the photoemission probability obtained by them is valid at any field strength of light. The correction factor preceding the exponential term of expression (5.11) follows automatically from the calculation.

## B. Nonperturbative Theory

An alternative to the quasi-classical theory of photoemission is the nonperturbative theory of the momentum translation method of Reiss (1970a,b) for solving the wave equation. The transition probability derived using this method is valid at arbitrary high light intensities, but this transition probability shows a peculiar intensity dependence, strongly deviating from the usual transition probability of the perturbation theory.

### 1. Momentum Translational Method

The Schrödinger equation given as follows

$$i \, \partial \psi/\partial t = [\tfrac{1}{2}(\mathbf{p} - (1/c)\mathbf{A})^2 + {}^{(\mathrm{p})}H(\mathbf{r})]\psi(\mathbf{r}, t) \tag{5.26}$$

is solved by making a unitary transformation, shifting the momentum of the electron. ${}^{(\mathrm{p})}H(\mathbf{r})$ is the potential energy of the electron in the atom. The transformed wavefunction is expressed as

$$\psi'(\mathbf{r}, t) = \exp[-(i/c)\mathbf{A}\mathbf{r}]\psi(\mathbf{r}, t). \tag{5.27}$$

The Schrödinger equation (5.26) is transformed into the following form:

$$i \, \partial \psi'(\mathbf{r}, t)/\partial t = (H_0 + V')\psi(\mathbf{r}, t), \tag{5.28}$$

where

$$H_0 = \mathbf{p}^2 + {}^{(\mathrm{p})}H(\mathbf{r}) \tag{5.29}$$

is the Hamiltonian of the field-free atom and

$$\begin{aligned} V'(\mathbf{r}, t) = \tfrac{1}{2}(\partial A_i/\partial t)r_i &+ (1/c)(\partial A_j/\partial x_i)r_j p_i \\ &+ (i/2c)(\nabla^2 A_i)r_i + (1/2c^2)(\partial A_j/\partial x_i)(\partial A_k/\partial x_i)r_j r_k \end{aligned} \tag{5.30}$$

is the new interaction Hamiltonian. If the new interaction Hamiltonian $(V')$ can be neglected in Eq. (5.28) the solution of the transformed Schrödinger equation is the field-free wavefunction $\Phi(\mathbf{r}, t)$.[*] Inverting relation (5.27), the solution of the original Schrödinger equation (5.26) is given as

$$\psi(\mathbf{r}, t) = \exp(i\mathbf{A}\mathbf{r}/c)\Phi(\mathbf{r}, t). \tag{5.31}$$

[*] The validity of the method is questioned in the work of Herman (1973).

The $T$ matrix of the transition (Roman, 1965) can be expressed as

$$\tau_{f,g} = 2\pi i\, \delta(\mathscr{E}_f - \mathscr{E}_g)T_{f,g}$$

$$= -i \lim_{t_0 \to \infty} \int_{t_0}^{\infty} dt_0\, e^{iH_0 t_0}\psi_f(t_0)e^{-iH_0 t_1}V'(t_1)\psi_g(t_1)$$

$$= i \int_{-\infty}^{\infty} dt\Phi_f(t)V(t)e^{i\mathbf{A}\mathbf{r}/c}\Phi_g(t), \qquad (5.32)$$

where $V(t)$ is the interaction Hamiltonian (2.10).

$$1/\lambda_p \ll 1 \qquad (5.33)$$

in the long wavelength approximation; consequently the dependence on the space coordinates can be neglected in the vector potential, i.e.

$$\mathbf{A} = A\mathbf{e}_{\lambda p}\cos \omega_p t, \qquad (5.34)$$

and the expression (5.32) will have the form

$$\tau_{f,g} = 2\pi i\, \delta(\mathscr{E}_f - \mathscr{E}_g)T_{f,g}$$

$$= -i(\mathscr{E}_{ag} - \mathscr{E}_{af}) \cdot \left(\varphi_f, \int_{-\infty}^{\infty} dt\, \exp\{i(\mathscr{E}_{af} - \mathscr{E}_{ag})t\right.$$

$$\left. + i(A/c)(\mathbf{r}, \mathbf{e}_{\lambda p})\cos \omega_p t\}\varphi_g\right). \qquad (5.35)$$

$\varphi_f(\mathbf{r})$, $\varphi_g(\mathbf{r})$ are the stationary wavefunctions of the unperturbed atoms. The expression (5.32) is transformed into the form

$$\tau_{f,g} = \sum_{k=-\infty}^{\infty} 2\pi i\, \delta(\mathscr{E}_{af} - \mathscr{E}_{ag} - k\omega_p) \cdot i^k(\mathscr{E}_{ag} - \mathscr{E}_{af})$$

$$\cdot (\varphi_f(\mathbf{r}),\ J_k((A/c)(\mathbf{r}\mathbf{e}_{\lambda_p}))\varphi_g(\mathbf{r})) \qquad (5.36)$$

after expanding the function in the time integral into a Fourier series and performing the integration. The $T^{(k)}$ matrix is defined as belonging to the absorption of $k$ photons and has the form

$$T_{f,g}^{(k)} = i^k(\mathscr{E}_{ag} - \mathscr{E}_{af})(\varphi_f(\mathbf{r}),\ J_k((A/c)(\mathbf{r}\mathbf{e}_{\lambda p}))\varphi_g(\mathbf{r})), \qquad (5.37)$$

$J_k(x)$ is the Bessel function. This definition is slightly different from the usual definition of the transition $T$ matrix involving the absorption or emission of $k$ photons (Roman, 1965). Then the total transition probability can be expressed as the sum of the absorption probabilities of $k$ photons with $k$ going from zero to infinity, i.e.

$$P = 2\pi \sum |T_{f,g}^{(k)}|^2\, \delta(\mathscr{E}_{af} - \mathscr{E}_{ag} - k\omega_p). \qquad (5.38)$$

The omitted interaction term $V'(\mathbf{r})$ in the expression (5.28) can be further taken into account using the formalism of the perturbation theory with the perturbation Hamiltonian $V'(\mathbf{r})$.

## 2. *Hydrogen and Hydrogen-like Ions*

The momentum translational method was applied to the calculation of $k$-photon bound–bound transition probabilities in the hydrogen atom (Rahman and Reiss, 1972) and to the ionization probabilities of hydrogen atom and hydrogen-like ions by Parzynski (1972), Tewari (1972), Faisal (1972a), Mohan and Thareja (1972, 1973), and De Witt (1973). The calculation of the above authors differs depending on their choice of plane wave or Coulomb continuum state wavefunction.

The transition probability of the even photon $(2k)$ ionization of the hydrogen atom or hydrogen-like ions is given (Parzynski, 1972) as follows

$$P^{(2k)} = 2k\omega_p(2k\omega_p - \mathscr{E}_0)^{1/2}h(u, 2k), \qquad (5.39)$$

where

$$h(u, 2k) = \{[(1 + u^2)^{1/2} - 1]^{4k}/u^{4k}(1 - u^2)^3\}[2k(1 + u^2)^{1/2} + u^2 + 2]^2 \qquad (5.40)$$

and

$$u = A/\omega_p \qquad (5.41)$$

is the "intensity parameter."

The plot of the function $h(u, 2k)$ can be seen in Fig. 31. The parameter is the number of photons $(2k)$ absorbed in the ionization process.

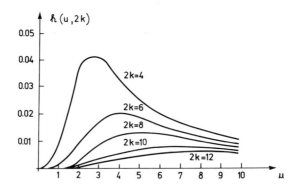

FIG. 31. $2k$-photon ionization probability $P \sim h(u, 2k)$ dependence on the "intensity parameter" $u = A/\omega_p$ for different numbers of photons $(2k)$ absorbed (Parzynski, 1972).

As is apparent, the transition probability dependence on the light power density $I \sim A^2$ strongly deviates from the result of the perturbation theory $P \sim A^{2 \cdot k}$. The transition probability has a maximum and the position of this maximum depends on the number of photons absorbed. This result is in qualitative agreement with the behavior of the bound–bound transition probability of the hydrogen atom calculated by Reiss (1970a) and Rahman and Reiss (1972).

The transition probability coincides with the result of the perturbation theory in the weak field limit.

It should be noticed that the " transition matrix element "

$$\langle \varphi_f(\mathbf{r}) \,|\, \exp(i\mathbf{A}\mathbf{r}/c) \,|\, \varphi_g \rangle$$

in expression (5.35) contains the dependence of the transition probability on the final state angular momentum quantum number $(l_f)$. This dependence is plotted in Fig. 32 at two different " intensity parameters $D$ " (De Witt, 1973). The number of absorbed photons is the parameter of the curves. It is apparent that the greater the orbital quantum number at which the transition probability has a maximum value, the greater the number of photons absorbed. Unfortunately the results of the nonperturbative theory of Reiss have yet to be confirmed experimentally and it is not clear how the distortion of the atomic structure influences the strong field photoemission described by this theory.

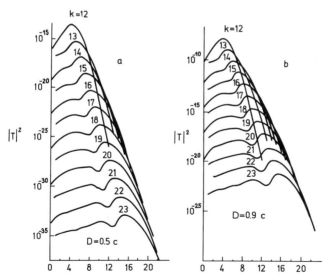

FIG. 32. The dependence of the $k$-photon ionization probability of the hydrogen atom $P^{(k)} \sim |\, T^{(k)}\,|^2$ on the orbital quantum number of the final continuum state at different " intensity parameter " values. (a): $D = 0.5c$; (b): $D = 0.9c$ (De Witt, 1973).

## VI. POLARIZATION, FIELD STATISTICS

From the very beginning of the discussion of the properties of multiphoton ionization, linear polarization of the light and only one field mode of
excitation have been supposed, with the exception of Section V,A,1,b. The
light beam of the real laser is characterized by many parameters in addition
to the light power density, the frequency, and the spatial and temporal
distribution of the light power density which were discussed in the preceding
sections.

First and foremost, the above-mentioned two suppositions on polarization and monochromaticity have to be rejected. The very statistical nature of
the light beam of the laser also involving polarization (Born and Wolf, 1964)
and monochromatic properties has to be realized. Also the strong influence
of the statistical properties of light on the multiphoton ionization probability must be recognized.

### A. Light Polarization

The dependence of the multiphoton ionization properties on the state of
polarization of the light is investigated with the assumption of one field
mode of excitation (monochromaticity). The statistical nature of the polarization state is disregarded in this section for the time being.

### 1. Linear Polarization

The transition matrix element of $k_0$th order [expression (2.21)] contains
the product of one-photon matrix elements of the form

$$\langle a_k \,|\, \mathbf{e}_{\lambda p}\mathbf{r} \,|\, a_{k-1}\rangle. \tag{6.1}$$

The coordinate system is directed with the $z$ axis in the direction of the
polarization unit vector, i.e.

$$\mathbf{e}_{\lambda p} = (0, 0, 1). \tag{6.2}$$

Then

$$\mathbf{e}_{\lambda p}\,\mathbf{r} = r \cos \theta = (4\pi/3)^{1/2} Y_{10}(\theta, \varphi) \tag{6.3}$$

with

$$\mathbf{r} = (x, y, z) = (r, \theta, \varphi), \tag{6.4}$$

where $r$, $\theta$, $\varphi$ are the polar coordinates. The wavefunction of the bound state
has the form

$$|a_k\rangle = R_{v_k l_k} Y_{l_k m_k}(\theta, \varphi). \tag{6.5}$$

The final state of multiphoton ionization is given in the form of the partial wave expansion

$$|a_f\rangle = \frac{1}{\kappa} \sum_{l_f} \sum_{m_f} (i)^{l_f} \exp(i\,\Delta_{l_f}) R_{\kappa l_f}(\mathbf{r}) Y_{l_f m_f}(\theta,\varphi) Y^*_{l_f m_f}(\theta_\kappa,\varphi_\kappa); \qquad (6.6)$$

$v$ and $\Delta_{l_f}$ are given by the expressions

$$v_\kappa = (-2\mathcal{E}_\kappa)^{-1/2}; \qquad (6.7a)$$

$$\Delta_{l_f} = \delta_{l_f} + \eta_{l_f}; \qquad (6.7b)$$

$$\eta_{l_f} = \arg\Gamma(l_f + 1 + v_\kappa); \qquad (6.7c)$$

and

$$\boldsymbol{\kappa} = (\kappa, \theta_\kappa, \varphi_\kappa). \qquad (6.8)$$

Using expressions (6.3) and (6.5) the angular dependent part of the matrix element is given (Bethe and Salpeter, 1957; Rose, 1957) as follows

$$\langle Y_{l_k m_k}(\theta,\varphi)\,|\,(4\pi/3)^{1/2} Y_{10}(\theta,\varphi)\,|\,Y_{l_{k-1} m_{k-1}}(\theta,\varphi)\rangle \neq 0 \qquad (6.9)$$

only if

$$l_k - l_{k-1} = \pm 1 \qquad \text{and} \qquad m_k - m_{k-1} = 0 \qquad (6.10)$$

and the transition $l_{k-1} = 0 \to l_k = 0$ is strictly forbidden. These are the well-known selection rules of the electric dipole transition.

Consequently the orbital angular momentum changes by unity in every step to the next virtual state of multiphoton ionization. The "form" of the ionization can be characterized by the array of consecutive values of the orbital angular momentum

$$l_g, l_1, l_2, \ldots, l_f = \{l_i\}_g^f \qquad (6.11)$$

and

$$m_f = m_g \qquad (6.12a)$$

$$l_f \leq k_0 + l_g. \qquad (6.12b)$$

The parity of the final state is the same as the initial state if $k_0$ is even and the contrary holds if $k_0$ is odd.

The angular distribution of the photoelectrons of the $l_f$ partial wave is determined by the orbital quantum number of the final state in the array and is given by the square of the function $Y^*_{l_f m_f}(\theta_\kappa,\varphi_\kappa)$.

The differential cross section is given as the square of the sum of the contribution of the different "form" of the ionization characterized by the array $\{l_i\}_g^f$ and the function $Y^*_{l_f m_f}(\theta_\kappa,\varphi_\kappa)$. The angular dependence of the differential cross section contains the contribution of the angular distribution of the ionization of the $l_f$ partial wave and the interference terms between the functions $Y^*_{l_f m_f}(\theta_\kappa, \varphi_\kappa)$ of different partial waves.

## 2. Circular Polarization

The wave vector of the light $\mathbf{K}$ is oriented in the direction of the $z$ axis of the coordinate system. Then the polarization vector is found to be

$$\mathbf{e}_{\lambda_p} = (1/\sqrt{2})(\mathbf{e}_x \pm i\mathbf{e}_y). \tag{6.13}$$

The ambiguous sign refers to the case of right and left circular polarization. Therefore the matrix element of the operator

$$\mathbf{r} \cdot \mathbf{e}_{\lambda_p} = (1/\sqrt{2})(x \pm iy) \tag{6.14}$$

has to be calculated. This operator can be expressed by spherical harmonics in the form

$$\mathbf{re}_{\lambda_p} = -r(4\pi/3)^{1/2} Y_{1\pm1}(\theta, \varphi). \tag{6.15}$$

The angular part of the matrix element

$$\langle Y_{l_k m_k}(\theta, \varphi) | Y_{1\pm1}(\theta, \varphi) | Y_{l_{k-1} m_{k-1}}(\theta, \varphi) \rangle \neq 0 \tag{6.16}$$

only if

$$l_k - l_{k-1} = \pm 1 \qquad \text{and} \qquad m_k - m_{k-1} = \pm 1. \tag{6.17}$$

The electron can only continuously increase or decrease its magnetic quantum number in the step subsequent to the next virtual state of multiphoton ionization.

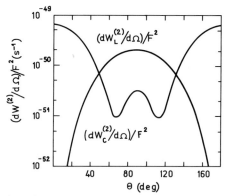

FIG. 33. Angular dependence of the generalized cross section of the two-photon ionization of potassium atoms at the wavelength $\lambda_p = 5295$ Å (Mizuno, 1973).

The only possible final state is that of magnetic quantum number $m_f = m_g + k_0$.

The angular distribution of the photoelectrons is determined only by the square of one $Y^*_{k_0, k_0}(\theta_\kappa, \varphi_\kappa)$ function and the interference terms are absent (Lambrapoulos 1972a). It is supposed that $l_g$ and $m_g$ are zero. The angular dependence of the two-photon generalized differential cross section of a potassium atom is plotted in Fig. 33 for the cases of linearly and circularly polarized light.

## 3. *Resonances*

Resonances of $k$th order can be realized only with levels corresponding to the selection rules of the linearly or circularly polarized light used. The number of resonances is generally smaller in the case of circularly polarized light.

The generalized cross section of the three-photon ionization of the metastable He atoms is plotted in Fig. 34, dependent on the frequency of the

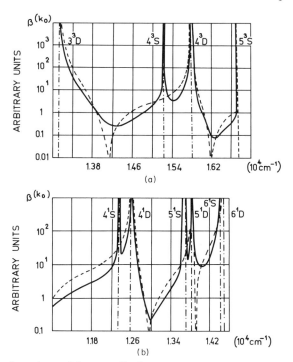

FIG. 34. The dependence of the generalized cross section of the three-photon ionization of the triplet (a) and the singlet metastable He atom (b) on the frequency of light using linearly polarized (solid line) or circularly polarized light (dashed line) given by Manakov *et al.* (1971b).

light (the unit is the wavenumber), for the case of linearly and circularly polarized light (Manakov *et al.*, 1971b). The different behavior and resonances of the two curves are clearly seen.

The total cross section of the ionization obviously also depends on the polarization state of the light used for the case of an unpolarized atom, as seen in Fig. 34. This feature of the total cross section is characteristic only for multiphoton ionization.

The difference in the total cross section of circularly and linearly polarized light is naturally the function of the frequency used.

The dependence of the total cross section on the polarization state of the light was measured by Fox *et al.* (1971) for the case of three-photon ionization of a Cs atom using a ruby laser, and by Kogan *et al.* (1971) for the case of two-photon ionization of a Cs atom using the second harmonics of ruby laser light. The measured ratios $\sigma_c^{(3)}/\sigma_1^{(3)}$ and $\sigma_c^{(2)}/\sigma_1^{(2)}$ of the total cross sections using circularly and linearly polarized light, $2.15 \pm 0.4$ and $1.28 \pm 0.2$, respectively, adequately coincide with the values calculated theoretically by Lambrapoulos (1972b).

The dependence of the ratio $\sigma_c^{(k_0)}/\sigma_1^{(k_0)}$ on the order of multiphoton ionization ($k_0$) was calculated by Klarsfeld and Maquet (1972) and an increase with the multiphoton order dominance of the cross section of circularly polarized light was found over that of the cross section of the linearly polarized light. The dominance of the cross section of the ionization leading to the partial wave $l_f = k_0$ was supposed to exist in the case of linear polarization.

This dependence of the ratio $\sigma_c^{(k_0)}/\sigma_1^{(k_0)}$ on the multiphoton order was calculated exactly by Reiss (1972a) and Parzynski (1973) using the momentum translation method. The dominant contribution of the final partial wave of $l_f < k_0$ to the cross section of ionization with linearly polarized light is demonstrated (Fig. 32; see also De Witt, 1973). Consequently a larger cross section of ionization with linearly polarized light is concluded, contrary to the result obtained by Klarsfeld and Maquet (1972), and Faisal (1972b). Reiss' result is in agreement with that of the calculation performed using the quasi-classical approximation (Section V).

## 4. *Electron Polarization*

The spin-orbit coupling can be comparable to or larger than the linewidth of the laser radiation. Then the fine structure splitting has to be taken into account in the calculation of the multiphoton ionization cross section. The final partial wave of definite spin quantum number is preferable when choosing the frequency of the light relative to the resonances with the fine structure levels. The result is the polarization of the photoelectrons (Lambrapoulos, 1973).

Lambrapoulos (1973) calculated the dependence of the degree of polarization of the photoelectrons in the case of alkaline atoms on the frequency of the light. The degree of polarization versus light frequency function is plotted in Fig. 35.

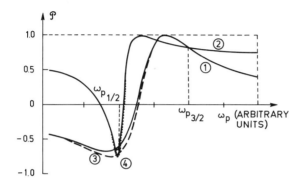

FIG. 35. Polarization degree of the photoelectrons and its dependence on the frequency of the light. The position of the fine-structure levels is also indicated ($\omega_{p_{1/2}}$, $\omega_{p_{3/2}}$) Curve 1: the ratio of the radial matrix element

$$\rho = \left(nP_{3/2}, nP_{1/2}\right)/ \left(nP_{1/2}, nP_{1/2}\right) = 1$$

Two-photon ionization takes place. Curve 2: $\rho = 5$ Curve 3: $\rho = 1$ and three-photon ionization takes place. Curve 4: $\rho = 5$ (Lambrapoulos, 1973).

As is apparent, the degree of polarization $\mathscr{P}$ is equal to unity for the case of a frequency lying between the fine structure levels.

## 5. Strong-Field Selection Rules

Laplanche et al. (1973) calculated the possible final angular momentum states in the theory of the momentum translation using expression (5.37) for the $T^{(k)}_{f, g}$ matrix.

The possible final angular momentum $l_f < k_0$ is of the same parity as $k_0$; $m_f = 0$ in the case of linearly polarized light. $l_f > k_0$ in the case of circularly polarized light and $l_f$ is the same parity as $k_0$. It is supposed that the angular momentum of the initial state is 0.

It is apparent that these selection rules strongly deviate from those of the perturbation theory. These selection rules of the nonperturbative theory coincide with those of the perturbation theory in the weak field limit.

## B. Light Statistics

The very simple model of the light beam is accepted in the previous discussions. The light beam is a plane wave of definite amplitude and frequency [Sections V,VI; e.g. expression (5.34)] in the semiclassical treatment. Otherwise it is supposed that only one field mode is excited in the quantum field theoretical treatment (Section II).

In fact the electric field strength of the light is a random variable either because of intrinsic quantum mechanical features of the light field or because of the classical statistical properties of the real light sources (Jánossy, 1957, 1959). Consequently the electric field strength of the light can be regarded as a random function of the time rather than the classical function as supposed until now.

It is the moments of the statistics of the electric field strength of the light that are accessible for observation. Actually the power density of the light is expressed as

$$I = c \langle |E|^2 \rangle$$

and the visibility of the interference fringes in the interferometer of Michelson can be given as

$$I_{max} - I_{min}/I_{max} + I_{min} = 2 \operatorname{Re}\langle E_1^*(\tau) E_2(0) \rangle / (\langle |E_1|^2 \rangle + \langle |E_2|^2 \rangle)$$

in the case of linear polarization. The angular bracket means expectation value (Born and Wolf, 1964).

The light beam can be fully characterized, including also its polarization properties, by the family of moments and correlation functions of the statistics (Glauber, 1963; Mandel and Wolf, 1965):

$$\Gamma_{j_1, \ldots, j_N, j_{N+1}, \ldots, j_{N+M}}^{N, M}(x_1, \ldots, x_N; x_{N+1}, \ldots, x_{N+M})$$

$$= \operatorname{Tr}[\rho A_{j_1}^{(-)}(x_1) \cdots A_{j_N}^{(-)}(x_N) A_{j_{N+1}}^{(+)}(x_{N+1}) \cdots A_{j_{N+M}}^{(+)}(x_{N+M})], \qquad (6.18)$$

where

$$\rho = \sum_i \rho_{ii} |i\rangle\langle i| \qquad (6.19)$$

is the density matrix of the light field. $j_1, \ldots, j_{N+M}$ are the polarization indexes. $A_{j_i}^{(-)}(x_i)$ and $A_{j_i}^{(+)}(x_i)$ are the parts of the vector potential containing only the emission and the absorption operators, respectively [see expression (2.12) and the similar definition in Section IV,E,1]. $x_i$ represent the space–time points.

The simplest detectors of the light measure the second moment of the statistics, i.e. the light power density. This simplest detector of the light can be an atom which makes a one-photon transition under the influence of the light beam (Glauber, 1965).

The higher than second moments and the correlation functions can be measured with the above mentioned detectors by using only two or more detectors and coincidence or correlation circuits (Brown and Twiss, 1956a,b,c; Adam et al., 1955), by measuring the photocount distribution (Arecchi, 1965; Arecchi et al., 1966), and by using interferometers.

The higher order processes in which more than one quantum is absorbed afford the possibility of measuring the higher order moments of the light statistics (Shen, 1967) using only one detector (Jaiswal and Agarwal, 1969; Teich and Diamant, 1969; and others). Conversely the cross sections measured by these processes cannot be compared with theory without knowing the statistical properties of the light with which the measurement is performed.

The special case of the multiple quantum transition, the two-photon absorption, is widely discussed in the literature (e.g., Lambrapoulos et al., 1966; Meadors, 1966; Teich and Wolga, 1966; Carusotto et al., 1967; Guccione-Gush et al., 1967; Lambrapoulos, 1968; and others). The only experimental work is related to the two-photon photoemission of electrons from a semiconductor thin layer (Shiga and Imamura, 1967) showing the increased probability in the case of using thermal light in comparison with the probability in the case of using coherent light.

### 1. Multiple Quantum Transitions

The $k_0$-photon transition probability is given by expression (2.15), i.e.

$$P_{f,g}^{(k_0)}(t) = |\langle f | U_I^{(k_0)}(t) | g \rangle|^2 \qquad (2.15)$$

which is valid at any light statistics. Here the time evolution operator $U_I^{(k_0)}(t)$ is given by expression (2.9). Substituting the interaction operator (2.10) into expression (2.9) and retaining the term containing only the product of $k_0$ absorption operators, the time development operator can be expressed (Agarwal, 1970) as follows:

$$U^{(k_0)} = (i/c)^{k_0} \int_0^t \cdots \int dt_1 \cdots dt_{k_0} \theta(t_1 - t_2) \cdots \theta(t_{k_0-1} - t_{k_0})$$

$$\cdot \mathbf{p}(t_1)\mathbf{A}^{(+)}(t_1) \cdots \mathbf{p}(t_{k_0})\mathbf{A}^{(+)}(t_{k_0}). \qquad (6.20)$$

The quadratic term is neglected in the interaction operator.

Let the initial state of the field be characterized by density matrix (6.19). The total probability of the $k_0$-photon absorption can be obtained using expression (6.20) and averaging it over the possible final states of the field.

The result is given as

$$P_{f,g}^{(k_0)}(t) = \sum_{\{i_{k(0)}\}\{j_{k(0)}\}} \int \cdot \frac{1}{0} \cdot \int dt_1 \cdots dt_{k_0} \, dt'_1 \cdots dt'_{k_0}$$

$$\cdot \, \Gamma_{i_1, \ldots, i_{k(0)}, j_1, \ldots j_{k(0)}}(t'_1, \ldots, t'_{k_0}; t_1 \cdots t_{k_0}) \zeta^*_{i_1, \ldots, i_{k(0)}}(t'_1, \ldots, t'_{k_0})$$

$$\cdot \, \zeta_{\{j_{k(0)}\}}(t_1, \ldots, t_{k_0}); \tag{6.21}$$

where

$$\zeta_{\{i_{k(0)}\}}(\{t_{k_0}\}) = (i/2\pi)^{k_0-1}(1/c)^{k_0} \int_{-\infty}^{\infty} \cdots \int d\{\omega_{k_0-1}\}$$

$$\cdot \exp[it_1(\omega_f - \omega_1) + it_2(\omega_1 - \omega_2) + \cdots + it_{k_0}\omega_{k_0-1}]$$

$$\cdot \sum_{a_{k(0)-1}, \ldots, a_1} \frac{\langle a_f | p_{i_1} | a_1 \rangle \cdots \langle a_{k_0-1} | p_{i_{k(0)}} | a_g \rangle}{(\omega_1 - \omega_{a_1} + i\gamma_{a_1}) \cdots (\omega_{k_0-1} - \omega_{a_{k(0)-1}} + i\gamma_{a_{k(0)-1}})}. \tag{6.22}$$

$\{j_{k_0}\}$ means $j_1, \ldots, j_{k_0}$.

As is apparent, the transition probability is proportional to the $k_0$ th-order correlation function of the field. The feature of the atom, the $k_0$ th-order transition matrix element, is contained in the $\zeta$ function.

If the field is stationary, i.e.

$$\Gamma_{\{i_{k(0)}\}\{j_{k(0)}\}}^{(k_0, k_0)}(\{t'_{k_0} + \tau\}; \{t_{k_0} + \tau\}) = \Gamma_{\{i_{k(0)}\}\{j_{k(0)}\}}^{(k_0, k_0)}(\{t'_{k_0}\}; \{t_{k_0}\}), \tag{6.23}$$

the transition probability is proportional to the time of the interaction, and the rate of the transition can be defined.

If the field is quasi-monochromatic with center frequency $\omega_p$ such that the scattering in the frequency of the light $\Delta\omega_p \ll \omega_p$ and the center frequency is near the resonance $\omega_f - k_0\omega_p < \Delta\omega_p < \gamma_f$, the transition rate is proportional to the correlation function of the field at $\{t'_{k_0}\} = \{t_{k_0}\} = \{0\}$, i.e.

$$W_{f,g}^{(k_0)} \sim \Gamma^{(k_0, k_0)}(\{0\}; \{0\}). \tag{6.24}$$

It is known (Mandel and Wolf, 1965) that the correlation functions of the thermal and coherent light are given, respectively, by

$$\Gamma_{th}^{(k_0, k_0)} = k_0! \langle I \rangle^{k_0} \tag{6.25a}$$

and

$$\Gamma_{coh}^{(k_0, k_0)} = \langle I \rangle^{k_0}. \tag{6.25b}$$

The ratio of the transition rates using thermal and coherent light, respectively, is expressed as

$$^{(\text{th})}W_{f,g}^{(k_0)}/^{(\text{coh})}W_{f,g}^{(k_0)} = k_0! \qquad (6.26)$$

which is used often in work dealing with multiphoton ionization in order to explain the deviation of the experimental data for the generalized cross sections from the theoretical results.

It was mentioned in Section III,D that the measured generalized cross section of the six-photon ionization of hydrogen atoms is higher than the theoretical value. The deviation is explained by the use of a multimode laser, the light of which is regarded to be thermal. Therefore the measured cross section should be 6! higher than that theoretically expected, according to expression (6.26).

Inversely the statistical properties of the light can be determined by measuring the $k_0$-photon transition rate. This can be done naturally only after agreement has been found between the result of the theoretical calculation (Section II) and the result of the measurement in which a single mode laser was used.

There has been no systematic experimental investigation of the influence of the light statistics on the multiple-quantum transition rate but it is felt that this is nonetheless highly desirable.

## 2. Resonances

The relation (6.26) changes if the bandwith of the amplitude stabilized laser is larger than the level width of the final state $\Delta\omega_p = 2b > \gamma_f$. Then

$$^{(\text{th})}W_{f,g}^{(k_0)}/^{(\text{coh})}W_{f,g}^{(k_0)} = \begin{cases} k_0(k_0\ !) & \text{if} \quad k_0\omega_p - \omega_f \ll k_0 b, \\ (k_0 - 1)! & \text{if} \quad k_0\omega_p - \omega_f \gg k_0 b. \end{cases} \qquad (6.27)$$

The resonance increase of the $k_0$-photon transition rate is higher in the case of thermal light than laser light, while the resonance is narrower in the case of thermal light than in the case of laser light (see also Mollow, 1968).

## 3. Conclusions

The light field statistics strongly influence the multiphoton transition rate. The fact that the comparison of the experimentally measured generalized cross sections with the theoretical result is not possible without knowing the light statistics notwithstanding, no systematic experimental investigation of the subject exists.

The influence of light statistics on the resonance multiphoton ionization has not been investigated theoretically or experimentally.

The Stark shift of levels taking place during the process of multiphoton ionization at different light statistics also presents an interesting topic for future research.

The method of taking into account the influence of the light field statistics on multiphoton ionization naturally differs from that outlined above for a very strong field. There has been no work in this direction as yet.

## VII. CONCLUDING REMARKS

The phenomenon of multiphoton ionization is one of the most thoroughly investigated multiphoton elementary processes. The results of these investigations provide information on the general properties in common of the different multiphoton transitions, i.e. multiphoton absorption, multiphoton induced emission, and scattering with simultaneous absorption and emission of many quanta of different frequencies. The latter processes take place in the atom simultaneously with multiphoton ionization but with increasing dominance of the ionization if the light power density increases.

The first, nonvanishing term of the perturbation theory which satisfies the conservation of energy can be applied in the description of the multiphoton ionization process at medium light power density and without any resonances. The process of absorption of the quanta travels through the virtual energy states of the atom. It is supposed that the states of the atom are not perturbed by the field of the light at this medium light power density.

The result of this perturbation theoretical treatment is confirmed by experiment for up to five of the absorbed photons of the Nd glass and ruby lasers and of that of the first harmonics of these laser radiations. The agreement between theory and experiment is satisfactory in both qualitative and quantitative respects.

The distortion of the bound states of the atom, except the ground state in the light field, has to be taken into account in the case of resonance multiphoton ionization, i.e. when the energy of some quanta is near the energy of one atomic level. This distortion is observable even at medium light power density if resonance occurs. The distortion of states can be calculated up to the second order of perturbation theory and results in the higher order approximation for the multiphoton ionization. This approximation appeared to be adequate for explanation of almost all the experimental results.

The distortion of the level structure of the atom has to be taken into account in the higher than second-order approximation in a strong light field. The summed up contribution of the approximations of different orders of the perturbation theoretical series leads to the appearance of new phenomena (anomalous behavior of the transition rate versus light power density function, the shift of energetic levels, appearance of new satellite resonances, etc.).

These new phenomena are not adequately confirmed by experiment and demand further investigation. Conversely, some phenomena observed experimentally, such as the decrease of the nonlinearity of the multiphoton ionization of rare gases, the satellite maximum on the rate versus frequency function of the Cs atoms, and the observed large shift of levels of the He atoms, may possibly find their explanation in these new phenomena.

The continuum state of the atom is disturbed most heavily by the field of the light wave. It is curious that the distortion of the continuum state leading to the tunnel effect, the acceleration of the photoelectron in the simultaneously present fields of the light wave and the residual atom, has not, as yet, been observed experimentally.

New experimental and theoretical investigations are consequently desirable in the field of strong field multiphoton photoemission with and without resonance in order to understand the interaction process also in that region of the light power density where the perturbation theory is not applicable.

The statistical properties of the light wave have as much influence on the features of the multiphoton photoemission as the structure and the distortion of the structure of the atom. These statistical properties of the light beam are not adequately known in most of the experiments. The unique system of the light sources is the single mode laser, the statistical properties of which are regarded as known. Unfortunately the multiphoton ionization experiments were performed using multimode lasers and therefore the statistics of light were not determined. Further experimental investigations are also highly desirable in this field.

## ACKNOWLEDGMENTS

The author wishes to express his gratitude to G. A. Delone, N. B. Delone, Y. Gontier, V. A. Kovarski, and L. P. Rapoport for stimulating discussions, to all those colleagues who helped him by furnishing copies of their own work, and to all his co-workers who enabled him to assemble the material for this review.

## REFERENCES

Abella, I. D. (1962). *Phys. Rev. Lett.* **9**, 453.
Adam, A., Jánossy, L., and Varga, P. (1955). *Acta Phys.* **4**, 301. (In Russ.)
Agarwal, G. S. (1970). *Phys. Rev. A* **1**, 1445.
Agostini, P., Barjot, G., Bonnal, J. F., Mainfray, G., Manus, C., and Morellec, J. (1968a). *IEEE J. Quantum Electron.* **4**, 667.
Agostini, P., Bonnal, J. F., Mainfray, G., and Manus, C. (1968b). *C. R. Acad. Sci., Ser. B* **266**, 1034.

148     J. S. BAKOS

Agostini, P., Barjot, G., Mainfray, G., Manus, C., and Thebault, J. (1970a). *Phys. Lett. A* **31**, 367.
Agostini, P., Barjot, G., Mainfray, G., Manus, C., and Thebault, J. (1970b). *C. R. Acad. Sci., Ser. B* **270**, 1566.
Agostini, P., Barjot, G., Mainfray, G., Manus, C., and Thebault, J. (1970c). *IEEE J. Quantum Electron.* **6**, 782.
Alimov, D. T., Berezhetskaya, N. K., Delone, G. A., and Delone, N. B. (1971). *Kratk. Soobshch. Fiz.* No. 11, p. 21. Lebedev Phys. Inst., Moscow.
Alimov, D. T., Berezhetskaya, N. K., Delone, G. A., and Delone, N. B. (1973). *Zh. Eksp. Teor. Fiz.* **64**, 1178.
Arecchi, F. T. (1965). *Phys. Rev. Lett.* **15**, 912.
Arecchi, F. T., Gatti, E., and Sona, A. (1966). *Phys. Lett.* **20**, 27.
Askar'yan, G. A., and Rabinovich, M. S. (1965). *Sov. Phys.—JETP* **21**, 190.
Bakos, J. S., Kantor, J., and Kiss, A. (1970a). *Pis'ma Zh. Eksp. Teor. Fiz.* **12**, 371.
Bakos, J. S., Kantor, J., and Kiss, A. (1970b). *Kratk. Soobshch. Fiz.* No. 11, p. 18. Lebedev Phys. Inst., Moscow.
Bakos, J. S., Kiss, A., Nagaeva, M. L., Petrosiyan, K. B., and Rozsa, K. (1971). *Int. Conf. Phenomena Ionized Gases, 10th, Contrib. Pap., Oxford* (R. N. Franklin, ed.), p. 43. Parson, Oxford.
Bakos, J. S., Kiss, A., Szabo, L., and Tendler, M. (1972a). *Phys. Lett. A* **39**, 283.
Bakos, J. S., Kiss, A., Szabo, L., and Tendler, M. (1972b). *Phys. Lett. A* **39**, 317.
Bakos, J. S., Kiss, A., Szabo, L., and Tendler, M. (1972c). *Phys. Lett. A* **41**, 163.
Bakos, J. S., Nagaeva, M. L., Ovchinykov, V. L., and Rubin, G. (1973a). *Krat. Soobshch. Fiz.* No. 9, p. 3. Lebedev Phys. Inst., Moscow.
Bakos, J. S., Kiss, A., Szabo, L., and Tendler, M. (1973b). *Pis'ma Zh. Eksp. Teor. Fiz.* **18**, 403.
Bakos, J. S., Nagaeva, M. L., Ovchinykov, V. L., and Rubin, G. (1973c). *Int. Conf. Phenomena Ionized Gases, 11th* (I. Stoll et al., eds.), p. 19. Inst. Physics, Prague.
Bakos, J. S., Kiss, A., Szabo, L., and Tendler, M. (1973d). *Int. Conf. Phenomena Ionized Gases, 11th* (I. Stoll et al., eds.), p. 18. Inst. Physics, Prague.
Baravian, G., Benattar, R., Bretagne, J., Godart, J. L., and Sultan, G. (1970). *Appl. Phys. Lett.* **16**, 162.
Baravian, G., Benattar, K., Bretagne, J., Callede, G., Godart, J. L., and Sultan, G. (1971). *Appl. Phys. Lett.* **18**, 387.
Barhudarova, T. M. (1969). *Fiz. Chim. Obr. Mat.* No. 4, p. 10.
Barhudarova, T. M. (1970). Thesis, Univ. of Moscow.
Barhudarova, T. M., Voronov, G. S., Delone, G. A., Delone, N. B., and Martakova, N. K. (1967). *Int. Conf. Phenomena Ionised Gases, Contrib. Pap., Vienna* p. 266.
Barjot, G. (1971). *J. Appl. Phys.* **42**, 3641.
Bates, D. R., and Damgaard, A. (1950). *Phil. Trans. Roy. Soc. London, Ser. A* **242**, 101.
Baz, A. I., Zel'dovich, Y. B., and Perelomov, A. M. (1971). "Scattering, Reactions and Decays in the Nonrelativistic Quantum Mechanics." Nauka, Moscow. (In Russ.).
Bebb, H. B. (1966). *Phys. Rev.* **149**, 25.
Bebb, H. B. (1967). *Phys. Rev.* **153**, 23.
Bebb, H. B., and Gold, A. (1965). In "Physics of Quantum Electronics" (P. L. Kelley, B. Lax, and P. E. Tannenwald, eds.), p. 489. McGraw-Hill, New York.
Bebb, H. B., and Gold, A. (1966). *Phys. Rev.* **143**, 1.
Benattar, R. (1971). Thesis, No. 847. Univ. Paris-Sud, Cent. Orsay, Paris.
Bethe, H. A., and Salpeter, E. E. (1957). In "Handbuch der Physik" (S. Flügge, ed.), p. 88. Springer-Verlag, Berlin and New York. B 35.
Bistrova, T. B., Voronov, G. S., Delone, G. A., and Delone, N. B. (1967). *Pis'ma Zh. Eksp. Teor. Fiz.* **5**, 223.

Bonch-Bruevich, A. M., and Khodovoi, B. A. (1965). *Usp. Fiz. Nauk* **85**, 3.
Bonch-Bruevich, A. M., and Khodovoi, B. A. (1967). *Usp. Fiz. Nauk* **93**, 71.
Bonch-Bruevich, A. M., Kostin, N. N., Khodovoi, V. A., and Khromov, V. V. (1969). *Zh. Eksp. Teor. Fiz.* **56**, 144.
Bonch-Bruevich, A. M., Khodovoi, V. A., and Przhibel'skii, S. G. (1973). *Conf. Interaction Electrons Strong Electromagn. Field, Invited Pap., Balatonfuered, Hungary, 1972* (J. S. Bakos, ed.), p. 255. Cent. Res. Inst. Phys., Budapest.
Born, M., and Wolf, E. (1964). " Principles of Optics." Pergamon, Oxford.
Brossel, J., Cagnac, B., and Kastler, A. (1953). *C. R. Acad. Sci.* **237**, 984.
Brown, R. H., and Twiss, R. Q. (1956a). *Nature (London)* **177**, 27.
Brown, R. H., and Twiss, R. Q. (1956b). *Nature (London)* **178**, 1046.
Brown, R. H., and Twiss, R. Q. (1956c). *Nature (London)* **178**, 1447.
Bunkin, F. V., and Prokhorov, A. M. (1964). *Zh. Eksp. Teor. Fiz.* **46**, 1090.
Burgess, A., and Seaton, M. J. (1960). *Mon. Notic. Roy. Astron. Soc.* **120**, 121.
Carusotto, S., Fornaca, G., and Polacco, E. (1967). *Phys. Rev.* **157**, 1207.
Chan, F. T., and Tang, C. L. (1969). *Phys. Rev.* **185**, 42.
Chang, C. S., and Stehle, P. (1971). *Phys. Rev. A.* **4**, 641.
Chang, C. S., and Stehle, P. (1973). *Phys. Rev. Lett.* **30**, 1285.
Chin, S. L., and Isenor, N. B. (1970). *Can. J. Phys.* **48**, 1445.
Chin, S. L., Isenor, N. B., and Young, M. (1969). *Phys. Rev.* **188**, 7.
Cohen-Tannoudji, C. (1962). *Ann. Phys. (Paris)* **7**, 423.
Cohen-Tannoudji, C., and Kastler, A. (1966). *Progr. Opt.* **5**, 3.
Dalgarno, A., and Lewis, J. T. (1956). *Proc. Roy. Soc., Ser. A* **233**, 70.
Davidkin, V. A., Zon, B. A., Manakov, N. L., and Rapoport, L. P. (1971). *Zh. Eksp. Teor. Fiz.* **60**, 124.
Delone, G. A. (1973). *Conf. Interaction Electrons Strong Electromagn. Field, Invited Pap., Balatonfuered, Hungary, 1972* (J. S. Bakos, ed.), p. 77. Cent. Res. Inst. Phys., Budapest.
Delone, G. A., and Delone, N. B. (1968). *Zh. Eksp. Teor. Fiz.* **54**, 1067.
Delone, G. A., and Delone, N. B. (1969). *Pis'ma Zh. Eksp. Teor. Fiz.* **10**, 413.
Delone, G. A., Delone, N. B., Donskaya, N. P., and Petrosiyan, K. B. (1969). *Pis'ma Zh. Eksp. Teor. Fiz.* **9**, 103.
Delone, G. A., Delone, N. B., and Piskova, G. K. (1971a). *Int. Conf. Phenomena Ionised Gases, 10th, Contrib. Pap., Oxford* (R. N. Franklin, ed.), p. 40. Parson, Oxford.
Delone, G. A., Delone, N. B., Kovarski, V. A., and Perelman, N. F. (1971b). *Kratk. Soobshch. Fiz.* No. 8, 37. Lebedev Phys. Inst., Moscow.
Delone, G. A., Delone, N. B., and Piskova, G. K. (1972). *Zh. Eksp. Teor. Fiz.* **62**, 1272.
Delone, G. A., Delone, N. B., Zolotarev, V. K., Manakov, N. L., Piskova, G. K., and Tursunov, M. A. (1973). *Zh. Eksp. Teor. Fiz.* **65**, 481.
DeMichelis, C. (1969). *IEEE J. Quantum Electron.* **5**, 188.
De Witt, R. N. (1973). *Proc. Phys. Soc., London (At. Mol. Phys.)* **6**, 803.
Evans, R. G., and Thonemann, P. C. (1972). *Phys. Lett. A* **39**, 133.
Evans, R. G., and Thonemann, P. C. (1973). *Phil. Mag.* **27**, 1387.
Faisal, F. H. M. (1972a). *Proc. Phys. Soc., London (At. Mol. Phys.)* **5**, L196.
Faisal, F. H. M. (1972b). *Proc. Phys. Soc., London (At. Mol. Phys.)* **5**, L233.
Farkas, G., Náray, Z., and Varga, P. (1967). *Phys. Lett. A* **24**, 572.
Farkas, G., Horváth, Z., and Kertész, I. (1972). *Phys. Lett. A* **39**, 231.
Feyman, R. P. (1949). *Phys. Rev.* **76**, 749.
Fox, R. A., Kogan, R. M., and Robinson, E. J. (1971a). *Phys. Rev. Lett.* **26**, 1416.
Franken, P. A., Hill, A. E., Peters, C. W., and Weinreich, G. (1961). *Phys. Rev. Lett.* **7**, 118.
Geltman, S. (1963). *Phys. Lett.* **4**, 168.

150     J. S. BAKOS

Giordmaine, J. A., and Howe, J. A. (1963). *Phys. Rev. Lett.* **11**, 207.
Glauber, R. J. (1963). *Phys. Rev.* **130**, 2529.
Glauber, R. J. (1965). *In* "Quantum Optics and Electronics," Les Houches Lectures, 1964 (C. DeWitt, A. Blandin, and C. Cohen-Tannoudji, eds.), p. 65. Gordon & Breach, New York.
Göppert-Mayer, M. (1929). *Naturwissenschaften* **17**, 932.
Göppert-Mayer, M. (1931). *Ann. Phys.* **9**, 273.
Gold, A., and Bebb, H. B. (1965). *Phys. Rev. Lett.* **14**, 60.
Gontier, Y., and Trahin, M. (1968a). *Phys. Rev.* **172**, 83.
Gontier, Y., and Trahin, M. (1968b). *C. R. Acad. Sci., Ser. B* **266**, 1177.
Gontier, Y., and Trahin, M. (1971). *Phys. Rev.* **4**, 1896.
Gontier, Y., and Trahin, M. (1973a). *Int. Conf. Phenomena Ionized Gases, 11th* (I. Stoll *et al.*, eds.), p. 21. Inst. Phys., Prague.
Gontier, Y., and Trahin, M. (1973b). *Phys. Rev.* **7**, 1899.
Gordon, W. (1928). *Z. Phys.* **48**, 180.
Guccione-Gush, R., Gush, H. P., and van Kranendonk, J. (1967). *Can. J. Phys.* **45**, 2513.
Hall, J. L., Robinson, E. J., and Branscomb, L. M. (1965). *Phys. Rev. Lett.* **14**, 1013.
Heitler, W. (1954). "The Quantum Theory of Radiation." Oxford Univ. Press, London and Newew York.
Held, B., Mainfray, G., Manus, C., and Morellec, J. (1971). *Phys. Lett. A* **35**, 257.
Held, B., Mainfray, G., and Morellec, J. (1972). *Phys. Lett. A* **39**, 57.
Held, B., Mainfray, G., Manus, C., Morellec, J., and Sanghez, F. (1973). *Phys. Rev. Lett.* **30**, 423.
Henneberger, W. C. (1968). *Phys. Rev. Lett.* **21**, 838.
Herman, M. (1973). *Int. Conf. Phys. Electr. Atom. Collisions,* Abstr. Paps., (B. C. Čobić and M. V. Kurepa, eds.), p. 588. *Inst. Phys.,* Beograd.
Jaiswal, A. K., and Agarwal, G. S. (1969). *J. Opt. Soc. Amer.* **59**, 1446.
Jánossy, L. (1957). *Nuovo Cimento* **6**, 111.
Jánossy, L. (1959). *Nuovo Cimento* **12**, 370.
Kaiser, W., and Garrett C. G. B. (1961). *Phys. Rev. Lett.* **7**, 229.
Karule, E. (1971). *Proc. Phys. Soc., London (At. Mol. Phys.)* **4**, L67.
Keldish, L. V. (1964). *Zh. Eksp. Teor. Fiz.* **47**, 1945.
Kertis, L. (1970). "Coulomb Wavefunctions." Publ. Acad: USSR, Moscow (In Russ).
Klarsfeld, S. (1970). *Lett. Nuovo Cimento* **3**, 395.
Klarsfeld, S., and Maquet, A. (1972). *Phys. Rev. Lett.* **29**, 79.
Kogan, R. M., Fox, R. A., Burnham, G. T., and Robinson, E. J. (1971). *Bull. Amer. Phys. Soc.* **16**, 1411.
Korobkin, U. V., Leontovich, A. M., Popova, M. H., and Stchelev, M. I. (1967). *Zh. Eksp. Teor. Fiz.* **53**, 16.
Kotova, L. P., and Terentev, M. V. (1967). *Zh. Eksp. Teor. Fiz.* **52**, 732.
Kovarski, V. A. (1969a). *Zh. Eksp. Teor. Fiz.* **57**, 1613.
Kovarski, V. A. (1969b). *Int. Conf. Phenomena Ionized Gases, 9th, Contrib. Pap.,* Bucharest (G. Musa, I. Ghica, A. Popescu, and L. Nestase, eds.), p. 38. Editura Academiei, Rep. Soc. România.
Kovarski, V. A. (1973). *Conf. Interaction Electrons Strong Electromagn. Field, Invit. Pap.,* Balatonfuered, Hungary, 1972 (J. S. Bakos, ed.), p. 125. Cent. Res. Inst. Phys., Budapest.
Kovarski, V. A., and Perelman, N. F. (1971a). *Zh. Eksp. Teor. Fiz.* **61**, 1389.
Kovarski, V. A., and Perelman, N. F. (1971b). *Zh. Eksp. Teor. Fiz.* **60**, 509.
Kramers, H. A., and Heisenberg, W. (1925). *Z. Phys.* **31**, 681.
Lambrapoulos, P. (1968). *Phys. Rev.* **168**, 1418.
Lambrapoulos, P. (1972a). *Phys. Rev. Lett.* **28**, 585.
Lambrapoulos, P. (1972b). *Phys. Rev. Lett.* **29**, 453.

Lambrapoulos, P. (1973). *Phys. Rev. Lett.* **30**, 413.
Lambrapoulos, P., Kikuchi, C., and Osborn, S. K. (1966). *Phys. Rev.* **144**, 1081.
Landau, L. D., and Lifshitz, E. M. (1958). "Quantum Mechanics." Pergamon, Oxford.
Laplanche, G., Jaquen, M., and Rachman, A. (1973). *Opt. Commun.* **8**, 37.
Lebedev, I. V. (1971), *Opt. Spektrosk.* **30**, 381.
Low, F. (1952). *Phys. Rev.* **88**, 53.
Lu Van, M., Mainfray, G., Manus, C., and Tugov, I. (1973). *Phys. Rev. A* **7**, 91.
Mainfray, G. (1973). *Conf. Interaction Electrons Strong Electromagn. Field, Invited Pap., Balatonfuered, Hungary, 1972* (J. S. Bakos, ed.), p. 155. Cent. Res. Inst. Phys., Budapest.
Maker, P. D., Terhune, R. W., and Savage, C. M. (1964). *In* "Quantum Electronics" (P. Grivet and N. Bloembergen, eds.), p. 1559. Columbia Univ. Press, New York.
Manakov, N. L. (1971). Thesis, Voronezh State Univ.,
Manakov, N. L., Rapoport, L. P., and Zon, B. A. (1971a). *Int. Conf. Phenomena Ionized Gases, 10th, Contrib. Pap., Oxford* (R. N. Franklin, ed.), p. 38. Parson, Oxford.
Manakov, N. L., Rapoport, L. P., and Zon, B. A. (1971b). *Int. Conf. Phenomena Ionized Gases, 10th, Contrib. Pap., Oxford* (R. N. Franklin, ed.), p. 46. Parson, Oxford.
Mandel, L., and Wolf, E. (1965). *Rev. Mod. Phys.* **37**, 231.
Meadors, J. G. (1966). *IEEE J. Quantum Electron.* **2**, 638.
Messiah, A. (1965). "Quantum Mechanics," Vol. 2. North-Holland Publ., Amsterdam.
Meyerand, R. G., and Haught, A. F. (1963). *Phys. Rev. Lett.* **11**, 401.
Minck, R. W. (1964). *J. Appl. Phys.* **35**, 252.
Mizuno, J. (1973). *Proc. Phys. Soc., London (At. Mol. Phys.)* **6**, 314.
Mohan, M., and Thareja, R. K. (1972). *Proc. Phys. Soc., London (At. Mol. Phys.)* **5**, L134.
Mohan, M., and Thareja, R. K. (1973). *Proc. Phys. Soc., London (At. Mol. Phys.)* **6**, 809.
Mollow, B. R. (1968). *Phys. Rev.* **175**, 1555.
Mollow, B. R. (1972). *Phys. Rev. A* **5**, 1827.
Morton, V. M. (1967). *Proc. Phys. Soc., London* **92**, 301.
Nikishev, A. I., and Ritus, V. I. (1966). *Zh. Eksp. Teor. Fiz.* **50**, 255.
Nikishev, A. I., and Ritus, V. I. (1967). *Zh. Eksp. Teor. Fiz.* **52**, 223.
Oppenheimer, J. R. (1928). *Phys. Rev.* **13**, 66.
Papoular, R., and Platz, P. (1973). *Conf. Interaction Electrons Strong Electromagn. Field, Invited Pap., Balatonfuered, Hungary, 1972* (J. S. Bakos, cd.), p. 211. Cent. Res. Inst. Phys., Budapest.
Parzynski, R. (1972). *Acta Phys. Pol. A* **42**, 745.
Parzynski, R. (1973). *Opt. Commun.* **8**, 75.
Perelomov, A. M., and Popov, V. S. (1967). *Zh. Eksp. Teor. Fiz.* **52**, 514.
Perelomov, A. M., Popov, V. S., and Terentev, M. V. (1966a). *Zh. Eksp. Teor. Fiz.* **50**, 1393.
Perelomov, A. M., Popov, V. S., and Terentev, M. V. (1966b). *Zh. Eksp. Teor. Fiz.* **51**, 309.
Perelomov, A. M., Popov, V. S., and Kuznetsov, V. P. (1968). *Zh. Eksp. Teor. Fiz.* **54**, 841.
Peressini, R. E. (1966). *Phys. Quantum Electron. Conf. Proc., San Juan, P.R., 1965* p. 499.
Pert, G. J. (1972). *IEEE J. Quantum Electron.* **8**, 623.
Peticolas, W. L., Goldsborough, J. P., and Rieckhoff, K. E. (1963). *Phys. Rev. Lett.* **10**, 43.
Phelps, A. V., and Molnár, J. P. (1953). *Phys. Rev.* **89**, 1209.
Platz, P. (1968). *Phys. Lett. A* **27**, 714.
Platz, P. (1969). *Appl. Phys. Lett.* **14**, 168.
Platz, P. (1970). *Appl. Phys. Lett.* **16**, 70.
Popov, V. S., Kuznetsov, V. P., and Perelomov, A. M. (1967). *Zh. Eksp. Teor. Fiz.* **53**, 331.
Rahman, N. K., and Reiss, H. R. (1972). *Phys. Rev. A* **6**, 1252.
Raizer, Y. P. (1965). *Usp. Fiz. Nauk* **87**, 29.
Rapoport, L. P. (1973). *Conf. Interaction Electrons Strong Electromagn. Field, Invited Pap., Balatonfuered, Hungary, 1972* (J. S. Bakos, ed.), p. 99. Cent. Res. Inst. Phys., Budapest.

Rapoport, L. P., and Zon, B. A. (1968). *Phys. Lett. A* **26**, 564.

Rapoport, L. P., Zon, B. A., and Manakov, L. P. (1969). *Zh. Eksp. Teor. Fiz.* **56**, 400.

Reiss, H. R. (1970a). *Phys. Rev. Lett.* **25**, 1149.

Reiss, H. R. (1970b). *Phys. Rev. A* **1**, 803.

Reiss, H. R. (1971). *Phys. Rev. D* **4**, 3533.

Reiss, H. R. (1972a). *Phys. Rev. Lett.* **29**, 1129.

Reiss, H. R. (1972b). *Phys. Rev. A* **6**, 817.

Robinson, E. J., and Geltman, S. (1967). *Phys. Rev.* **153**, 4.

Roman, P. (1965). "Advanced Quantum Theory," p. 281. Addison-Wesley, Reading, Massachusetts.

Rose, M. E. (1957). "Elementary Theory of Angular Momentum." Wiley, New York; Chapman & Hall, London.

Schwartz, C. (1959). *Ann. Phys.* (*New York*) **2**, 169.

Schwartz, C., and Thiemann, J. J. (1959). *Ann. Phys.* (*New York*) **2**, 178.

Seaton, M. (1958). *Mon. Notic. Roy. Astron. Soc.* **118**, 117.

Sestakov, A. F., Khristenko, S. V., and Vetchinkin, S. I. (1972). *Opt. Spektrosk.* **33**, 413.

Shen, Y. R. (1967). *Phys. Rev.* **155**, 921.

Shiga, F., and Imamura, M. (1967). *Phys. Lett. A* **25**, 706.

Sobelman, I. I. (1963). "Introduction to the Theory of Atomic Spectra." State Publ. Phys. Math. Lit., Moscow. (In Russ.)

Teich, M. C., and Diamant, P. (1969). *J. Appl. Phys.* **40**, 625.

Teich, M. C., and Wolga, G. J. (1966). *Phys. Rev. Lett.* **16**, 625.

Terhune, R. W. (1963). *Bull. Amer. Phys. Soc.* **8**, 359.

Tewari, S. P. (1972). *Phys. Rev. A* **6**, 1869.

Tozer, B. A. (1965). *Phys. Rev. A* **137**, 1665.

Vetchinkin, S. I., and Khristenko, S. V. (1968). *Opt. Spektrosk.* **25**, 650.

Volkov, D. M. (1935). *Z. Phys.* **94**, 250.

Voronov, G. S. (1966). *Zh. Eksp. Teor. Fiz.* **51**, 1496.

Voronov, G. S., and Delone, N. B. (1965). *Pis'ma Zh. Eksp. Teor. Fiz.* **1**, 42.

Voronov, G. S., and Delone, N. B. (1966). *Zh. Eksp. Teor. Fiz.* **50**, 78.

Voronov, G. S., Gorbunkov, V. M., Delone, G. A., Delone, N. B., Keldish, L. V., Kudrevatova, O. V., and Rabonovich, M. S. (1966a). *Proc. Int. Conf. Phenomena Ionized Gases, 7th,* (B. Perovic and D. Tosic, eds.), Vol. 1, p. 806. Gradevinska Kujiga, Belgrade.

Voronov, G. S., Delone, G. A., and Delone, N. B. (1966b). *Zh. Eksp. Teor. Fiz.* **51**, 1660.

Wright, J. K. (1964). *Proc. Phys. Soc., London* **84**, 41.

Zel'dovich, Y. B. (1966). *Zh. Eksp. Teor. Fiz.* **51**, 1492.

Zel'dovich, Y. B. (1973). *Conf. Interaction Electrons Strong Electromagn. Field, Invit. Pap., Balatonfuered, Hungary, 1972* (J. S. Bakos, ed.), p. 5. Cent. Res. Inst. Phys., Budapest.

Zel'dovich, Y. B., and Raizer, Y. P. (1965). *Sov. Phys.—JETP* **20**, 772.

Zernik, W. (1964). *Phys. Rev. A* **135**, 51.

Zernik, W. (1968). *Phys. Rev.* **176**, 420.

Zernik, W., and Klopfenstein, R. W. (1965). *J. Math. Phys.* **6**, 262.

Zon, B. A., and Rapoport, L. P. (1968). *Pis'ma Zh. Eksp. Teor. Fiz.* **7**, 70.

Zon, B. A., Manakov, N. L., and Rapoport, L. P. (1968). *Zh. Eksp. Teor. Fiz.* **55**, 924.

Zon, B. A., Manakov, N. L., and Rapoport, L. P. (1969). *Dokl. Akad. Nauk. SSSR* **188**, 560.

Zon, B. A., Manakov, N. L., and Rapoport, L. P. (1971a). *Zh. Eksp. Teor. Fiz.* **61**, 968.

Zon, B. A., Manakov, N. L., and Rapoport, L. P. (1971b). *Zh. Eksp. Teor. Fiz.* **60**, 1264.

# Recent Advances in the Hall Effect: Research and Application

## D. MIDGLEY

*Department of Electronic Engineering,*
*The University of Hull,*
*Hull, United Kingdom*

## I. INTRODUCTION

The Hall effect is fundamental in the solid state. It is the expression of the $\mathbf{v} \times \mathbf{B}$ law for a charge moving with velocity $\mathbf{v}$ in a magnetic field $\mathbf{B}$. The form taken by the same law in the study of electrical machines leads to results for the induced emf and the force on each conductor. Why have all the many and varied applications of electrical machines not been replicated in the solid state by Hall effect devices? The explanation is that the effect is negligibly small in most materials and that losses occur as heat in the resistances of the few materials which have a significant Hall coefficient.

Stricker (1) has reviewed the Hall effect and its applications, with references up to 1967. He gives a short outline of the theory for charge carriers of one polarity and also for two polarities; materials and their properties are listed; the geometry of plates, their efficiency and residual voltages are considered; applications to amplifiers, oscillators, analog multipliers, field measurement, gyrators, isolators, and circulators are discussed and anticipated.

Weiss (2) has written about the structure and application of "galvanomagnetic" devices with references up to 1968. The Hall effect and magnetoresistance effects are both embraced by the term galvanomagnetic, which is linguistic currency throughout continental Europe. He emphasizes the design and fabrication of raster plate magnetoresistors, known also as field plates, with applications to contact-free signal generation, contact-free variable resistors, and potential dividers.

The present review places some stress on the self-magnetic field of Hall currents. References up to 1973 have been collected in part by a computer profile of keywords and parts of words such as *magnet*. The profile is amended continuously if significant references are found to be not retrieved by the existing profile. No attempt is made to include measurements of Hall constants, where this has been done for its own sake as part of pure physics and the routine process of collecting the properties of elements and compounds as such. The Hall effect is certainly important in that context as a means of establishing the mobility and sign of charge carriers, but the point of view taken here is that it represents just one of the applications.

## II. ANALYTICAL THEORIES

### A. Electronic Motion and the Hall Angle

In free space an electron immersed in a constant uniform electric field $E_x$ moves in a straight line with constant acceleration. An electron with initial velocity $v_x$ entering a constant uniform magnetic field $B_z$ moves in a circle. The motion is a combination of straight line and circle into a cycloidal path, if $E$ and $B$ exist together as crossed fields. The result is well known in the study of magnetrons and magnetically deflected cathode-ray tubes. With no magnetic field the electron moves in the $x$ direction, but the net motion in that direction is reduced to zero by a magnetic field in free space; the average velocity is converted to the $y$ direction for every value of $B$.

When the atoms of a material are present, collisions are usually so frequent that the cycloidal motion of every electron is terminated before the first cusp has been reached as in Fig. 1. Suppose that a given electron is brought to rest at intervals equal to the mean time between collisions. The motion is a cascade of incomplete cycloidal paths. A straight line representing the average velocity is a good approximation to this motion, and the direction of the line is that of the resultant mean current density $J$. The angle $\theta$ between $E$ and $J$ is the Hall angle.

One result of the effect of a magnetic field, therefore, is that the current density $J$ is no longer in the direction of the electric field $E$. The current is deflected through a Hall angle, which increases with $B$ and the mean free path. Typically, the Hall angle exceeds 80° at room temperature in indium antimonide with a magnetic flux density of 1 T.

$$\tan \theta = \mu_e B, \tag{1}$$

where

$$\mu_e = e\bar{t}/m_e \tag{2}$$

is the electron mobility expressed in terms of the electronic charge $e$, effective mass $m_e$, and mean time $\bar{t}$ between collisions.

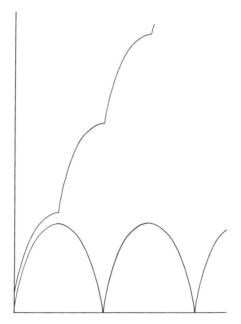

FIG. 1. Motion of an electron in crossed fields. Upper curve: Hall angle about 30° and frequent collisions. Lower curve: Hall angle, 90°.

### B. Magnetoresistance as an Aspect of Hall Effect

The discovery that a resistor can change its value in a magnetic field was attributed to Thomson (3) in 1856, more than twenty years before the Hall effect itself became known. They are, however, not truly separate effects. Magnetoresistance is frequently explained in terms of the longer path taken by electrons when they are deflected. It also emerges from an analysis (1) in which the scalar conductivity is replaced by a tensor, whose off-diagonal terms describe the Hall effect but the diagonal itself displays the magnetoresistance.

To understand easily why the square of the magnetic flux density controls magnetoresistance, we consider a Hall plate of indefinite length. This ensures that the only currents are the uniform source currents, flowing along the plate. There are transverse emf's but they are only able to deflect electrons to the side of the plate, where they set up an electrostatic field. They do not flow continuously, and there are no Hall currents. Thus we have the simplest first-order effect with emf directly proportional to the magnetic flux density.

Let the two sides of the plate now be joined together by a short circuit or, equivalent to the short circuit, let the plate be deformed into a tube with the source currents along its axis and the applied magnetic flux density in a radial direction. The electrostatic components are obliterated. The greatest possible Hall current flows and it may be treated as a separate component of current from the source current, especially as the two are perpendicular to each other. Since these new currents are also flowing across the magnetic field, they must create a second-order Hall effect, whose electric field inserts a back emf to oppose the source. The "$IR$ voltage drop" of an ordinary resistor is sometimes regarded as a back emf. We have shown that the resistance therefore is increased and that the increase must be a function (as in Fig. 2) of the square of the magnetic flux density, because it is a second-order Hall effect.

For a material with electron mobility $\mu_e$, Hall constant $K_H$, and conductivity $\sigma$, let the original source current density and electric fields be $J_0$ and $E_0$.

$$J_0 = \sigma E_0, \qquad E_H = K_H J_0 B, \qquad J_H = \sigma E_H.$$

The final electric field strength $E$ in the source circuit is made up of $E_0$ and $\Delta E$, where $\Delta E = K_H J_H B$. Combining these equations and solving for $J_0$, we see that the conductivity $\sigma_B$ in the presence of $B$ is:

$$\begin{aligned}
\sigma_B &= J_0/E, \\
&= \sigma/[1 + (\sigma K_H B)^2], \\
&= \sigma/[1 + (\mu_e B)^2].
\end{aligned} \tag{3}$$

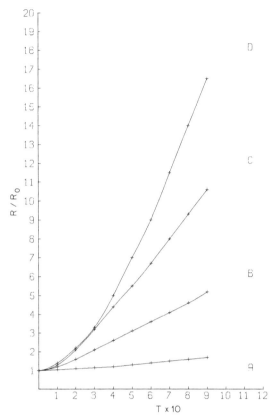

FIG. 2. Magnetoresistance against magnetic flux density. Ratio of resistance in magnetic field to initial resistance for: Curve A, a Hall plate with length 10 times width, Curve B, a Hall plate of square shape. Curve C, a Hall plate with width 3 times the length. Curve D, a Corbino disk in an axial field or a tubular magnetoresistor in a radial field with axial current.

In rectangular coordinates $x$, $y$, $z$ with $\mathbf{B} = 0, 0, B_z$ the tensor conductivity may be written as a $2 \times 2$ matrix:

$$\begin{vmatrix} \sigma_{xx} & \sigma_{xy} \\ \sigma_{yx} & \sigma_{yy} \end{vmatrix} = \begin{vmatrix} \sigma_B & \mu_e B_z \sigma_B \\ -\mu_e B_z \sigma_B & \sigma_B \end{vmatrix}. \tag{4}$$

These results apply to isotropic materials with charge carriers of only one polarity and to an isothermal state. It is also implied that no contribution to the magnetic field is made by the source currents or the Hall currents. The argument does not lead to an infinite regression of $n$th-order Hall effects, if it is supposed that the original source current density is maintained after the application of the magnetic field by a suitable increase in the applied voltage.

All of the quantities discussed so far have been unvarying in time; the transients have been allowed to decay and the currents have been able to penetrate the materials uniformly throughout at zero frequency.

## C. Helicon Waves as an Aspect of Hall Effect

Whatever bulk conducting solid state devices are devised, to be effective they must be penetrated by electric or magnetic fields or by a current density. But it is the nature of conducting and semiconducting materials to exhibit skin effects at high frequencies. Perfect conductors exclude all fields even at zero frequency; the propagation of electromagnetic waves through plasmas, sea water, and conductors in general appears to be an almost insuperable problem. Yet, seemingly impenetrable materials can be pierced, as it were, by a type of corkscrew, namely, the helicon wave.

The helicon wave requires a steady magnetic field in the direction of propagation. In general the electric and magnetic fields are transverse, but they are circularly polarized with one component corkscrewing in a right-hand direction and the other in a left-hand direction. One component is rapidly attenuated, but the other may penetrate to some distance.

A significant advance in the understanding of helicon waves was made when Chambers and Jones (4) showed that they could be explained entirely as a Hall effect. The following analysis is based upon their treatment, particularly in the unusual use of the square root of minus one. Normally, we denote this by $j$ and attach it to the exponent of the time factor, when considering waves at a single frequency. For the purposes of this analysis, we shall denote root minus one by $i$ to distinguish $y$ components from $x$ components in the three-dimensional field. This simplifies the wave equations.

It is convenient to introduce another constant into the analysis in the form of a constant current density in the same direction as the magnetic steady field. This current must be set to zero for the ordinary results of plain helicon waves. The intention is to show that growing helicon waves can occur in the presence of that current. They can never grow in the constant magnetic field alone, because there is no mechanism for the transfer of energy to the wave in order to make good its losses in the conductivity of the material through which it travels.

Let the $z$-directed steady magnetic field intensity be denoted by $H_0$ and the $z$-directed constant current density be $J_0$.

The transverse components of electric field strength in the wave are $E_x$ and $E_y$ but these are neither constant with respect to time nor with respect to $z$. The same applies to transverse components of the magnetic field intensity $H_x$, $H_y$ and current density $J_x$, $J_y$. All of these transverse components are combined into single entities as complex transverse vectors by the following definitions:

$$E = E_x + iE_y,$$
$$H = H_x + iH_y, \tag{5}$$
$$J = J_x + iJ_y.$$

It is supposed that the steady component of the magnetic field caused by $J_0$ is negligible compared with both $H_0$ and $H$. A solution is pursued in which nothing shall vary with respect to $x$ or $y$: $\partial/\partial x = \partial/\partial y = 0$. The plane $z = 0$ is the boundary at the surface of a semiinfinite slab of material with given conductivity and Hall constant.

$$E_x = J_x/\sigma - K_H H_0 J_y + K_H H_y J_0, \tag{6}$$
$$E_y = J_y/\sigma + K_H H_0 J_x - K_H H_x J_0. \tag{7}$$

These equations, which give effect to the conductivity and Hall fields, may be combined into a single equation if the second is multiplied through by $i$ and is then added to the first, thus:

$$E = J/\sigma + iK_H H_0 J - iK_H J_0 H. \tag{8}$$

The constants may be simplified by defining an effective resistivity $\rho_e$ and a quantity $k$:

$$\rho_e := 1/\sigma + iK_H H_0 \tag{9}$$

where $\rho_e$ is complex, and

$$k := K_H J_0. \tag{10}$$

Equation (8) now takes the form

$$E = \rho_e J - ikH. \tag{11}$$

If Maxwell's equations are written out in component form as was done for Eqs. (6) and (7), the same action of combining pairs of equations using $i$ yields these results:

$$\text{curl } E = -\mu \, \partial H/\partial t \tag{12}$$
$$\text{curl } H = J \tag{13}$$

in which, as is usual in conducting materials below X-ray frequencies, displacement current is negligible compared with conduction current.

$$i \, \partial E/\partial z = -\mu \, \partial H/\partial t, \tag{14}$$
$$i \, \partial H/\partial z = J. \tag{15}$$

Here, the nonzero components of curl $E$ and curl $H$ have been written. Those results are now combined with Eq. (11) to give a differential equation in $H$

$$\frac{\partial^2 H}{\partial z^2} - \frac{k}{\rho_e}\frac{\partial H}{\partial z} + \frac{\mu}{\rho_e}\frac{\partial H}{\partial t} = 0. \tag{16}$$

This is the wave equation for helicon waves, when the constant current density $J_0$, and hence $k$, is zero. Solutions for a single angular frequency $\omega$ may now be found by the conventional use of root minus one:

$$\partial/\partial t = j\omega. \tag{17}$$

The absolute magnetic permeability $\mu$ is not to be confused with $\mu_e$, the electron mobility. Particular solutions of the equation take the form:

$$H = A\varepsilon^{mz} \tag{18}$$

where $A$ is an arbitrary constant and $m$ is given by

$$m = \frac{k}{2\rho_e} \pm \left\{\frac{1}{2}\left[\left(\frac{k^4}{16\rho_e^4} + \frac{\omega^2\mu^2}{\rho_e^2}\right)^{1/2} + \frac{k^2}{4\rho_e^2}\right]\right\}^{1/2}$$
$$\pm i\left\{\frac{1}{2}\left[\left(\frac{k^4}{16\rho_e^4} + \frac{\omega^2\mu^2}{\rho_e^2}\right)^{1/2} - \frac{k^2}{4\rho_e^2}\right]\right\}^{1/2} \tag{19}$$

The upper sign represents a wave traveling in the negative $z$ direction and suffering attenuation. The lower sign means a wave with travel in the positive $z$ direction but again attenuated. Because the quantity $k/2\rho_e$ appears under each square root sign, the most that can be done by the introduction of steady current is to make the attenuation approach zero for one of the two waves. The expectation of a growing helicon wave is not realized.

In physical terms the transfer of energy from the source of the direct current into the helicon wave has taken the form of increasing its spin. The phase is changed ever more rapidly as the constant current is increased. Whitehouse has performed experiments on bounded helicons in indium antimonide in this context of uniform properties and currents (5). The conclusion is that amplification cannot be obtained.

Vikulin and his collaborators (6), however, have reported amplification during studies of helical instabilities in oscillistors. The significant new factor appears to be that the distribution of plasma along the sample is no longer uniform and this is achieved by the connection of a $p$ junction to one end and an $n$ junction to the other.

Suppose, then, that the effective conductivity or the Hall coefficient or the current density varies with $z$. The quantity $k$ becomes a function of $z$, which may be expanded into a Taylor series with a linear first term. Let this term be denoted $Kz$ and examine the change in Eqs. (10) to (16). A linear term now precedes the first derivative and $j\omega$ replaces $\partial/\partial t$:

$$\frac{\partial^2 H}{\partial z^2} - K\frac{z}{\rho_e}\frac{\partial H}{\partial z} + j\omega\frac{\mu}{\rho_e}H = 0. \tag{20}$$

By a change of variable and redefinition of the constants the equation is recognized as a Hermite differential equation of imaginary order. Let

$$r := Kz/2\rho_e \tag{21}$$

$$M := 2\omega\mu/K. \tag{22}$$

Equation (20) now takes the Hermite form:

$$\frac{\partial^2 H}{\partial r^2} - 2r\frac{\partial H}{\partial r} + jMH = 0. \tag{23}$$

The general solution of this may be written in terms of arbitrary constants $A_0$ and $A_1$:

$$H = A_0 H(r) + A_1 H(r). \tag{24}$$

The Hermite functions are expressible in terms of gamma functions and the hypergeometric function $\Phi$ (7).

$$H(r) = [2^{-jM/2}\Gamma(1/2)/\Gamma(1/2 + jM/4)]\Phi(jM/4, 1/r; r^2)$$
$$+ [2^{-jM/2}\Gamma(-1/2)/\Gamma(jM/4)]r\Phi(1/2 + jM/4, 3/2; r^2). \tag{25}$$

Except when $K$ is zero, the solutions always become asymptotic to $\exp(r^2)$. Thus a semiinfinite slab with a current density in the same direction as the steady magnetic field at its surface must eventually sustain a growing helicon wave if the current density increases steadily with $r$, and therefore with $z$, until a point is reached where the rate of increase in source energy is sufficient to compel growth of the wave.

The helicon wave is one example of an effect caused by the self-magnetic field of Hall currents; it arises naturally. Other examples can be contrived and they also share the feature that the field is liable to decay if source currents do not exceed a critical value.

## D. Skin Effect and Hall Effect in Cylinders

The penetration of fields and currents in cylindrical wires is a topic of particular emphasis in the study of the skin effect. Special names have been given to Bessel functions of real and imaginary argument and fractional order, to express the results, e.g., ber, bei, ker, kei.

Just as the Hall effect brings about deep penetration in the form of helicon waves for a plane geometry, similarly waves can penetrate or even grow as they radiate through a cylinder. Instead of a magnetic field along the direction of propagation, aided by a superimposed direct current, the latter takes precedence in the cylindrical case. This follows from the impracticability, not to say impossibility through div $B = 0$, of having the magnetic lines of force converge at the axis.

The Corbino disk is the archetype of the Hall effect in the round, especially at zero frequency. It is a hollow cylinder of material with, say, high electron mobility; a good conductor fills the central hole and a second electrode occupies the outer curved surface. A source of radial current density is passed between these electrodes in the presence of an axial magnetic field. Hall currents flow in circles around the axis; their own magnetic field at the axis aids or opposes the applied field according to the direction of the source current. There are Hall emfs but no potential differences exist because the emfs are used to drive currents against the resistivity of the materials as fast as they are created; as with tubular plates and axial currents, the magneto-resistance is maximized. The resultant current flow lines are tight or loose spirals according to the source current direction, and a partial rectification takes place in that circuit as a result of self-magnetic field action.

We now extend this theory to include alternating currents. The direct current remains as the source of energy with its radial flow, and alternating signal currents are also imposed on the central wire. Interest then focuses on the impedance presented to that wire and on the magnetic field, Hall currents, and electric fields surrounding it. These all exist as cylindrical radial waves.

An infinitely long cylinder is considered with axial symmetry so that in a coordinate system $r$, $\theta$, $z$ nothing varies with respect to $z$ or $\theta$. The source current in unit axial length is radial:

$$J_s = I/2\pi r, \tag{26}$$

where $I$ is the total current in amps providing a current density $J_s$ which increases toward the center. The magnetic field $B$ of the signal currents is $\theta$-directed, so that the Hall currents flow axially. The zero frequency magnetic field of the source current is not taken into account; it is zero in magnitude over a central plane and could in principle be cancelled by a superimposed field if necessary.

The resultant electric field $E$ is made up of a Hall component $E_{\text{Hall}}$ and a component $E_{\text{trans}}$ by transformer action from the changing magnetic field.

$$E = E_{\text{Hall}} + E_{\text{trans}}, \tag{27}$$

$$\text{curl}(E_{\text{trans}}) = -\partial B/\partial t = -j\omega B, \tag{28}$$

$$E_{\text{Hall}} = K_H I B/2\pi r. \tag{29}$$

Again recognizing that displacement currents in a material of conductivity $\sigma$ are negligible, the relevant Maxwell equation is

$$\text{curl } B = \mu J. \tag{30}$$

The only nonzero component after an expansion of the curl is

$$(1/r)(\partial/\partial r)[rB_\theta] = \sigma\mu E_z . \tag{31}$$

Omitting the unambiguous directional suffices and expanding,

$$dB/dr + (1/r)B = \mu\sigma E. \tag{32}$$

Differentiating with respect to $r$ and combining with Eqs. (27)–(29):

$$\frac{d^2B}{dr^2} - \frac{1}{r^2}B + \frac{1}{r}\frac{dB}{dr} = j\mu\sigma\omega B + \frac{k}{r}\frac{dB}{dr}, \tag{33}$$

where the constant $k$ has absorbed the following constants:

$$k := \mu\sigma K_H I/2\pi, \tag{34}$$

$$\frac{d^2B}{dr^2} + \frac{(1-k)}{r}\frac{dB}{dr} - \left(j\omega\mu\sigma + \frac{1}{r^2}\right)B = 0. \tag{35}$$

A change of variable is needed before this differential equation can be solved in terms of known functions

$$U := r^{k/2}B. \tag{36}$$

The transformation of Eq. (35) requires results for the first and second derivatives of $B$:

$$dB/dr = (-k/2)r^{(-k/2-1)}U + r^{-k/2}\,dU/dr, \tag{37}$$

$$d^2B/dr^2 = r^{(-k/2)}\,d^2U/dr^2 - kr^{(-k/2-1)}\,dU/dr$$
$$+ (k/2)(k/2 + 1)r^{(-k/2-2)}U. \tag{38}$$

Expressed as a function of the new variable $U$ Eq. (35) is

$$d^2U/dr^2 + (1/r)\,dU/dr + (-j\mu\omega - \eta^2/r^2)U = 0, \tag{39}$$

where

$$\eta^2 := 1 - 3k^2/4. \tag{40}$$

A further simple substitution produces a form of Bessel's equation:

$$p := r(\mu\omega\sigma)^{1/2}, \tag{41}$$

$$\frac{d^2U}{dp^2} + \frac{1}{p}\frac{dU}{dp} + \left(-j - \frac{\eta^2}{p^2}\right)U = 0. \tag{42}$$

The solution of the first kind is

$$U = \text{ber}_n(p) + j\,\text{bei}_n(p), \tag{43}$$

where $\text{ber}_n$ and $\text{bei}_n$ are Thompson functions of order $\eta$. Similarly the solution of the second kind is

$$U = \text{ker}_n(p) + j\,\text{kei}_n(p). \tag{44}$$

Returning to the original variables, a solution may be expressed in terms of arbitrary constants $A_0$ and $A_1$:

$$\begin{aligned}
B = r^{(-k/2)}\{A_0[&\text{ber}_n(r(\mu\omega\sigma)^{1/2}) + j\,\text{bei}_n(r(\mu\omega\sigma)^{1/2})] \\
+ A_1[&\text{ker}_n(r(\mu\omega\sigma)^{1/2}) + j\,\text{kei}_n(r(\mu\omega\sigma)^{1/2})]\}.
\end{aligned} \tag{45}$$

Again the Thompson functions are included in the functions that can be computed from an existing algorithm for the confluent hypergeometric function (8).

It is useful to examine the wave impedance at a small radius. This can be done without detailed computation by taking known small argument approximations for the Thompson functions.

Let a linear combination of the ker and kei functions of order $\eta$ be denoted by $F(p)$:

$$F(p) = \text{ker}_n(p) + j\,\text{kei}_n(p). \tag{46}$$

The wave impedance calculation requires $E$ to be calculated from curl $B$.

$$B = r^{(-k/2)}F(r(\mu\omega\sigma)^{1/2}), \tag{47}$$

$$d(rB)/dr = (1 - k/2)r^{(-k/2)}F + r^{(1-k/2)}(\omega\mu\sigma)^{1/2}F'. \tag{48}$$

Now

$$E = (1/\mu\sigma)(1/r)\,d(rB)/dr. \tag{49}$$

Therefore the wave impedance is

$$E/H = [(1 - k/2)/r]\sigma + (\mu\omega/\sigma)^{1/2}F'/F. \tag{50}$$

This result applies for all combinations of the Thompson functions at any radius, but we choose the ker and kei functions since they will dominate the solution at small radius.

An approximation for $F$ when $p$ is small is

$$F = (0.5)\varepsilon^{(-\eta\pi j/2)}\Gamma(\eta)[(p/2)\varepsilon^{(j\pi/4)}]^{-\eta}. \tag{51}$$

It follows from this that the ratio of the derivative to the function is given by

$$F'/F = -\eta/p + \cdots. \tag{52}$$

Thus an approximate expression for the wave impedance looking out along a radius from a point near the origin is

$$E/H = (1/r)\{(1 - k/2)/\sigma \pm [(\mu\omega/\sigma)(1 - 3k^2/4)]^{1/2}\}  \qquad (53)$$

This results suggests that the real part of the wave impedance can become negative for values of $k$ greater than 2. Now $k$ is proportional to the Hall constant, permeability, conductivity, and source current and a negative wave impedance indicates a wave which is acquiring energy from the source. So again there is a critical source current above which a regenerative amplifying action should become possible, and this is brought about by the presence of a self-magnetic field of the Hall currents.

## III. COMPUTATIONAL THEORIES

### A. Rectangular Plates and Equipotential Calculations

In common with most branches of mathematical science, the number of examples of wholly analytical solutions is limited to idealized situations of the kind described earlier. When realistic boundaries confine the solid state device, or separate it from dissimilar materials, computation may be needed. At first this often takes the form of replacing the differential equations by finite difference approximations so that the field vectors **E, B, J** or the potentials **A,** $V$ are estimated at a number of separated points. The points generally are the nodes of a rectangular net in two-dimensional problems or an array of such nets in three dimensions.

It is elegant to express solutions in terms of characteristic functions, when the boundaries permit this. Computation of a Fourier expansion of the functions is still required, but the size of the largest matrix may be reduced by an order of magnitude. The field can be computed at any chosen point, which need not coincide with a network node (9).

The rectangular plate continues to be the preferred shape for measurements of mobilities in materials which can be so fabricated. It is important to know how the potential varies near the point of attachment of the electrodes and to know how results depend on the aspect ratio of the rectangle.

De Mey (10) has described a new method for calculating the potentials without spreading a network of nodes through the body of the plate (11). The plate is immersed in a uniform magnetic field $\mathbf{B} = 0, 0, B_z$ and $\partial/\partial z = 0$. A point $(x, y)$ in the body of the plate is distant $r = (x^2 + y^2)^{1/2}$ from an arbitrary origin $(0, 0)$; a point $(x_1, y_1)$ on the open edges of the plate is at $r_1$ from the origin. Perfect conductors fill the other two edges and provide a connection to the source currents. The interest in this problem centers on these conductors, which short-circuit the Hall emf and allow Hall currents to flow at the ends of the plate. The resultant current is perturbed into a skewed pattern and the potentials on the open edges are reduced. Correction factors are needed when the application is mobility measurement. But the effects are useful and are deliberately enhanced by choice of aspect ratio or by the insertion of transverse short circuits when the application is to an effective magnetoresistor.

At any point $(x, y)$ the following equations hold:

$$\mathbf{E} = \mathbf{J}/\sigma + K_H(\mathbf{J} \times \mathbf{B}). \tag{54}$$

The currents are steady; therefore curl $\mathbf{E} = 0$ and it is permissible to let $\mathbf{E}$ be the gradient of a scalar:

$$\mathbf{E} = -\text{grad } V, \tag{55}$$

$$\text{div } \mathbf{B} = 0. \tag{56}$$

The plate has conductivity and a steady state has been reached; therefore no charges can remain or be accumulating within the plate and the equation of continuity applies to the current density.

$$\text{div } \mathbf{J} = 0. \tag{57}$$

It follows by taking the divergence of Eq. (54) that

$$\text{div grad } V = \nabla^2 V = 0. \tag{58}$$

The two-dimensional Laplace equation is satisfied by the potential function. The boundary conditions are mixed: $V$ takes the values $V_0$ and $V_1$ of the edge conductors, but its value is to be found on the open edges. On these the boundary condition applies to $\mathbf{J}$ which must remain parallel to the edge. Deflected charges, whose values are also to be calculated, occupy the open edges.

To perform the calculation the distributed edge charges are replaced by a finite number of charges at equally spaced nodes. The potential at an interior point in terms of the edge distributed charge density $q(r_1)$ is

$$V(r) = \frac{1}{2\pi\varepsilon_0} \int q(r_1)\ln \frac{1}{|\mathbf{r} - \mathbf{r}_1|} \, ds, \tag{59}$$

where $ds$ is an element of the perimeter and the integration is around the whole perimeter.

Equation (59) may be adapted to express the potential that each equivalent discrete charge produces at the site of every one of the nodes. This generates a set of simultaneous equations for the unknown charges; the number of equations is one order of magnitude less than the number that would be needed if a two-dimensional net had been spread. The equations are solved by standard computer routines to give $q(r_j)\Delta s$, the effective charges at $j$ nodes spaced at $\Delta s$ along the open edges of the perimeter.

The potential at any internal point is then given by

$$V(r) = (1/2\pi\varepsilon_0) \sum_j q(r_j)\, \Delta s\, \ln(1/|\mathbf{r} - \mathbf{r}_j|). \tag{60}$$

The time taken to compute a set of equipotentials is greatly reduced compared with a two-dimensional method and the accuracy can be checked by increasing the number of nodes.

The pattern of equipotentials and lines of electric field strength in the absence of a magnetic field is a rectangular grid. In the presence of the magnetic field, as shown in Fig. 3, there is a skew symmetric pattern in which the Hall potential difference is reduced to zero at the corners of the plate; it is reduced significantly at the center of the plate if the length does not exceed the breadth by a factor of four. The Hall currents form vortices around the source–current connections. The vortices are skewed from the center line by an increasing distance as the magnetic field or the mobility increases.

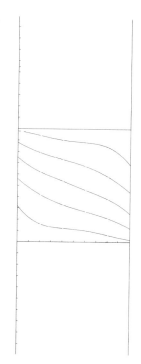

FIG. 3. Pattern of equipotentials for a square Hall plate with source current perfect conductors filling two sides. The Hall angle is about 45°.

### B. *Plates with Internal Electrodes and Surface Printed Circuits*

It is not always practical or convenient to manufacture a rectangular block of material with electrodes for the source currents in contact with the whole of two opposite faces. Mobility measurements can be made on materials in extended thin layers if, for instance, point electrodes are inserted in a square configuration; finite electrodes, as shown in Fig. 4, can replace the points. A printed circuit on the surface of the Hall material can provide the electrodes, but this takes the problem outside the scope of a two-dimensional analysis.

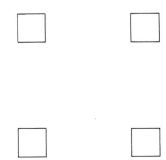

FIG. 4. Electrodes for a square type of Hall plate on the surface of a material or through a thin film.

White, Knotek, and Ritchie (*13*) have computed potentials, geometrical correction factors, and magnetoresistance for square Hall plates with such electrodes. They apply finite difference equations to a two-dimensional grid and they limit the value of $\tan(\theta_H)$ to 0.5, because it is known that extremely fine grid spacing is needed for reasonable convergence by this method at higher values.

With point electrodes at the outside corners of a square Hall plate, results for the potential difference on the diagonal are virtually identical with those for an ideal infinitely long plate. With internal electrodes arranged at the corners of a square of side 13/15 of the length of side of the plate itself, there is a reduction in the computed Hall voltage of less than 2% of the ideal value.

Finite electrodes of the order 1/15 of the side of the square plate cause a geometrical correction factor which reduces the Hall voltage significantly by more than 14% in one example. In general, magnetoresistance effects are smaller than with the full-width electrodes considered previously. The short-circuit effect from these small electrodes is slight and no external Hall currents are drawn from them; the resistance is never changed by more than 10% for $\tan(\theta_H) < 0.5$.

## C. Plates with Output Current and Various Electrodes

Newsome and Silber (*12*) have studied a complementary set of electrodes to those at the corners of a plate. They consider symmetrical plates of rectangular shape; electrodes extend an assigned distance from the center line along each of the four edges, as shown in Fig. 5 so that the corners can be free of electrode material; source currents flow between one pair of opposite edges and Hall currents are drawn from the other pair.

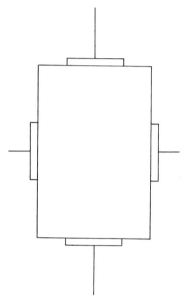

Fig. 5. Electrodes for source current and output current on a rectangular Hall plate. The proportion of each side occupied by an electrode is adjustable.

The method is to compute the potentials by finite differences, by a two-dimensional rectangular mesh, and by the relaxation method, in which direct inversion of a large matrix is avoided. Successive approximations are made to the potentials at the nodes; each adjustment at one node is usually allowed to affect only the four nearest other nodes. The adjustment in this case has the special feature of a dynamic relaxation factor; this is varied automatically to assist rapid convergence of the solution. The iterative process continues until no node experiences an adjustment greater than an assigned small value, which is a measure of the accuracy for that computation. It is not a measure of the error that can be ascribed to having replaced the continuous material by an equivalent net. There is an interesting connection between the fineness of a mesh and the number of terms that ought to be retained in a Fourier expansion to achieve comparable accuracy. The link between these opposite approaches is the sampling theorem of information theory. There should be two nodes or samples within the period of the highest term that is retained in the Fourier series (9).

Computations that include the effect of output current taken from suitably large electrodes allow a complete two-port representation of a given plate to be reached. The efficiency, maximum output, and linearity of the device can be predicted. Although the Hall voltage is maximized with re-spect to the source electrodes by giving them the full width of the plate, linearity is improved by reducing the fraction of the width occupied by them. As in the discussion of magnetoresistance, these electrodes short-circuit some of the linear first-order Hall emf's. The short-circuit currents react again with the magnetic field to become the source of second-order effects dependent upon the square of $B$, the magnetic flux density. In consequence the output current is a nonlinear function of $B$; the extent of the nonlinearity depends on the length/breadth ratio and on the values of resistances in the source circuit. Optimum values can be selected and the problem is of interest mainly to designers, who seek linearity factors better than 95%.

The maximum efficiency of this type of Hall plate is shown to be

$$\eta_{\max} = (2^{1/2} - 1)/(2^{1/2} + 1) = 17\%. \tag{61}$$

This figure is approached only as the magnetic flux density becomes infinitely large. The low efficiency arises in part from the use of Hall currents at a load which is separated from the place where they are generated. In magnetoresistors and Corbino disks they are used at source, and higher efficiencies become possible.

## IV. Applications of Existing Devices

### A. Hall Mobility Measurements

A well established and widespread application of the Hall effect is to the measurement of electron or hole mobility in materials that are to be built into devices for other branches of electronic technology. If the material can be shaped into a rectangular block, and if electrodes can be connected to it, measurement is straightforward. If the material cannot be maintained at a uniform temperature, the electrodes can be made massive and extensive to suppress thermal differences of potential.

With some materials it is not possible to attach electrodes or to elaborate the shape of the specimen material. In this case it is possible to deduce the mobility by passing helicon waves through a slice of the material. The waves are reflected at each surface, resonances occur, and these may be shock excited by sudden changes in the applied field. The frequency and rate of decay of the oscillations determine the Hall coefficient and resistivity of the sample (4). Alternatively, an alternating magnetic field is applied continuously to the sample and an emf is induced in a secondary coil; the $Q$-factors and frequencies of resonance are then recorded. For instance, a 1 cm slice of indium antimonide at room temperature resonates at about 100 MHz with a $Q$-factor of 2.5.

Shewchun and his co-workers (14) have developed an automatic system for measuring conductivity and the Hall effect in semiconductors. The development of high performance silicon, discrete, and monolithic devices in particular depends upon complete and detailed knowledge of mobility over a range of temperatures. Thin film samples are connected to heat sinks and are surrounded by a copper shroud to assist the maintenance of a constant temperature and to give shielding from liquified and convected gases. A process control computer puts the samples through a sequence of events for up to fifty selected temperatures between 4.2°K and 400°K. The sample is brought to an assigned temperature by a heater winding, if necessary. The current through the sample is set and recorded, reversed, and recorded again. These actions are repeated with an applied magnetic field and are repeated again with a reversed magnetic field, while the Hall voltage is recorded throughout.

The results of these four combinations of source current and applied field and the close control of temperature allow the Hall effect to be isolated from the other thermomagnetic effects and from spurious potential differences of geometrical misalignments. The results are processed by digital voltmeter, paper tape. digital computer, and computer graph plotter to display mobility against temperature fully automatically.

## B. Measurement of Magnetic Fields

Either a Hall plate with its four leads or a magnetoresistor with only two leads may be applied to magnetic field measurement. In weak fields, typically less than 0.3 T, the resistance of ordinary magnetoresistors is proportional to $B^2$ and so is insensitive to the direction of $B$. Yamada (15) has shown that a form of magnetodiode can be two or three orders of magnitude more sensitive to a magnetic field than a Hall plate. The device resembles a plate with no output electrodes; instead one side is highly polished, and the other side is arranged to have a surface which creates many recombination centers for electrons and holes. The earliest techniques for this type of diode were to sandblast that side, but more effective methods probably include irradiation or neutron bombardment. The direction of $B$ determines to which side of the plate the carriers will be deflected, and the recombination rate determines the conductivity for that direction. Therefore, in this instance, the magnetoresistance is a function of both magnitude and direction of $B$.

The Hall plate as a probe for magnetic field measurement has the advantage of good linearity and sensitivity to the direction of the field. Small probes capable of being inserted into the air gap of electrical machines are useful. Unlike the calibrated search coils of a fluxmeter, they define a precise location in the field, and do not have to be turned over and reinserted in order to measure the field in the air gap of a permanent magnet.

Instruments which combine a Hall plate with a dc amplifier can be sensitive to fields of the order $10^{-7}$ T. The addition of shaped pole pieces to concentrate the field can change this to $10^{-9}$ T. It is usual to tolerate the lower sensitivity of indium arsenide relative to antimonide in order to gain more constant performance against temperature, and accuracies of 1 part in $10^4$ are attainable in the measurement of magnetic fields.

## C. Contact-Free Switches, Variable Resistors, and Position Transducers

These devices are a consequence of the magnetic field measurement capability. A magnet moves into a new position and so activates the Hall voltage or the change in resistance. The Hall generator actuated by a moving permanent magnet has the advantage that no energy is required for the magnet, but an emf is generated for transmitting a signal. Such generators have found applications in the telemetry for remote reading of domestic electricity meters, in which a Hall generator is activated by magnets on the circumference of a rotor to create ten pulses per turn.

For a volume control or variable resistor that does not wear out, the magnetoresistor is appropriate. The resolution is not limited by the fineness of wire windings or carbon granules. Contact noise and bounce are eliminated, and self-inductances and capacitances are small enough to permit high frequencies.

The dependence of resistance on the square of $B$ does not affect the linearity that is possible between resistance and the displacement of the magnet. Given a uniform field into which the resistor is partly inserted, it is easy to achieve linearity of 1% from $R_0$, the initial resistance, to $R_B$, a maximum resistance for full insertion with $R_B/R_0 = 10$. By this means linear potential dividers can be constructed with the magnetically controlled resistor in series with other resistors or, preferably, thermistors to achieve temperature compensation.

A rotary potential divider can have a permanent magnet with an air gap holding a magnetoresistor and a rotating soft iron cam, the shape of which can be elegantly chosen (2) (p. 272) so that the ratio of potential division is a chosen function of the angle of rotation. If a linear function is required the linearity can be within 0.3%.

In one respect ordinary potential dividers are superior in their ability to deliver an output voltage down to zero volts at zero output impedance.

One position transducer uses a single magnetoresistor passing constant current; its voltage drop is then a known function of the distance from a magnet. Precise location of position is possible if two magnetoresistors are inserted in diametrically opposite air gaps of a toroidal winding, which passes a constant current. The resistors are arranged in a bridge circuit which remains balanced if the whole toroidal assembly is in zero external magnetic field. Balance also occurs if the external field is symmetrical about the line of the two air gaps, such that the two air-gap flux densities are equally affected. A slight displacement of a strong external magnet from this central position causes one magnetoresistor to experience a stronger total field and the other a weaker one; the bridge is sharply unbalanced and an output voltage appears at the detector. The curve of this voltage against displacement of the magnet resembles the derivative of the magnitude of its field strength against displacement.

The meander path raster plate is the outstanding device for these and related applications; the usual substances are indium antimonide with nickel antimonide inclusions: InSb–NiSb.

### D. InSb–NiSb Magnetoresistors

Steidle (16) describes the recent status of the device. Recalling the effect óf length-to-breadth ratio in a plate and the short circuit effected by the input leads, we conclude that the plate should be as short as possible to maximize the magnetoresistive effect. To draw short-circuit lines across a long plate as frequently as possible in the manner of a television raster is what is required (see Fig. 6). Such raster plates can be made by evaporating, say, silver films onto an indium antimonide plate and etching the pattern after a photographic exposure.

This approach is carried to a limit by the inclusion of less than 2% of NiSb in the original material. The inclusions take the form of crystalline whiskers or needles of diameter below 1 $\mu$ and length about 50 $\mu$. They are aligned along one direction and have more than 100 times the conductivity of the matrix. Thus the result resembles the raster plate, but the short circuits are distributed through the interior and the elaborate fabrications on the surface are avoided.

This composite material is then deposited on a substrate to a depth of some 25 $\mu$ as a meander path; it is insulated from the substrate. If the latter is nonmagnetic, the ratio of resistances in and out of the magnetic field $R_B/R_0$

FIG. 6. A raster plate modification to convert a Hall plate into an effective magnetoresistor.

is proportional to $B^2$ up to 0.3 T; after this the relationship is linear up to 10 T. The substrate can also be iron or ferrite, in which case there is similar behavior apart from saturation around 1.5 T.

By varying the structure of the meander path and by doping the normally intrinsic InSb, for instance, with tellurium, it is possible to vary the resistance $R_0$ from a few ohms to several kilohms. At 25°C and a flux density 1 T applied along a normal to the surface of the substrate, intrinsic InSb has a conductivity of $2 \times 10^4$ Siemens/m and $R_B/R_0 = 15$. A material with $n$-type doping to bring the conductivity to $8 \times 10^4$ S/m suffers a reduction in the mean free path caused by the concentration of donor atoms, and the corresponding ratio $R_B/R_0$ is approximately 6.

The intrinsic material is more sensitive to temperature changes than the doped. A change of 10°C can lead to 25% changes in the resistance of the former compared with 2% in the doped material. A rise in temperature above 25°C always reduces the ratio $R_B/R_0$ and this is especially true of the doped material. Thus, it is not easy to prescribe one material or degree of doping for all applications; $R_0$ itself and $R_B/R_0$ are separate functions of the doping and the temperature.

The inclusions of InSb–NiSb confer mechanical stability to the extent that etchings of meander paths can be transferred during manufacture from one substrate to another. They also ensure that the current density in a magnetic field is uniformly distributed throughout the material and this is quite unlike the extended Hall plate with its areas of high current density.

Consequently, temperatures can rise to 150°C before irreversible damage occurs, and this is likely to be to the adhesives and solder associated with the magnetoresistor. Adhesives and procedures are carefully prescribed for mounting the iron bases in order to mitigate the effects of alternating mechanical forces caused by an alternating magnetic field.

## V. New, Potential, and Unrealized Applications

### A. Alternator Excitation and Nonreciprocity

The gyrator is the archetypal nonreciprocal device. It has a conductivity matrix with the sign configuration of Eq. (4), and therefore the Hall plate is a natural gyrator. Grutzmann (17) has shown that gyrators and the related devices, isolators and circulators, can be made efficient by the presence of multiple electrodes on the edges of an ordinary Hall plate. Efficiencies above 80% are possible, but the rival products based upon gyromagnetic effects in ferrites are usually to be found in microwave systems; a permanent magnet is needed in each case, but the plain ferrite devices are simple to construct.

The widespread application of the Hall effect device as a gyrator requires future study and this is true of the majority of the possible applications. There is a good possibility of it being used in the design of alternators, particularly for aerospace and the autocar industry. Progress toward reliability and simplicity has led to the progressive elimination of commutators, moving brushes, and the apparatus of dc generators. The next steps could well be the elimination of slip rings, rotating windings, and rectifiers. Controlled self-excitation of a simple piece of iron as the rotor is made possible by two stator windings connected to each other through a gyrator (18, 19).

Alternatively, the same result is achieved if magnetoresistors are inserted in the stator. This version (see Fig. 7) is easy to understand. Let a two-pole

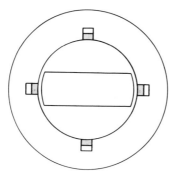

Fig. 7. A magnetoresistive self-excited alternator. Shaded area: magnetoresistive coil sides. Unshaded area: ordinary conductor coil sides.

soft magnetic rotor be already magnetized and turning between two coils. It helps to name these the "emf coil" and the "mmf coil." At the instant considered the first is in the position to experience maximum induced emf. The second is in the position to supply mmf to maintain or increase the existing magnetization. In other words, the current in the mmf coil must be looped by the magnetic field of the rotor. If the circuits are overwhelmingly resistive and if the rotor speed is above a critical minimum, the emf coil can supply the necessary current to an mmf coil at exactly 90 electrical degrees simply by putting the two coils in series.

The correct direction for this series connection is the crux of the problem. It can easily be found for one position of the rotor, but as the rotor advances by 90° it will be found that the previous connections now lead to a destruction of the magnetization. The connections need to be reversed and that is exactly the function of a gyrator.

If the circuits contain significant inductance, the angle between the mmf coil and the emf coil can with advantage be changed from 90° so that, when the peak of current is finally reached after the phase lag, the mmf coil is exactly positioned to maximize the magnetic flux.

The magnetoresistive version requires the active conductors of the mmf coil to contain a magnetoresistor. This has low resistance when the rotor is adjacent to the emf coil and allows the mmf coil to operate. After a 90° shift of the rotor, the former mmf coil experiences a strong magnetic field, the magnetoresistor acquires high resistance, and demagnetizing currents are much reduced. The former mmf coil is thus inhibited from acting as a new emf coil.

Stator windings are usually arranged in a double layer. Continuous build-up and control of the magnetization requires the pairing of emf and mmf coils around all the slots of the stator. The mmf coil sides are located in the layer nearer to the rotor, where their magnetoresistors sense the approach of the rotor and effectively switch the induced currents always to aid the magnetization (20).

Permanent magnet rotors are too simple in that they do not allow control of the generated emf, but they are to be found in pilot exciters for two-stage alternators in aircraft. The same protection from accidental or thunderstorm demagnetization could be provided by inserting a small permanent magnet into the magnetoresistive alternator rotor.

Self-excited inductor alternators are known to exist, but they are somewhat dependent on the phase angle of the load. The cost size and weight of capacitors to guarantee a leading power factor is a deterrent for aerospace application. There is therefore some likelihood that the next generation of ultra high speed solid rotor alternators may require the Hall effect or magnetoresistive control of excitation (21).

## B. Brushless Motors

An unusual type of dc motor is one in which the commutator is replaced by contact-free sensors. The magnetoresistor can perform this function. Let the rotor be a simple permanent magnet. The stator contains at least three equally spaced coils around its circumference. If these are energized in turn, the magnet is turned.

In one version the magnetoresistors report the present position of the rotor and their signals cause the energization of the next set of stator coils. The magnet turns, but energization again moves ahead and the rotor goes on being accelerated up to the maximum torque, like a donkey chasing its rider's carrot.

Such motors can have speeds of 20,000 rpm with good speed control, high efficiency, and durability. They can be less noisy electrically and acoustically than comparable machines with commutators, but unlike the self-excited alternator some additional electronic apparatus is needed to energize and control the stator.

## C. Multipliers, Wattmeters, and Modulators

Not only is the Hall plate a natural gyrator, it is also a four-quadrant multiplier. This means that the product is computed with correct magnitude and sign for positive and negative values of both input quantities; the output is the Hall voltage. One input is the source current and the second input is the current in a coil to provide the magnetic field.

A major application of multipliers of this type has been in analog computers. There are disadvantages, for instance, if hysteresis and nonlinearity are encountered in a ferromagnetic magnetic circuit. The two inputs are likely to have dissimilar input impedances and special drive circuits may be needed to pass source currents of the order of one hundred milliamps through the plate's resistance of the order of one ohm.

A particular difficulty has been the determination of the exact location of the output electrodes at the center of each side of the plate, to achieve zero misalignment voltage. If they are not so centered, a residual voltage remains when the magnetic field is zero. An early remedy was to adjust a potential divider between two grossly misaligned electrodes, both on the same side of the plate. Strand (22) has proposed a method which improves the signal-to-noise ratio by about one hundred, regarding the misalignment voltage as an ingredient of noise. The plate is at least three times longer than it is wide and these dimensions are, as usual, much greater than the misalignment factor. This factor arises from the mechanical tolerances in imperfect masks, when devices are constructed by printing and etching. The method is to have several pairs of output electrodes with separate amplifiers and to sum their outputs; three pairs lead to the improvement of one hundred. The basis of the improvement is that the mechanical misalignments are randomly distributed, so that summation of the signals is not accompanied by full summation of the noise.

With microwaves, multiturn coils and magnetic circuits are replaced by direct immersion of the Hall device in the electric and magnetic fields of the wave. There is no failure of the Hall effect itself below infrared frequencies. Since the power in the microwave is proportional to the product of its electric and magnetic fields, the Hall multiplier becomes a direct indicator of the power flux density. Its source current is determined by the electric field at the same time as it is immersed in the magnetic field. It has even been suggested that radiation pressure itself is in a related way a manifestation of Hall effect. Barlow (23) has proposed several microwave wattmeters and has applied the multiplicative property also to microwave mixers (24).

The magnetoresistor is also capable of similar applications, but it is not a four-quadrant device. As a multiplier the output can be the voltage drop across the magnetoresistor. One input can be the current into it; the other input can be the current in a coil providing the magnetic field. The output is thus not dependent on the sign of the second input, unless double-injection magnetodiodes are included in the term "magnetoresistor." For reasonable linearity the outer, linear part of the resistance curve as a function of magnetic field is needed; a separate bias magnetization can provide that. Instead of a small misalignment voltage with zero second input, there is a massive voltage drop in a single magnetoresistor; but, a bridge configuration with two such voltage drops opposed to each other provides a solution to that problem.

Nalecz and Rijnsdorp (25) propose a multiplier with three inputs. The third of these is an application of the position transducer, in which the Hall plate is moved through a nonuniform magnetic field so that its output becomes a function of its displacement. Examples of constructions by Nalecz include transducers for seismographs, accelerometers, pendulum, microphone, and hand-tremor measurement. The function of displacement can be linear with zero output in the center of the range; this is achieved by tapered pole pieces starting with N–S in the air gap and ending with S–N. Other functions may be chosen by shaping the pole pieces; a change from analog to digital action requires, for instance, a staircase function.

If three currents are to be multiplied, one method is to fix the Hall plate on the arm of a moving coil instrument, so that it moves through the shaped pole pieces. The output of the plate is then proportional to: first the source current, secondly the current in a coil magnetizing the pole pieces, and thirdly the current in the moving coil. The multiplier is sensitive to the eight possible sign combinations of the three inputs.

Three-input multipliers have possible applications in process control. They are capable of output voltages larger than those of thermocouples. Mass-flow measurement is one possibility: fluid flow is converted by an orifice plate into a differential pressure across the orifice, which actuates a diaphram to move the Hall plate. The other inputs can be actuated by temperature and pressure transducers with refinements to allow the density to be computed from the gas law. The multiplier finally indicates the mass-flow rate by taking the product of density and velocity.

The Hall plate is a natural choice for applications such as the three-input multiplier and the wattmeter immersed in the fields that operate it. Otherwise, four-quadrant multipliers based upon the diode function generator as a means for taking the sum and difference of squares are serious competitors. They require no magnetic circuit and the precise alignment or the duplication of output electrodes is avoided.

### D. Magnetoresistive Oscillators

Baibakov (26) has constructed Corbino disks of indium antimonide to oscillate at the temperature of liquid nitrogen. A power output of 20 W is reported with an efficiency of 20%, which compares with a theoretical maximum for Class A operation of 37.5%. Continuous oscillation occurs from 100 Hz to 20 MHz, coherent oscillation up to 400 MHz, and incoherent output up to 4 GHz; see Fig. 8.

A self-excited Corbino disk is observed to oscillate from 50 MHz to 400 MHz in the sense that no external capacitors or inductors are connected. There is, however, an externally provided axial magnetic field. Nevertheless, the structure is one of elegant simplicity. There is a disk and a coaxial

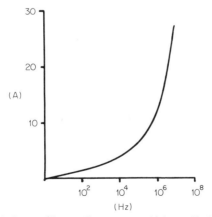

FIG. 8. Magnetoresistive oscillator. Current at which oscillation starts against log frequency.

cylindrical magnet. A ring electrode on the outer curved surface and a central electrode pass a radial source current. Circulating rf currents flow around the center, and their magnetic fields modulate the axial imposed field, which combines with the source current to create these rf currents through the Hall effect. In this way, a positive feedback or bootstrap action may be argued. The final product is like an electrical analog of a watch spring natural escapement mechanism.

Some questions of theoretical interest remain about the self-exciting action in a Corbino disk. Cowling (27) reaffirms that his results on the impossibility of self-maintaining magnetic fields in axisymmetric cases are applicable. The original studies applied to sunspots and the earth's magnetic field with self-excitation in the sense of a self-excited dc generator. The difficulty with the unencumbered Corbino disk is that a ring of Hall current can be arranged to create a magnetic field inside itself to support similar rings, by way of providing them with a nonzero magnetic field. If so, the magnetic field outside the ring is counterproductive. The outermost ring requires a current that is unjustified in terms of the self-generated magnetic field. It may be speculated that the impossibility theorems will be circumvented, for instance, by the existence of viscosity in the electron circulating fluid; otherwise, two axisymmetric disks might supply axial fields to each other. Or a multiply-connected structure might guarantee the feedback (28, 29). Mathews (30) has stated the case against self-maintenance of the Corbino disk at zero frequency. These questions are about the internal fields and currents and do not affect the theory of Baibakov's oscillator, which starts from the premise that a magnetoresistor is operating under the square law

$$R = R_0[1 + S^2(I + I_1)^2], \tag{62}$$

where the final resistance $R$ is expressed in terms of an initial resistance $R_0$, a steady current $I_1$ in the controlling coil, and a current $I$ which flows in both the coil and the disk and the loop of a series tuned circuit. There is also a steady current $I_0$ flowing in the disk. The constant $S$ is a function of the carrier mobility, the turns of the coil, and the magnetic circuit. The frequency and damping in terms of the inductance $L$ and capacitance $C$ and resistance $r$ of the tuned circuit are given by

$$d^2I/dt^2 + (1/L)[r + R_0 + R_0S^2(I + I_1)^2 + 2S^2R_0(I - I_0)(I + I_1)] \, dI/dt$$
$$+ (1/LC)I = 0. \tag{63}$$

At the start of oscillation, $I \ll I_0$ and $I \ll I_1$ and the equation reduces to

$$d^2I/dt^2 + (1/L)[R + r - 2(R - R_0)I_0/I_1] \, dI/dt + (1/LC)I = 0. \tag{64}$$

It follows that oscillation begins when the coefficient of the middle term becomes negative. This may be expressed as a minimum value for $I_0$, the steady current in the disk:

$$I_0 > [(R + r)/2(R - R_0)]I_1. \tag{65}$$

The disks are no larger than a transistor of similar rating; they have 0.5 mm thickness and inner radius 0.5 mm and outer radius 2 mm. Indium electrodes are fused at the inner and outer radii. The disks are of $n$-type indium antimonide with carrier density $2 \times 10^{14}$ cm$^{-3}$ and mobility $5.5 \times 10^5$ cm$^2$ V$^{-1}$ sec$^{-1}$ at 77°K.

The effective value of circuit resistance $r$ increases at high frequencies, since it takes into account the eddy current and hysteresis losses in the magnetic circuit. The minimum current $I_0$ therefore also increases and to prevent overheating continuous operation has to be replaced by pulsing. Radio frequency voltages in the oscillatory circuit reach a magnitude of 400 V and the radio frequency component of the magnetic field reaches 0.04 T. In other forms of oscillator it is often necessary to limit the magnitude of the oscillation by inserting a thermistor or other nonlinear device to reduce the loop gain as the limit is reached. With this oscillator such a process occurs naturally, since the resistance $R$ is already a function of the square of the magnetic field.

It is anticipated that coherent oscillations will be extended to the microwave range and that a complete explanation of the theory of the device will include the strong electric fields arising in a nonequilibrium electron–hole plasma and the propagation of a type of helicon wave inside the Corbino disk.

The significance of this oscillator relative to microwave transistors and other bulk solid-state devices, which depend on space-charge wave propagation, could be compared with the early magnetron and its rivals. The structure can be large, the power output high, and the efficiency high. Similar problems of modulating a transmitter with high peak power arise as with the magnetron and there are other similarities.

### E. Magnetoresistive Bistable Circuits

If the unencumbered Corbino disk were capable of oscillation down to zero frequency, it would be a bistable device of great simplicity. But this remains to be demonstrated in theory and in practice. Just as water draining from a bathtub can be triggered into a stable vorticity of either sign, so the current pouring into the central electrode might enter a self-preserving spin. An array of pin electrodes piercing a sheet of semiconductor is a speculative form of computer binary store, based upon the principle.

At present, the magnetoresistive bistable circuit can be constructed from two coils and two magnetoresistors (31). The inductance of the coil $L$ slows down the speed of transition, but the resistance $R$ in the complete circuit can be so high that the time constant $L/R$ is acceptable. With Hall plates the resistance is usually so low that feedback-coupled plates as bistable circuits are unacceptably slow.

The principle is that each magnetoresistor controls the current in the coil that cuts off the other; at switch-on they try to strangle each other.

Let the two currents be denoted as $x$ and $y$; otherwise the conditions of Eq. (62) apply. A common potential $V$ across each coil-plus-magnetoresistor is then given by

$$V = xR_0 + xS^2y^2, \tag{66}$$

$$V = yR_0 + yS^2x^2. \tag{67}$$

Equation (66) is multiplied by $x$, (67) by $y$; a subtraction leads to the cancellation of $x - y$ from $x^2 - y^2$ and a simple condition holds:

$$V = (x + y)R_0 . \tag{68}$$

A substitution of (68) into (66) or (67) leads to further cancellations of cubic terms, since $x \neq 0$. It is found that

$$x^2 - Vx/R_0 + R_0/S^2 = 0,$$
$$y^2 - Vy/R_0 + R_0/S^2 = 0. \tag{69}$$

The solutions of these quadratic equations are

$$x, y = (1/2)\{V/R_0 \pm [(V^2/R_0^2) - 4R_0/S^2]^{1/2}\}. \tag{70}$$

The critical value for $V$ above which two solutions occur for $x$ and $y$ is

$$V_{\text{crit}} = 2[(R_0^3)/(S^2)]^{1/2}. \tag{71}$$

It is most unusual to be able to analyze a bistable circuit so simply and so completely. This depends upon the magnetoresistor remaining within the square-law of its dependence on the magnetic field. The resistance of the coil has been included in $R_0$.

It is also unusual that the bistable circuit operates without regard to the polarities of the supply voltages. Both halves can have positive supplies, and both can have negative supplies. There can be mixed polarities, which effectively puts the four elements in series with a single supply (see Fig. 9) In this series case, the former reference point at the center of the circuit can be allowed to float to convert the circuit into an astable configuration, whose relaxation time is under the control of whatever capacitors are present in the output impedances of the power supplies.

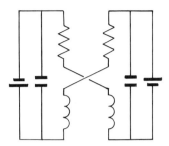

FIG. 9. Bistable magnetoresistor circuit. Each magnetoresistor is controlled by the coil on its axis. Also one supply voltage can be reversed and the upper and lower rails made common.

Experimental bistable circuits have given $x = 20$ mA with $y = 10$ mA, or $x = 10$ mA with $y = 20$ mA $(\pm 1$ mA$)$ as the stable states. With a 500 ohm magnetoresistor and a 50 ohm coil, $R_0$ can range from 550 ohms up to 9050 ohms if an initial bias is applied to increase the magnetoresistances. The idea of a time constant applies strictly only to linear systems, but allowing the notion that $L/R$ still represents a time constant, the notional time constant with a 1000 turn coil on a ferrite core can be 40 $\mu$sec. It is better to site the magnetoresistors in a gap between two of the outer arms which carry the return flux of a ferrite core than to use the central cylinder. The flux density is intensified by shaping those outer arms and removing all others.

When the bistable states exist, it is found that one of the currents plotted against the common applied voltage shows a negative resistance section. The other current continues to have a positive slope against the common applied voltage. The sum of these currents may not exhibit negative resistance, so that although the stable states are maintained once they have been entered, a truly cumulative action may not accompany the transition between states.

This leads to questions about creating a total negative resistance curve and, if possible, creating this and a bistable configuration with only one active element. Although bistable valve and transistor circuits conventionally are symmetrical and have two active elements, single element circuits are known. The tunnel diode also gives negative resistance and bistable action in a single element.

The change that is needed in the magnetoresistor is that the resistance, which now increases when current is passed, should henceforth decrease. However, this alone is not sufficient; in the product $(I + \Delta I) \cdot (R + \Delta R)$ it is implied that when the current changes the resistance will also change. An additional condition is that the decrease in resistance, if it occurs, should be great enough to compensate for the increase in voltage drop from the term $\Delta I \cdot R$.

### F. Magnetoresistive Amplifier

In order to create an amplifier and an active element with negative resistance, the essential step is to immerse the magnetoresistor in a constant bias magnetic field (32). Thus, high resistance is possessed initially. When current is passed a self-magnetic field appears and this must be directed against the initial bias. An increase in the applied current now leads to a reduction in the resultant field, a possibly large reduction in the resistance, and a possible fall in the potential difference.

If the self-magnetic field is weak, the self-controlling action can be enhanced by having a coil in series with the magnetoresistor; again, the field of this coil must oppose the initial bias field (see Fig. 10). It is simpler to think

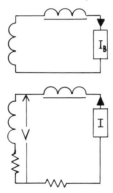

FIG. 10. Magnetoresistor with bias and self-control. $I$ against $V$ can show negative slope as in Fig. 11. The load resistor is included in dc amplifier applications. The chokes are included for oscillator applications with a capacitor across the $V$ terminals, but only if the sources do not possess high output impedance.

of two separate coils with the separate functions of providing first bias and secondly self-control, but in fact the two functions can be performed in one coil. A constant-current generator passes a bias current $I_B$ through the coil in one direction and at the same time a constant-voltage generator can supply a potential difference $V$ to the coil and magnetoresistor in series, with current $I$ in the opposite direction to $I_B$.

A slight modification to Eq. (62) applies to the single coil version, or to the two coils, if they have equal turns:

$$R = R_0[1 + S^2(I_B - I)^2]. \tag{72}$$

Recalling that $R_0$ includes the resistance of the coil and that $V$ is the voltage across the coil–magnetoresistor combination, $V$ is given by

$$V = IR_0 + S^2II_B^2 - 2S^2I^2I_B + S^2I^3. \tag{73}$$

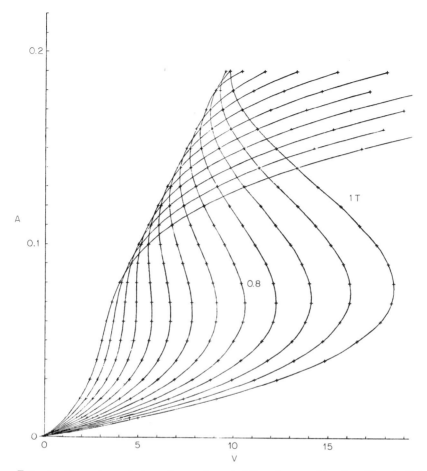

FIG. 11. Current against voltage characteristics for a biased self-controlled magnetoresistor.

The third term can operate to create negative resistance, if sufficient bias has been given. This may be illustrated (see Fig. 11) by plotting $I$ against $V$ for several fixed values of $I_B$. The drop in resistance becomes quite large from a starting point well removed from the origin in the quadratic dependence of $R$ on the total current.

A minimum value for the bias current can be found by differentiation of Eq. (73) with respect to $I$:

$$dV/dI = R_0 + S^2 I_B^2 - 4S^2 I I_B + 3S^2 I^2. \tag{74}$$

The incremental resistance becomes zero when

$$3S^2 I^2 - 4S^2 I_B I + S^2 I_B^2 + R_0 = 0. \tag{75}$$

This is solved as a quadratic equation in $I$ to give

$$I = [4S^2 I_B \pm (16 S^4 I_B^2 - 12 S^4 I_B^2 - 12 S^2 R_0)^{1/2}]/6S^2. \tag{76}$$

Two such places of zero incremental resistance just fail to appear when the square root is zero:

$$4 S^4 I_B^2 = 12 S^2 R_0. \tag{77}$$

Expressed as a minimum value for $I_B$ this is

$$I_B \geq (3 R_0)^{1/2}/S. \tag{78}$$

Experimental results confirm the shape of the cubic characteristics of this theory and an amplifier based upon 500 ohm magnetoresistors has recorded continuously at room temperature a power gain of 75 in Class A dc operation with an efficiency of 12%. This should rise to 21% if a permanent magnet were to provide the bias field. The operating point is $V = 20$ V, $I = 43$ mA on a load line of 210 ohms and with a bias of 60 mA.

The characteristics of the biased self-controlled magnetoresistor are of exceptional interest. They represent a family of curves comparable with those of vacuum tubes and transistors, but with a region in which negative resistance is already present. They are closely predictable by a simple theory. The tunnel diode possesses a single curve with negative resistance, but here we have an entire family of such curves.

The possible applications include amplifiers, oscillators, bistable and monostable and astable multivibrators, low impedance controlled voltage sources, controlled incremental resistance of infinite ratio, and all the existing applications of a magnetoresistor, with an enhanced incremental resistance change.

A disadvantage is shared with positive feedback schemes in general. It is that an indefinite increase, for instance, in the power gain of the dc amplifier by bringing the slope of the load line ever closer to the characteristic curve, must entail an increase in the sensitivity of the circuit to a variation in the component values.

## G. Magnetoresistors Compared with Thyristors

Thyristors cannot be switched off at their control electrode, and the rate of switch on is very fast and again is not under control. During conduction a voltage drop of about one volt can occur, which is not negligible in low voltage applications, like an electric small vehicle. The consequences are that a limited set of waveforms accompanies thyristor control. The waveforms have steep edges, are rich in harmonics, and create interference. When three-phase loads are controlled, the current in the neutral conductor can be as large as that in any other conductor at certain phase angles of firing at the control electrode; the neutral conductor must be redesigned (*33*).

The magnetoresistor with bias and self-control has possibilities which may solve some of these problems. It is only necessary to take those parts of the characteristics up to the bias, where negative resistance begins. The principle is to change the operating point from that which gives zero incremental resistance to a point of maximum resistance. The change is to be made rapidly, so that high dissipation in the magnetoresistor itself is avoided, but not with such rapidity that interference is excessive.

Let a power source at mains frequency be connected to a load in series with the primary winding of a transformer. The secondary winding contains a loop of elements in series, namely, a magnetoresistor, the coil which controls it, and a pulse generator. The turn ratio ensures that currents in the secondary are lower than those in the primary. The pulse generator has two output levels with variable mark-space and rise–fall rate. One level fixes the "on" state in which actually zero impedance is reflected into the primary, a steady current is established in the secondary and all the power dissipated in the magnetoresistor is supplied by the pulse generator. The other level fixes the "off" state, with maximum impedance reflected into the primary and some assignably small amount of power is transferred through the transformer from primary to secondary.

The system can be switched on or off at any point in the cycle and at a controlled rate. There is no residual voltage drop in the main circuit. For instance, harmonics can be minimized by arranging a symmetrical final waveform on each side of the peak, because the option to switch off at a chosen instant now exists. In applications of this type, transformers and magnetic circuits are already in use. The magnetoresistors are conceivably accommodated by incorporating them into modifications of the existing magnetic system.

If a special magnetic circuit is to be designed there is preliminary evidence that a loudspeaker configuration might have advantages. The air gap takes a tubular polycrystalline magnetoresistor with axial source currents. This has the same advantages as the Corbino disk in that the Hall currents flow in closed circular paths around the axis. It is the Corbino disk with an interchange between the directions of source current and applied magnetic field. The current densities are more uniform than in the disk and there is no special problem of collecting current from the central electrode. The coil controlling magnetoresistance can be substantially assisted by the self-magnetic field.

Large monocrystalline samples are notoriously fragile and liable to accidental cleavage. But there is no obvious reason why large powers should not be handled by polycrystalline structures.

Polycrystalline magnetoresistors are robust mechanically. They are not hypersensitive to electrical abuse. Thyristors on the other hand are known to be "unforgiving" of lapses from the rated electrical conditions. Switch-off has been known not to occur at all during a period of mains overvoltage of some 5%.

Therefore, an area of development and application for the magneto-resistor possibly exists in many of the functions which are at present performed by thyristors. The first necessity is the development of increased power handling, and indications are that the power of existing indium antimonide meander path devices can be increased above 0.5 W by at least two orders of magnitude in tubular constructions of about 1 cm diam and 1 cm length.*

## H. Unusual Applications and Recent Advances

New applications for the Hall effect are being proposed constantly and improvements in devices, techniques, and materials occur. Some of the proposals cannot expect to be implemented until these improvements have been consolidated. For instance, helicon waves could in principle be made to penetrate sea water to communicate with or detect submarines. Noncommunication with satellites during reentry is caused by a plasma which is similarly difficult to penetrate. The problems, which seem insuperable at present, include how to supply a sufficiently powerful magnetic field or direct current or to inject sufficient heavy ions into the fluid in order to make the Hall effect significant. A selection of the more noteworthy proposals and improvements is now given.

A typewriter keyboard has been proposed in which all wearing and moving rigid parts are eliminated; the keys would be printed on a flexible elastomer sheet (34). When a marked area is touched a permanent magnet is brought near to a Hall plate, which then delivers a unique output for that key.

Another use for the displacement transducer is a thermogravimetric balance (35). In this a V-shaped permanent magnet is hung from the arm of the balance near an indium arsenide Hall plate attached to the center post. The Hall voltage is calibrated in terms of the unbalance and the whole apparatus allows weighings to be continued during the heating of specimens.

A bladder motility detector is similar in principle but different in application (36). Permanent magnetic disks of about 2 mm diam are cut from a sheet, whose surfaces are the poles. A thin film indium arsenide Hall plate is made almost 3 mm square. The output voltage is calibrated in terms of the

*Electronic Materials Unit, Royal Radar Establishment, Malvern, U.K.

leakage flux of the permanent magnet as the two sheets are separated in the same plane. These light elements are attached to the bladder to become a low force, unidirectional, biological displacement transducer, which yields a topographical record of bladder movements.

A differential position transducer based on magnetoresistors uses two field plates, set side by side on the pole piece of a permanent magnet (37). When a soft iron toothed wheel is turned near this differential sensor, pulses can be generated. Applications include a tachometer, shaft digitizer, noncontacting limit switch, and noncontacting potential dividers such as sine and cosine function generators by shaping the soft iron parts.

Displacement transducers using a Hall plate and four pole pieces can be designed to have constant gradient of output against displacement over a significant range (38).

Tape recorders conventionally have magnetic heads with coils for electromagnetic pickup. The linearity of output voltage against flux density is improved if the coil is replaced by a Hall plate in the form of a thin film device (39). It is possible to combine such thin film Hall plates with thin film field effect transistors into a composite device. The control electrode of the transistor is supplied by the Hall output terminals and the composite device is a sensitive magnetoresistor (40).

Another tape reader is proposed, in which the tape runs close to the face of a cathode-ray tube (41). On the inside of the face is a thin wire magnetoresistor. This is influenced by the magnetic field of the tape and at the same time it is the target for an electron beam, which scans the length of the thin wire. The beam current as a function of time acquires the same signal that is present in the magnetized tape. As a video tape reader this instrument promises to outstrip versions which use a coil and an air gap for the pickup head in both bandwidth and speed. There is possibly some enhancement of the magnetoresistive effect, through the self-magnetic field, since a thin wire is employed (8). Such thin wires can also be the basis of Hall effect amplifiers, with the input current flowing along the wire and the source current leaving the surface of the wire to enter a surrounding material of high electron mobility. Feedback is inherent, since the magnetic field of the Hall currents coincides with that of the signal current. The thin wire has the advantage that in its vicinity there is a great concentration of both current densities and magnetic fields, but the material in the bulk away from the wires is not used to full advantage.

Modulators, pulse generators, and amplifiers are all possible in thin wire configurations at liquid helium temperature with bismuth as the active material. New phenomena of quantized flux jumps and the Josephson effect occur at this temperature (42). The flux jumps are manifest as voltage pulses of uniform height and width. The pulse position is under the control of the source current, so that modulation can occur. Large changes in the Hall currents cause correspondingly large changes in the numbers of pulses that occur, so that the pulse count becomes a measure of the Hall effect.

Tubular bulk semiconductors permit a more uniform current density than the thin wire pattern; they also allow the self-magnetic field to take effect. The tubular shape is chosen often for resistors, diodes, capacitors, and nuclear particle counters. With the object of measuring Hall coefficients for such components, correction factors have been found in the case where the magnetic field is uniform and perpendicular to the axis of the tube (43). For Hall currents to circulate around the tube, the magnetic field would be radial.

Special effects occur at the surface of semiconducting materials. In particular there is a special version of the helicon surface wave (44) which is axisymmetric. There is also a connection between galvanomagnetic and acoustomagnetic surface waves with strong excitation currents (45). Propagation of submillimeter waves along thin slabs of indium antimonide inside a waveguide is a subject which might lead to useful variable attenuators, if not traveling wave amplifiers eventually (46).

Refinements in the application of the Hall effect to magnetic field measurement continue to be made. Varying fields can be handled and even pulsed magnetic fields (47–49). The Corbino disk is used in place of a Hall plate for the measurement of mobility. This it achieves directly without need of a separate measurement of conductivity. Also there is no need to provide the voltage contacts of the Hall plate (50).

Refinements in the Hall plate itself include the optimizing of geometric ratios and the fitting of self-screening magnetic systems (51). The same attention to geometric ratios combined with appropriate linear matching resistors enhances the linearity of indium arsenide plates (52); consequently highly sensitive power transducers are produced (53). Attention is also given to the thermal properties and the concentration of fields by ferrite pole pieces (54, 55).

Indium antimonide and indium arsenide continue to be the major materials used; as true chemical compounds these occupy a special place in the list of possible materials (56). Germanium, silicon, and gallium arsenide are also in use; silicon magnetodiodes are chosen, for instance, in a bridge configuration (57, 58). The true test of materials for many applications is the product of conductivity with Hall constant and also with magnetic permeability. Very pure iron is worth considering but it has anomalous characteristics such as a reversal of the Corbino effect with increasing applied field. Magnetic semiconductors are in principle ideal subjects for investigation as Hall elements; they are usually II–VI ternary compounds with an inconveniently low Curie temperature. There is much interest in the band structure of the materials and in their ability to make transitions from high to low conductivities (59–62). A specific interaction of helicon and spin waves is reported in cadmium chromium selenide with p-type doping (63).

Studies of the Hall effect in unusual situations include measurement in the semiconducting molten state (64). Plates of arbitrary shape should be capable of gyrator action from a study of the antireciprocity condition in such plates (65). Small Hall voltages and amorphous germanium films have been studied to produce a sensitive measurement method (66). The Hall mobility in sintered powders of cadmium sulfide has been measured (67). Pulse counters have been used in the measurements of Hall and magnetoresistive effects in semiconductors and metals (68). A virtual cascade of Hall generators, in which all share the same magnetic field but the output of one becomes the source for the next in the cascade, exhibits an output voltage of the $n$th power of the magnetic flux density (69). Magnetic bubble domains have been detected by the evaporation of thin films of indium antimonide (70).

Throughout these studies of Hall research and applications charge carriers of high mobility are essential. High mobility leads to exceptional performance, but it must be remembered that to consider only the average mobility in a material may lead to a neglect of the exceptionally few electrons with much more than average mobility. On such a basis, Cullwick hypothesized that magnetic energy itself could be equated with the kinetic energy of electrons, even though the drift velocity in a conductor is low. Integration over carrier speeds is considered in one exposition of the theory of galvanomagnetic effects in thin slabs (71, 72).

## VI. CONCLUSION

The number of actual applications of the research on the Hall effect is a small fraction of what is possible. There has even been a decline in what was formerly a most widespread application to analog multipliers; manufacturers have turned to diode function generators. These may have many components but the components are cheap, coils are avoided, impedances are not different for each input, and there are no acute problems of balance to secure zero output for zero input.

The lack of mass production on the scale of that for diodes and transistors prevents a comparison of reliability and stability on a statistical basis. However, the InSb–NiSb magnetoresistor has been in production for several years, and it is known that the device is not subject to aging (16). This is because the change in resistance is a bulk effect and is not dependent on surface properties. Experiments have confirmed the freedom from aging in the semiconductor itself, but allowance must be made for changes that can arise if epoxy resins are used in potting, since the resins are subject to aging. Noise is not normally a problem in the magnetoresistors, provided that mechanical strains have not induced microscopic cracks through failure to observe the mounting instructions. Temperature dependence is a complex but known function of the doping in InSb and of the magnetic field strength. A reasonable independence of temperature can be had in exchange for some loss of sensitivity, either by selecting the doping or by changing to indium arsenide. Therefore in these respects there is not an unfavorable comparison with diodes and transistors. In other respects, the Hall and magnetoresistive devices are superior to their monocrystalline competitors.

Apart from the obvious vulnerability of large monocrystals to fracture through cleavage, the action of even small diodes and transistors depends upon carrier lifetime, surface effects, and diffusion. The electrical characteristics are vulnerable to overloads of very short duration indeed. They are vulnerable to neutron bombardment, radiation, the ingress of impurities, and temperature and pressure changes. Reliable engineering is firmly based on annealed, polycrystalline or homogeneous materials. It is to the latter category that most Hall devices belong.

The electrical noise of bulk polycrystalline Hall devices is a low, mainly thermal Johnson noise; current noise is apparently absent. No other static device has comparable ability to convert a magnetic flux density input into a proportional voltage output. Coils require a changing magnetic field and have sensitivity proportional to area. Ferromagnetic probes disturb the field in which they are inserted, but the Hall plate retains its sensitivity down to minute probe dimensions.

The Hall plate is a unique embodiment of a passive, essentially linear, yet nonreciprocal element. It is the archetypal gyrator.

The fabrication of polycrystalline plates and magnetoresistors does not entail all the elaborate processes of transistor technology. $p$-$n$ junctions are sensitive to the presence of recombination centers and therefore need to have material of higher purity than suffices for Hall devices. Dislocations in crystal structure are equally unimportant in the Hall device. Minority carriers do not feature in the action of Hall devices; therefore inhomogeneous doping to control diffusion is not required, and high uniform doping can be given.

Given these properties and advantages, why then have Hall devices not yet made their proper impact in practical applications? It may be that the comparability of magnetoresistors with transistors as devices with families of amplifier characteristics has remained unknown until recently (32). It may be that coils, inductors, permanent magnets, and soft magnetic circuits are avoided at all costs by those responsible for printed circuits. But, for instance, the Corbino disk microwave oscillator needs no coil (26). Again, ferrite core stores, magnetic bubble domains, and even magnetostrictive delay lines have found their niche in computer technology.

A renewal of interest in the application of Hall devices is now due. It is likely to center on polycrystalline magnetoresistors. Single crystals are most painfully inappropriate in large sizes, so that the polycrystalline magneto-resistive amplifier will probably first find acceptance for high powers and low frequencies. Recall that, unlike the thyristor, it can be switched off or on or be controlled smoothly through any intermediate state. There are many other exciting possibilities including microwave oscillators, low attenuation helicon wave structures, distributed amplifiers, nonreciprocal devices, bistable circuits, contact-free position transducers, solid rotor alternators, and a whole class of electronic engineering for hostile environments near radiation, accelerated particles, or extreme temperatures, where other techniques fail.

## REFERENCES

1. S. Stricker, *Advan. Electron. Electron. Phys.* **25**, 98–143.
2. H. Weiss, "Structure and Application of Galvanomagnetic Devices." Pergamon, Oxford, 1969.
3. W. Thomson, *Phil. Trans. Roy. Soc. London* **146**, 649 (1856).
4. R. G. Chambers and B. K. Jones, *Proc. Roy. Soc. Ser. A* **270**, 417 (1962).
5. A. C. D. Whitehouse, *Brit. J. Appl. Phys.* **1**, 1637 (1968).
6. I. M. Vikulin, L. L. Lyuze, and V. A. Presnov, *Sov. Phys.—Semicond.* **2**, 1073 (1969).
7. D. Midgley, *Electron. Lett.* **6**, 497 (1970).
8. D. Midgley, *IEEE Trans. Electron Devices* **19**, 375 (1972).
9. D. Midgley and S. W. Smethurst, *Proc. Inst. Elec. Eng.* **112**, 1945 (1965).
10. G. De Mey, *Electron. Lett.* **9**, 264 (1973).
11. D. Midgley, *Ind. Electron.* **1**, 383 (1963).
12. J. P. Newsome and W. H. Silber, *Solid-State Electron.* **12**, 631 (1969).
13. D. J. White, M. L. Knotek, and M. H. Ritchie, *J. Appl. Phys.* **44**, 1870 (1973).
14. J. Schewchun, K. M. Ghanekar, and R. Yager, *Rev. Sci. Instrum.* **42**, 1797 (1971).
15. T. Yamada, *Proc. Int. Conf. Phys. Semicond., Moscow*, p. 672 (1968).
16. H. G. Steidle, *Siemens Rev.* **40** (4), 177 (1973).
17. S. Grutzmann, *Proc. IEEE* **51**, 1584 (1963).
18. D. Midgley, *Electron. Lett.* **4**, 101 (1968).
19. D. Midgley, *Elec. Times* **154**, 640 (1968).
20. D. Midgley, Brit. Pat. No. 1,171,541 (1970).
21. J. C. Shenton and T. D. Mason, *Roy. Aircraft Estab.* (*Farnborough*), *RAE Tech. Memo* **IEE 200** (1968).
22. R. J. Strand, *IBM Tech. Disclosure Bull.* **14**(10), 2928 (1972).
23. H. M. Barlow, J. C. Beal, and H. G. Effemey, *IEEE Trans. Instrum. Meas.* **14**, 238 (1965).
24. H. M. Barlow and K. V. G. Krishna, *Proc. Inst. Elec. Eng., Part B* **109**, 131 (1962).
25. M. Nalecz and J. E. Rijnsdorp, *Rozprawy Elektrotech.* **18**, 3 (1972).
26. V. I. Baibakov, *Sov. Phys.—Semicond.* **5**(12), 2070 (1972).
27. T. G. Cowling, *Mon. Notic. Roy. Astron. Soc.* **94**, 39 (1933).

28. D. Midgley, *Nature (London)* **186**, 377 (1960).
29. D. Midgley, *Electron. Eng.* **38**, 728 (1966).
30. I. H. Mathews, *Nature (London)* **188**, 651 (1960).
31. D. Midgley, *Nature (London)*, *Phys. Sci.* **240**, 22 (1972).
32. D. Midgley, *Nature (London)* **243**, 514 (1973).
33. A. Holland and D. Midgley, *Electron. Power* **19**(6), 129 (1973).
34. C. P. Ludeman, *IBM Tech. Disclosure Bull.* **14**(10), 2924 (1972).
35. R. E. Jensen and R. P. Swenson, *J. Chem. Educ.* **49**(9), 648 (1972).
36. J. A. Woltjen, G. W. Timm, F. M. Waltz, and W. E. Bradley, *IEEE Trans. Bio-Med. Eng.* **20**(4), 295 (1973).
37. K. Wetzel, *Elektronicker* **1**, 21 (1972); also U. Von Borcke, *Siemens Electron. Components Bull.* **8**(2), 53 (1973).
38. J. Oledzki and Z. L. Warsza, *Arch. Elektrotech. (Warsaw)* **20**(2), 479 (1971).
39. M. Murai, *J. Electron. Eng. (Tokyo)* **70**, 58 (1972).
40. J. Cervenak, *Slaboproudy Obz.* **34**(5), 207 (1973).
41. C. R. Weidmann, U.S. Pat. No. 318,299 (1965).
42. R. D. Parks, "Superconductivity." Dekker, New York, 1969.
43. N. I. Pavlov, V. L. Konkov, and A. S. Kukui, *Ind. Lab. (USSR)* **37**, 559 (1971).
44. V. I. Baibakov and V. N. Datso, *Sov. Phys.—Solid State* **13**(10), 2637 (1972).
45. L. E. Gurevich, I. V. Ioffe, and S. L. Kuliev, *Sov. Phys.—Solid State* **13**(2), 459 (1971).
46. Y. G. Altshuler, L. I. Hats, and R. M. Revzin, *Radio Electron. Commun. Syst.* **15**(8), 938 (1972).
47. N. S. Babenko, *Avtometriya* No. 5, 117 (1971).
48. K. W. Bonfig and A. Kazamalikis, *Arch. Tech. Mes. Ind. Messtech.* No. 439, 141 (1972).
49. N. S. Babenko, *Avtometriya* No. 2, 73 (1972).
50. G. P. Carver, *Rev. Sci. Instrum.* **43**(9), 1257 (1972).
51. D. S. Rusev and D. D. Aleksandrov, *Elektroprom. Priborostr.* **7**(3), 102 (1972).
52. R. Maniewski, *Pomiary, Automat., Kontr.* **18**, 361 (1972).
53. I. Zawicki, *Pomiary, Automat., Kontr.* **19**(3), 115 (1973).
54. A. Kobus, *Electron. Technol.* **5**(3/4), 73 (1972).
55. A. Kobus, A. Milczarek, and A. Dajna, *Electron. Technol.* **5**(3/4), 31 (1972).
56. E. D. Sisson, *Anal. Chem.* **43**(7), 67A (1971).
57. V. Husa, I. Benc, J. Kriz, and J. Ladnar, *Elektrotech. Obz.* **60**(4), 179 (1971).
58. P. Kordos, *Slaboproudy Obz.* **33**(8), 354 (1972).
59. P. Wachter, *Crit. Rev. Solid State Sci.* **3**(2), 189 (1972).
60. J. M. Langer, *Postepy Fiz.* **24**(1), 57 (1973).
61. R. Wadas, *Electron. Technol.* **5**(2), 49 (1972).
62. G. Heber, *Wiss. Z. Tech. Univ. Dresden* **20**(2), 315 (1971).
63. L. J. Abella and B. Vural, *IEEE Trans. Electron Devices* **19**(8), 928 (1972).
64. A. A. Andreev and M. Mamadaliev, *Ind. Lab. (USSR)* **37**(12), 1905 (1971).
65. C. Sora, *Rev. Roum. Sci. Tech., Ser. Electrotech. Energ.* **16**(4), 679 (1971).
66. R. A. Lomas, M. J. Hampshire, and R. D. Tomlinson, *J. Sci. Instrum.* **5**(8), 819 (1972).
67. K. Takahagi, Y. Kashiwaba, and H. Saito, *Technol. Rep. Iwate Univ.* **5**, 37 (1971).
68. B. Vichev, *Izv. Inst. Elektron.* **5**, 97 (1971).
69. N. Masuda, U.S. Pat. No. 3,684,997 (1972).
70. S. Yoshizawa, T. OI., J. Shigeta, I. Mikami, and G. Kamoshita, *Amer. Inst. Phys. Conf. Proc., Chicago* **5**, Part 1, 230 (1971).
71. M. Biermann, *Sci. Elec.* **18**(3), 73 (1972).
72. E. G. Cullwick, *Proc. Inst. Elec. Eng., Part C* **103**, 159 (1955).

# Research and Development in the Field of Walsh Functions and Sequency Theory

HENNING F. HARMUTH

*Department of Electrical Engineering,*
*The Catholic University of America,*
*Washington, D.C.*

## I. INTRODUCTION

Sinusoidal functions play a dominant role in electrical communications and, beyond that, in physics. Whenever the term frequency is used, one refers implicitly to these functions. Let us see how this dominant role came about and where its limitations are.

During the nineteenth century, the most important functions for communications were block pulses. Voltage and current pulses could be generated by mechanical switches, amplified by relays, and detected by a variety of magnetomechanical devices. Sine-cosine functions and the exponential

function were well known and so was Fourier analysis, although in a somewhat rudimentary form. Almost no practical use could be made of this
knowledge with the technology available at that time. Heinrich Hertz used
the exponential function to obtain his famous solution of Maxwell's equations for dipole radiation but he was never able to produce sinusoidal waves.
His experiments with wave propagation were done with what we would call
colored noise today.

Toward the end of the nineteenth century, more practical means to
implement capacitances than Leyden jars and metallic spheres were found.
The implementation of inductances through the use of coils had been known
long before. Practical resonant circuits for the separation of sinusoidal electromagnetic waves of different frequencies could thus be built around the
turn of the century. Low-pass and band-pass filters using coils and capacitors
were introduced in 1915 and a large new field for the application of sinusoidal functions was opened. Speaking more generally, the usefulness of sinusoidal functions in communications is intimately related to the availability of
linear, time invariant circuit components in a practical form.

The last sentence gives the first clue why a theory of communications
based on sine-cosine functions would eventually prove unsatisfactory. One
cannot transmit information if everything is (time) invariant. The telegra-
pher's key, the microphone, and the amplitude modulator are linear but time
*variable* devices. Making them time invariant by not operating the key, not
speaking into the microphone, or not feeding a time variable modulating
voltage into the modulator puts an end to the transmission of information.
The requirement of time variability for information transmission holds quite
generally. An atom with all orbital electrons in certain quantum states trans-
mits no information. A photon is emitted by a change of a quantum state,
and this time variation provides information about the energy difference of
the initial and the final state.

Sine-cosine functions are obtained as the eigenfunctions of systems
described by linear differential equations with constant coefficients. Hence,
sine-cosine functions are most convenient as long as one may ignore the time
variability. Increasing sophistication forces one to use equations, not necessarily differential equations, with variable coefficients; their eigenfunctions
are no longer sinusoidal functions.

We have so far discussed *time* variability. An obvious generalization is
*space* variability. A tunable generator for sinusoidal functions always produces time functions and not space functions. The theory of filters in electrical communications almost exclusively refers to the filtering of time signals
rather than space signals. A black-and-white photograph has various shades
of grey as a function of the two space variables $x$ and $y$. Hence, it is a space
signal with two variables. A television signal is a function of two space

variables and the time variable. The omnipresence of TV signals makes it safe to conclude that most information does not consist of functions of the time variable only. Why then do we hear so little about space signals and filters for space signals? There are two reasons. The concept of time invariance, meaning that something has always been as it is now and will always remain so, is acceptable to our thinking although we know it is unrealistic. Space invariance, on the other hand, is so unrealistic that we cannot accept it. A television image clearly has a left and a right edge, a top and a bottom, while the finite extension in time is much less obvious. There are $30 \times 3600 = 108,000$ images as a sequence of time per hour, but there are never more than some five or six hundred space points in the $x$ and $y$ directions. Hence, a theory of space filters must start with space variable filters and cannot consider space invariant filters as a first, simpler case. The second reason for not hearing much about filters for space signals is that filters for time signals are overwhelmingly implemented by inductances and capacitances, but this technology is not applicable to filters for space signals.

In addition to the concepts of time variability and space signals, there is a third basic reason for going beyond sine-cosine functions: the convergence of Fourier series and Fourier transform. Any practical signal can be approximated by the Fourier series or the transform in the sense of a vanishing mean-square-error. This mean-square-convergence is often sufficient but not always. Most of the solutions of the wave equation or Maxwell's equations cannot be approximated with uniform convergence by a superposition of sine-cosine functions. This fact has become of interest due to semiconductor technology. The typical on–off type switching functions preferred by semiconductor circuits do not permit an approximation of the transients due to the Gibbs phenomenon, and the transients are the important part of switching functions. The practical engineer has long recognized the difficulty of using sine-cosine functions for approximations. Nobody builds pulse generators that contain many amplitude-, frequency-, and phase-stable sinusoidal oscillators in order to produce block pulses according to the Fourier series. On the contrary, it is general practice to synthesize stable sinusoidal oscillations by means of block pulses generated by digital circuits.

To see in which way one may profitably generalize our theory of communications based on sine-cosine functions, let us consider Fig. 1. Block pulses, which were the historically first important system of functions, are shown on the right. The sine-cosine functions plus the constant function used in the Fourier series are shown on the left. One may readily see why we have an extensive theory based on sine-cosine functions but not one based on block pulses. The block pulses differ by a time shift only. In other words, they contain one free parameter only, which we call *delay*. The periodically continued sine-cosine functions contain the parameter delay too, which is

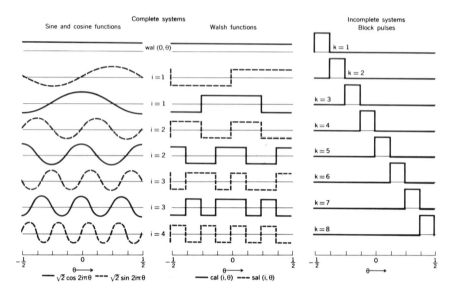

FIG. 1. Sine-cosine functions, Walsh functions, and block pulses.

called *phase* for these particular functions, but in addition they contain the parameter frequency. In essence, sine-cosine functions of different frequency have a different shape, while the block pulses have all the same shape. For a satisfying more general theory, one will thus have to look for nonsinusoidal functions that have at least as many parameters as the sinusoidal functions. Since sine-cosine functions are a particular system of complete orthogonal functions, one may replace them by general systems of complete orthogonal functions. The resulting more general theory is called sequency theory for reasons soon to be explained, while the theory based on sine-cosine functions is called frequency theory.

When looking for useful complete systems of orthogonal functions, one soon finds the Walsh functions. A few of them are shown in Fig. 1. These functions assume the two values $+1$ or $-1$ only, which makes them comparably simple as the block pulses and leads to simple equipment. On the other hand, the functions have different shapes like the sinusoidal functions, which assures that theory and equipment are not as restricted as for block pulses. Actually, the shape of the Walsh functions depends on two parameters, their sequency and their time base. Let us consider the sequency first. The frequency of the periodically continued sine and cosine functions $\sin 2\pi i\theta$ and $\cos 2\pi i\theta$ in Fig. 1 is the number $i$ of oscillations in the half-open time interval $-\frac{1}{2} \leq \theta < \frac{1}{2}$. The number $2i$ gives the zero crossings of sine and cosine functions in the same interval. Consider now the Walsh functions $\mathrm{sal}(i, \theta)$ and $\mathrm{cal}(i, \theta)$ in Fig. 1. The letters al allude to Walsh, while the letters s and c indicate the similarity with either the sine or the cosine functions. The functions $\text{sal}(i, \theta)$ and $\text{cal}(i, \theta)$ have $2i$ "jumps" or zero crossings in the half-open interval $-\frac{1}{2} \le \theta < \frac{1}{2}$, just like $\sin 2\pi i\theta$ and $\cos 2\pi i\theta$. The distance between the zero crossings is always the same for a particular sine or cosine function but this is in general not so for the Walsh functions. One may readily see from Fig. 1 that the distance between zero crossings for $\text{sal}(3, \theta)$ and $\text{cal}(3, \theta)$ is either $1/8$ or $1/4$.

The normalized sequency is defined as one-half the number of zero crossings in a half-open time interval of duration 1. The nonnormalized sequency is one-half the number of zero crossings per second; its unit is the zps in analogy with the old unit cps for the frequency. Sequency and frequency are identical for sine and cosine functions.

In order to see the meaning of the time base of Walsh functions, let us compare them with sine-cosine functions in their general form. The normalized functions $V \sin(2\pi i\theta - \alpha)$ and $V \cos(2\pi i\theta - \alpha)$ have amplitude $V$, normalized frequency $i$, phase angle $\alpha$, and normalized time $\theta$. The nonnormalized functions are obtained by the substitutions $i = fT$ and $\theta = t/T$, where $f$ and $t$ are the nonnormalized frequency and time, while $T$ is the time base. One obtains the nonnormalized functions $V \sin(2\pi ft - \alpha)$ and $V \cos(2\pi ft - \alpha)$. The time base $T$ drops out due to the relation $i\theta = fT(t/T) = ft$.

Let us repeat the same process for Walsh functions. The normalized functions $V \text{ sal}(i, \theta - \theta_0)$ and $V \text{ cal}(i, \theta - \theta_0)$ contain amplitude $V$, normalized sequency $i$, and normalized time $\theta$ just like the sine-cosine functions. The phase angle $\alpha$ is replaced by the normalized delay $\theta_0$; this is an insignificant difference in writing since one can substitute $\alpha = 2\pi i\theta_0$ in the argument of the sinusoidal functions. The substitutions $i = fT, \theta = t/T$, and $\theta_0 = t_0/T$ yield the nonnormalized functions $V \text{ sal}[fT, (t - t_0)/T]$ and $V \text{ cal}[fT, (t - t_0)/T]$. The time base $T$ does not drop out. The physical significance may readily be seen from Fig. 1. The periodically continued sine or cosine functions for $i = 3$ become identical to the functions for $i = 1$ if "stretched" by a factor of 3 $(T \to 3T)$. The Walsh functions for $i = 3$, on the other hand, cannot be transformed by stretching into the Walsh functions for $i = 1$.

The notation $\text{sal}(i, \theta)$ and $\text{cal}(i, \theta)$ stresses the similarity with sine and cosine functions. This is helpful for an introduction to sequency theory. The notation becomes cumbersome at a more advanced level, just as $\sin 2\pi i\theta$ and $\cos 2\pi i\theta$ are too cumbersome for most practical purposes and are replaced by the exponential function. The corresponding simpler notation for Walsh functions is provided by the following relations:

$$\text{cal}(i, \theta) = \text{wal}(2i, \theta), \quad \text{sal}(i, \theta) = \text{wal}(2i - 1, \theta).$$

The notation wal($j$, $\theta$), with $j = 0, 1, 2, \ldots$ thus replaces the notation wal($0$, $\theta$), sal($i$, $\theta$), and cal($i$, $\theta$) with $i = 1, 2, \ldots$. The resulting simplification is readily appreciated by considering Walsh functions with two variables $x$ and $y$. One may write them as wal($l$, $x$)wal($m$, $y$) with $l$, $m = 0, 1, 2, \ldots$. The old notation would require the writing of nine products.

The discussion is restricted here to Walsh functions. Information on other functions may be found in a comprehensive textbook (*1*), six Proceedings of symposiums (*2*, *3*), and two collections of references (*4*, *5*).

Three areas of applications will be discussed in the following sections: sequency multiplexing of digital signals, two-dimensional sequency filters for TV image processing and for image generation by sound waves, and electromagnetic Walsh waves applied to radar.

The most basic theoretical investigations of sequency theory are presently made in a rather abstract field of mathematics and theoretical physics. The sinusoidal functions are mathematically connected by way of group theory to the real numbers. The Walsh functions are similarly connected to binary numbers or the *dyadic group*, which is essentially a different name for the same thing. Differential calculus is also based on the real numbers. Whenever a differential or integral equation is used to describe a physical process in time and space, one introduces the real numbers which, in turn, introduce the sine-cosine functions. Expressed in a more mathematical fashion, the use of differential calculus introduces the topology of the real numbers or of the continuum into the physical space and time, and sine-cosine functions have certain shift-invariant features for this topology. The mathematical convenience of differential calculus created our present concept of the topology of time and space, just as the convenience of Euclidean geometry created the metric of Euclidean space–time, and the convenience of planar geometry the flat earth. We have learned that the shape of the earth and the metric of space–time can only be obtained from observation, and we are in the process of realizing that the topology of space–time must come from observation too, not from differential calculus.

The dyadic group also defines a topology of space–time, in which Walsh functions have certain shift invariant features. This topology creates phenomena similar to those obtained by the introduction of nuclear exchange forces in the real number topology. The similarity to the metric of Riemann geometries versus Euclidean geometry plus gravitational forces is evident. Of course, one cannot expect that the dyadic group yields *the* topology of space–time. Just as the metric of space is created by the property mass of the matter in space, one must expect that the topology is created by some other property of the matter. The contribution of Walsh functions and the dyadic group is that a model of topology for space–time other than that based on real numbers is investigated in detail. This work should help us toward a better understanding of the topology of space–time similar to the way the first models of non-Euclidean geometry furthered the development of Riemann geometries and the concept of metric of space–time. The interested reader will find much on the mathematical foundations in the papers by Butzer, Gibbs, Pichler, and Wagner.*

## II. SEQUENCY MULTIPLEXING

### A. Principle of Sequency Multiplexing

Consider the multiplication theorems of sine and cosine functions:

$$2 \cos i\theta \cos k\theta = +\cos(k - i)\theta + \cos(k + i)\theta,$$
$$2 \sin i\theta \cos k\theta = -\sin(k - i)\theta + \sin(k + i)\theta,$$
$$2 \cos i\theta \sin k\theta = +\sin(k - i)\theta + \sin(k + i)\theta,$$
$$2 \sin i\theta \sin k\theta = +\cos(k - i)\theta - \cos(k + i)\theta.$$

Let $\cos i\theta$ and $\sin i\theta$ stand for components of a Fourier expansion of a time signal $F(\theta)$, while $\cos k\theta$ and $\sin k\theta$ stand for carriers. The multiplication theorems then represent the amplitude modulation of these carriers by one component of the Fourier expansion. On the right side, one obtains the "lower" components with frequency $k - i$ and the "upper" components with frequency $k + i$. This is the reason for the phenomenon of double sideband modulation. If amplitude modulation is done by the signal $F(\theta)$ rather than the two components $\cos i\theta$ and $\sin i\theta$, one obtains a lower and an upper sideband in place of the lower and upper components.

Consider now the multiplication theorems of Walsh functions:

$$\mathrm{cal}(i, \theta)\mathrm{cal}(k, \theta) = \mathrm{cal}(r, \theta) \qquad r = i \oplus k,$$
$$\mathrm{sal}(i, \theta)\mathrm{cal}(k, \theta) = \mathrm{sal}(r, \theta) \qquad r = [k \oplus (i - 1)] + 1,$$
$$\mathrm{cal}(i, \theta)\mathrm{sal}(k, \theta) = \mathrm{sal}(r, \theta) \qquad r = [i \oplus (k - 1)] + 1,$$
$$\mathrm{sal}(i, \theta)\mathrm{sal}(k, \theta) = \mathrm{cal}(r, \theta) \qquad r = (i - 1) \oplus (k - 1).$$

The sign $\oplus$ indicates an addition modulo 2. The numbers $i$ and $k$ are written in binary form and added according to the rules $0 \oplus 1 = 1 \oplus 0 = 1$, $0 \oplus 0 = 1 \oplus 1 = 0$ (no carry).

Let now $\mathrm{cal}(i, \theta)$ and $\mathrm{sal}(i, \theta)$ stand for the components of a Walsh expansion of a time signal $F(\theta)$, while $\mathrm{cal}(k, \theta)$ and $\mathrm{sal}(k, \theta)$ stand for carriers. Note

---

*In order to limit the list of references, only the name of one author and sometimes the year of publication is given. The references and the names of the co-authors may be found in the bibliography by Bramhall (4), which is available free of charge.

that these carriers have the time variation of the Walsh functions in Fig. 1. The amplitude modulation of a carrier yields only one component, cal(r, θ) or sal(r, θ), not two as in the case of sinusoidal functions. As a result, the amplitude modulation of a Walsh carrier by the time signal F(θ) yields only one *sequency* band which is as wide as the sequency band of F(θ) and not twice as wide as in the case of sinusoidal carriers and *frequency* bands. Hence the need for single sideband filters is eliminated and so are the distortions caused by them.

For an explanation of the practical implementation of sequency multiplexing by means of Walsh carriers, refer to Fig. 2. The output voltages of

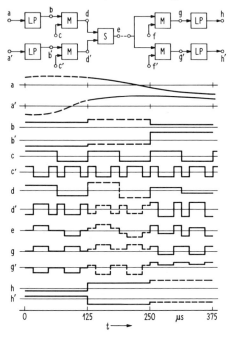

FIG. 2. Principle of a sequency multiplex system. LP—sequency low-pass filter; M—multiplier; S—adder.

two microphones are applied to points a and a'. They are passing through two sequency low-pass filters* LP. Step voltages appear at their outputs b and b'. These are fed to the multipliers M and amplitude modulate two periodic Walsh carriers applied to points c and c'. The modulated carriers d and d' are added in the summer S and the output voltage e is obtained. This voltage is multiplied at the receiver in two multipliers M by the same Walsh

---

* See Harmuth (1) for a discussion of sequency filters for time signals.

carriers as used at the transmitter. The two voltages appearing at the outputs g and g′ of the multipliers are fed through sequency low-pass filters LP, that are equal to those used in the transmitter. The step voltages at the outputs h and h′ are equal to those at b and b′. They may be fed directly into a telephone headset. The low-pass filters of the transmitter produce a delay of 125 μsec and those of the receiver produce another 125 μsec. The dashed sections of the time diagram of Fig. 2 indicate these delays.

### B. Features of Frequency, Time, and Sequency Multiplexing

In order to define areas of application for sequency multiplexing, let us list its characteristic features and compare them with those of time and frequency multiplexing.

1. Equipment can easily be implemented by the present (binary) semi-conductor technology, a feature shared with time multiplexing but not with frequency multiplexing.

2. The amplitude distribution of sequency multiplex signals is essentially a Gaussian distribution and very similar to that of frequency multiplex signals; this statement holds for analog as well as for digital signals. Hence, sequency multiplex signals have the amplitude distribution for which the existing frequency multiplex channels have been designed. Time multiplex signals generally do not have a Gaussian amplitude distribution; in particular, the amplitudes of time multiplexed binary digital signals assume two values only.

3. The average power of sequency multiplexed signals is independent of the activity factor* if automatic volume control is used. The average power per channel and thus the signal-to-noise ratio increase with a decreasing activity factor. The same holds true for frequency multiplexing. In the case of time multiplexing, the average power of an active channel is independent of the activity factor and the average power of the multiplex signal declines with the activity factor. In the transmission of peak power limited binary signals in the presence of thermal noise, sequency multiplexing definitely yields lower error rates than time multiplexing for activity factors of 50% or less while time multiplexing yields definitely lower error rates for activity factors close to 100%. Since the activity factor of a multiplex system rarely exceeds 25%, means have been developed to increase it for time multiplexing by methods usually referred to as "time assignment." These methods must yield activity factors well above 50%—not counting the time used for the transmission of additional information required by these methods—to make them worthwhile.

* The activity factor gives the fraction of channels actually busy.

4. Errors in digital signal transmission through the switched telephone network are mainly caused by burst-type interference. The error rates of digital time multiplex signals are orders of magnitude above those for frequency multiplex signals under these conditions. Sequency multiplexing yields about the same error rates as frequency multiplexing.

5. Sequency multiplexing presently yields some 40 dB crosstalk attenuation. This is sufficient for digital signals, most analog telemetry or TV signals, and low grade voice or scrambled voice signals. It is not sufficient for commercial voice channels that should have about 70 dB crosstalk attenuation. Although 70 dB apparent crosstalk attenuation can be attained by compandors, the more prudent approach is to restrict sequency multiplexing for the time being to applications for which 40 dB suffices.

6. Time multiplexing favors the time-shared use of one complex piece of equipment. Frequency and sequency multiplexing favor the simultaneous use of several but simpler pieces of equipment. Which approach is more economical depends on the state of technology. Time sharing communication lines and telegraphy equipment were universal before 1920. The introduction of frequency filters emphasized simultaneous use; e.g., several low speed teletypewriters replaced one high speed teleprinter. Time sharing again became attractive after 1950 due to the introduction of the transistor. The present trend to integrated circuits with high tooling-up costs and relatively low costs thereafter may again bring a shift from time sharing to simultaneous use.

7. Time multiplexing as well as sequency multiplexing are less suitable than frequency multiplexing for multiplexing of unsynchronized signals. Typically, such unsynchronized signals occur in networks where signals are generated at various points and multiplexed at other points.

8. Sequency and time multiplexing require no single sideband filters and thus avoid the costs as well as the distortions caused by them. The distortions are of little importance for voice transmission but are the main reason why the transmission rate of digital signals is generally one third or less of the Nyquist rate.

9. Individual extraction of a channel is done easily in time multiplexing but requires considerable expense in frequency multiplexing. Sequency multiplexing lies between these two extremes.

10. Frequency multiplexing requires individual tuning of filters. Time and sequency multiplexing do not require any individual tuning but use an accurate clock pulse generator instead that is shared by many circuits.

11. Synchronization errors can transfer digits from one channel to the other in time division while the different shape of carriers for different channels prevents this in frequency or sequency division. This problem is usually referred to as the *stuffing* and *spilling* problem in time division. We have to

elaborate on the stuffing and spilling problem in sequence division in some detail.

There are two distinct modes of transmission in sequence multiplexing. In the first mode, the characters are transmitted parallel while the digits of the characters are transmitted serially. This is shown in Fig. 3a for four characters, 1 to 4, each containing four digits, $A$ to $D$. One carrier is needed per character; hence, the carriers are denoted 1 to 4.

Figure 3b shows the second mode of sequence multiplexing. The four characters 1 to 4 are now transmitted serially while the four digits $A$ to $D$ of each character are transmitted in parallel. The four carriers needed for multiplexing are denoted by $A$ to $D$.

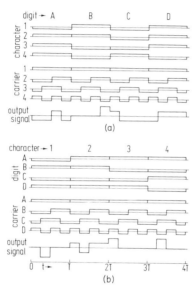

FIG. 3. The two modes of transmission of digital signals in sequence multiplexing. (a) Mode 1: Digits serially, characters parallel. (b) Mode 2: Digits parallel, characters serially.

The output signal of digital equipment usually consists of series of block pulses; that is, digits represented by block pulses and time multiplexed to form a character. This makes the first mode appear to be the more useful one. However, one should not conclude that digital equipment ranging from teletypewriters to computers *must* deliver digits serially to the output terminal. Consider an optical paper tape reader with eight digit characters. The digits of a character will be read simultaneously or in parallel, but successive characters would be read serially. This corresponds to the second mode of sequence multiplexing. One could actually build a sequence multiplexer right into the paper tape reader and provide serial characters with sequency

multiplexed parallel digits at the output of the equipment. The apparently
greater complexity of the second mode of sequence multiplexing is thus
caused by the use of time multiplexing for combining digits into a character;
the time multiplexed digits must be retransformed into their original simultaneous form to use sequence multiplexing to its best advantage.

Having recognized that in theory one needs none or few buffer storages,
we still have to face the fact that practical digital data sources use time
multiplexing and that large buffer storages are required to undo this time
multiplexing. However, semiconductor technology is a field with sharply
dropping prices. Hence, one does not have to be too concerned about complexity as long as the equipment can be implemented by semiconductor
technology. On the other hand, the number and the quality of radio channels is practically fixed by nature; the cost of coaxial cables and microwave
links has no tendency to decrease. This means that the trend must be toward
more complex terminal equipment that permits better use of the channels.
Sequence multiplexing as discussed here follows this trend.

Consider now stuffing and spilling for the two modes of sequence multiplexing. Figure 4 shows on top three characters, 1 to 3, with the digits
$a,b,...$. The three characters are not synchronized. For instance, the digits of
character 2 are delivered particularly slowly to the multiplexer and the digits
of character 3 particularly fast. The sample pulses transform the unsynchronized digits $a$, $b$, $c$, $d$ of character 1 into the synchronized digits $A$, $B$, $C$, $D$.
Since the digits of character 2 are delivered very slowly, the digit $c$ is sampled

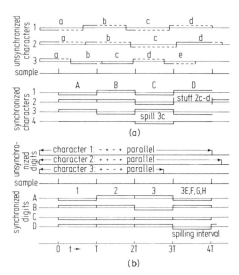

FIG. 4. Stuffing and spilling in the two modes of transmission of digital signals in sequence
multiplexing. (a) Mode 1: stuffing and spilling. (b) Mode 2: stuffing and spilling.

twice. For the second sample, one has to transmit a *stuffed digit* indicated by "stuff 2c-d." The digits of character 3, on the other hand, are delivered so fast that its digit $c$ is not sampled at all. Hence, the unsynchronized digits $a, b, d$ of character 3 become the synchronized digits $A, B, C$. The digit $c$ has to be transmitted as a *spilled digit* on the carrier for the not used character 4. This is indicated by "spill 3c." In addition to the spilled digit, the carrier for character 4 must also transmit from which character the spilled digit came. Similarly, if a digit is stuffed, one must transmit on this carrier information about which character contains in which location a stuffed digit.

Let us assume a spilled digit is not reinserted or a stuffed digit is not eliminated at the receiver. Although all the following digits transmitted by the respective Walsh carrier are detected properly, they will all be in the wrong position. All characters transmitted by that carrier will be decoded incorrectly. This is called a catastrophic error. In a particularly bad case, a digit spilled from a character transmitted by one carrier is reinserted into a character transmitted by another carrier and catastrophic errors occur in both channels.

In mode 2, all four digits of character 1 are transmitted in the first time interval according to Fig. 4b. Similarly, the digits of characters 2 and 3 are transmitted simultaneously in the following intervals. If one character has not yet arrived when its time for transmission has come, the zero character will be transmitted automatically and disregarded automatically at the receiver. In the usual alphabets the zero character consists of spaces only and no marks, or of zeros only. This solves the stuffing problem. If such a zero character is decoded incorrectly, one will receive one wrong character, but there will be no catastrophic accumulation of errors.

For the spilling operation, one must provide a spilling interval as shown in Fig. 4b. The four digits of the spilled character are transmitted in this interval. A second such interval is needed in which the number of the channel is transmitted from which the character was spilled. If the spilled character is decoded incorrectly, there will be no further error. If the number of the channel from which the character was spilled is decoded incorrectly, there will be two errors: one, because the character will be missing from the proper channel; the other, because an additional character appears in a wrong channel. The problem of catastrophic errors is thus avoided by transmitting the digits in parallel and the characters in series.

## C. Mixed Sequency–Frequency Multiplexing

A possible way to combine the strong points of sequency and frequency multiplexing is in the transmission of digital signals through the existing frequency multiplex plant. The multiplexing equipment discussed here is

designed for the standard of transmission of $75 \times 2^n$ digits per second.[*] Two
basic modules are used to build up multiplex systems of considerable
complexity. The first module, called basic sequency multiplexer, accepts
binary input signals. Eight Walsh carriers are used. This implies a time base
of $T = 1/75 \times 2^n \times 8$ or $T = 1/75 \times 2^{n+3}$ sec for the carriers wal$(j, t/T)$. In
the simplest case this multiplexer can be used either for the transmission
through a telephone channel with a nominal bandwidth of 4 kHz or through
a telephone group band with a nominal bandwidth of 48 kHz. The time
bases $T = 1/300$ sec, for low quality, and $T = 1/600$ sec, for high quality
telephone channels, can typically be chosen. The number of digits transmitted per second is then either $300 \times 8 = 2400$ or $600 \times 8 = 4800$. For
transmission through a telephone group band the proper time bases are
either $T = 1/4800$ or $T = 1/9600$ sec. The number of digits transmitted per
second becomes $4800 \times 8 = 38,400$ or $9600 \times 8 = 76,800$. The transition
between the four time bases requires only a change in the clock pulse rate.
The multiplexing circuit itself is not changed.

The basic sequency multiplexer accepts binary signals only. However,
the multiplexed signal can be binary, quarternary, etc. Hence, if the quality
of the channel permits, one may transmit 2, 3, ... bits per digit rather than
one bit per digit.

The second basic module is called the analog sequency multiplexer. It
combines two signals into one. The purpose of this module is to simplify
multiplexing of signals that arrive at different data rates. Since the accepted
standard permits all data rates of $75 \times 2^n$ digits per second, one must be able
to multiplex signals arriving at the rates of 75, 150, 300, 600, ... digits per
second.

Figure 5 shows the block diagram of the basic multiplexer for eight
channels. There are two significantly different modes of operation as mentioned before. One may feed binary digits at the rates of 300, 600, 4800, or
9600 per second—depending on the time base $T$ used—to each one of the
eight channels. This first mode requires that there are eight sources from
which the digits come. Assume now that there are $m$ sources from which
binary characters with eight digits each arrive. One can feed each character
to the eight channels in Fig. 5 and let each channel transmit one digit.
Having $m$ sources, each source may deliver characters at the rates $300/m$,
$600/m$, $4800/m$, or $9600/m$ to the basic sequency multiplexer. This is the
second mode.

Let us observe that the number of eight digits per character used to
explain the second mode was chosen for simplification only. Obviously one

[*] The development of this equipment was supported by the United Nations under
UNESCO Project IND-104 at the Regional Engineering College, Warangal, India.

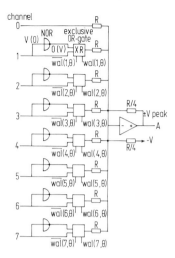

FIG. 5. Basic sequency multiplexer BSM for binary digital signals. Input voltage 0 or $V$.

could combine characters with three and five digits to yield eight digits. More generally, the second mode can be used as long as characters can be combined in such a way that the total number of digits equals or comes close to a multiple of four.

Binary digits $d_0, \ldots d_j, \ldots d_7$ are fed to the input terminals of the eight channels in Fig. 5. The inverters NOR produce the complements $\bar{d}_j$. The digits and their complements are fed to exclusive OR-gates XR. Walsh functions $\mathrm{wal}(j, \theta)$ and their complements $\overline{\mathrm{wal}}(j, \theta)$ are also fed to the exclusive OR-gates. At the output terminals one obtains the products $d_j \, \mathrm{wal}(j, \theta)$. The products are summed by the operational amplifier OA1 and the multiplex signal is obtained at point A.

Since the inverters NOR and the exclusive OR-gates XR operate between 0 and $V$ volts, rather than $-V$ and $+V$, the voltage $-V$ is fed through a resistor of value $R/4$ to the input of the operational amplifier. Its output voltage will then be between $-V$ and $+V$, having a Bernoullian amplitude distribution with mean value zero.

The summing resistors of all eight channels have the same value $R$. This implies that the digits of each channel are transmitted with equal energy. Using different values of the resistors, one may distribute the output signal energy unequally over the channels and thus provide channels with different error rates.

Let the output signal at point A of Fig. 5 be fed to point D of the demultiplexer in Fig. 6. The signal is multiplied with the same Walsh carrier. A typical circuit of the multipliers MU is shown in Fig. 7. If a bipolar FET

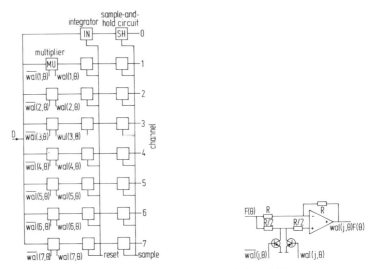

FIG. 6 (*left*). Basic sequency demultiplexer BSD.

FIG. 7 (*right*). Circuit of the multipliers MU in Fig. 6.

transistor is used instead of the pair of *npn* and *pnp* transistors, one needs only the carrier wal($j$, $\theta$) to produce the product wal($j$, $\theta$)$F(\theta)$ at the output terminal of the multiplier.

The products of the input signal with the eight Walsh functions wal(0, $\theta$)–wal(7, $\theta$) are fed to integrators IN and sample-and-hold circuits SH. The transmitted digits are obtained at the output terminals 0 to 7.

Most communication channels are so good that transmission capacity would be wasted by the transmission of binary signals. Sequency multiplexing makes it easy to use signals with 4 levels (quarternary), 8 levels (octonary), 16 levels, etc. To show what "level" means, refer to Fig. 8. Let a Walsh carrier wal($j$, $\theta$) be used to transmit the binary digit $a(j)$ with value $+1$ or $-1$. The amplitude $Va(j)$ of the modulated carrier $Va(j)$wal($j$, $\theta$) will thus be $+V$ for $a(j) = +1$ and $-V$ for $a(j) = -1$. These two amplitude values

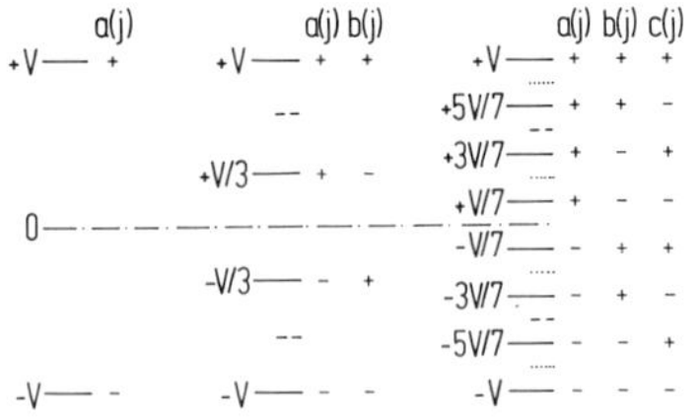

FIG. 8. Voltage levels of the carrier wal($j$, $\theta$) for binary signals $a(j)$, quarternary signals $a(j)$, $b(j)$, and octonary signals $a(j)$, $b(j)$, $c(j)$.

together with the sign of $a(j)$ are shown in Fig. 8 on the left. Detection of $a(j)$ is done by the integral

$$\int_{-1/2}^{1/2} Va(j)\text{wal}(j, \theta)\text{wal}(j, \theta) \, d\theta = (1/T) \int_{-1/2}^{1/2} Va(j)\text{wal}(j, t/T)\text{wal}(j, t/T) \, dt$$

$$= Va(j).$$

Hence, the amplitude $Va(j)$ may also be interpreted as the value of the integral produced by the respective integrator IN in Fig. 6, that is, the voltage which is sampled and held.

Consider now the modulated carrier $(2V/3)a(j)\text{wal}(j, \theta)$ and $(V/3)b(j)\text{wal}(j, \theta)$, with $a(j) = \pm 1$ and $b(j) = \pm 1$. The amplitude $(2V/3)a(j) + (V/3)b(j)$ of the sum can assume the following four values:

$$+(2V/3) + (V/3) = +V,$$
$$+(2V/3) - (V/3) = +V/3,$$
$$-(2V/3) + (V/3) = -V/3,$$
$$-(2V/3) - (V/3) = -V.$$

These four values of the amplitudes or sampled output voltages are shown in Fig. 8 in the middle, together with the signs of $a(j)$ and $b(j)$ which they represent.

Going one step further, let us consider the modulated carriers $(4V/7)a(j)\text{wal}(j, \theta)$, $(2V/7)b(j)\text{wal}(j, \theta)$, and $(V/7)c(j)\text{wal}(j, \theta)$. The amplitude of the sum can assume the following eight values which are shown in Fig. 8 on the right with the signs of $a(j)$, $b(j)$, and $c(j)$:

$$+(4V/7) + (2V/7) + (V/7) = +V,$$
$$+(4V/7) + (2V/7) - (V/7) = +5V/7,$$
$$+(4V/7) - (2V/7) + (V/7) = +3V/7,$$
$$+(4V/7) - (2V/7) - (V/7) = +V/7,$$
$$-(4V/7) + (2V/7) + (V/7) = -V/7,$$
$$-(4V/7) + (2V/7) - (V/7) = -3V/7,$$
$$-(4V/7) - (2V/7) + (V/7) = -5V/7,$$
$$-(4V/7) - (2V/7) - (V/7) = -V.$$

A circuit to produce quarternary signals from binary digits $a(j)$ and $b(j)$ is shown in Fig. 9a. Two basic multiplexers BSM1 and BSM2 as shown in Fig. 5 multiplex binary digits $a(j)$ and $b(j)$ with $j = 0, \ldots, 7$. The output voltage of the multiplexer BSM1 equals

$$(V/4) \sum_{j=0}^{7} a(j)\text{wal}(j, \theta),$$

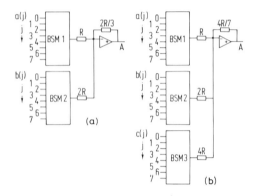

FIG. 9. Extension of the circuit of Fig. 5 for the transmission of quarternary signals (a) and of octonary signals (b). BSM basic sequency multiplexer of Fig. 5.

while the output voltage of the multiplexer BSM2 equals

$$(V/4) \sum_{j=0}^{7} b(j) \text{wal}(j, \theta).$$

The weighted sum of these two voltages at point A in Fig. 9a is given by

$$(V/4) \sum_{j=0}^{7} [(2/3)a(j) + (1/3)b(j)] \text{wal}(j, \theta).$$

Hence, the two binary digits $a(j)$ and $b(j)$ are multiplied by the proper factors 2/3 and 1/3 to be combined into one quarternary amplitude or digit.

The transformation of three binary digits into one octonary digit is shown by Fig. 9b. The output voltage at point A has the following value:

$$(V/4) \sum_{j=0}^{7} [(4/7)a(j) + (2/7)b(j) + (1/7)c(j)] \text{wal}(j, \theta).$$

Let us observe that the circuits of Fig. 9 readily permit the transmission of a mixture of binary, quarternary, and octonary signals. The different error rates of these signals thus provide channels with different quality. For instance, by choosing $a(0) = b(0) = c(0)$, and $c(1) = 0$, but all other digits $a(j)$, $b(j)$, $c(j) = \pm 1$, one may transmit $a(0)$ as a binary digit, $a(1)$ and $b(1)$ as essentially quarternary digits, and all others as octonary digits by means of the circuits of Fig. 9b.

In order to explain the detection of octonary signals, let us assume that the binary digits $a(0) = +1$, $b(0) = -1$, and $c(0) = +1$, have been transmitted. The amplitude of the carrier $\text{wal}(0, \theta)$ then equals

$$(V/4)[(4/7) - (2/7) + (1/7)] = (3/7)(V/8).$$

Choosing the time constant of the integrations in Fig. 6 equal to $T/3$ rather than $T$ thus yields the voltage $3V/7$ at the output terminals of the sample-and-hold circuit SH of channel 0. Let the circuit of Fig. 10 be connected to that terminal. This is a simple analog-to-digital converter. Let the saturation voltage of the three operational amplifiers be $\pm V$. The voltage $5V/7$ will drive the first amplifier CO1 to $-V$, since the resistance of $100R$ in the feedback loop provides one hundredfold amplification. The current flowing to the input of the second amplifier CO2 is thus $(3V/7)/R - V(7R/4) = -V/7R$. The voltage $+100V/7$ would be required at the output of CO2 to compensate this current. The amplifier CO2 will thus saturate at $+V$. The current flowing to the input terminals of the amplifier CO3 becomes $(3V/7)/R - V/(7R/4) + V/7R$. The voltage $-100V/7$ would be required at the output of CO3 to compensate this current, and CO3 will saturate at $-V$. Except for a change in sign the three transmitted binary digits are thus recreated.

For practical use, the circuit of Fig. 10 requires diodes to clamp the output voltages of the operational amplifiers exactly at $\pm V$. Furthermore, biasing is needed to place the threshold for the input voltage at exactly 0 V.

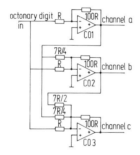

FIG. 10. Circuit for the transformation of one octonary digit into three binary digits.

## D. Combination of Basic Multiplexers

In order to combine basic sequency multiplexers and demultiplexers as shown in Figs. 5 and 6 into more complicated multiplexing systems, one needs two more basic modules. The output voltage of the multiplexer in Fig. 5 is practically a continuously varying voltage. To combine the output voltages of two such multiplexers one needs a circuit that can accept continuously varying voltages. Figure 11 shows such a circuit. Following general use, it is called an analog sequency multiplexer, even though the signals are digital. However, the circuit is capable of multiplexing genuine analog signals.

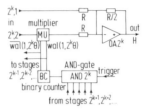

FIG. 11. Analog sequency multiplier ASM for repeated multiplexing of digital signals.

The channel $2^k$, 1 in Fig. 11 is like the channel 0 in Fig. 5. The channel $2^k$, 2 contains a multiplier as shown in Fig. 7 instead of the inverter NOR and the exclusive OR-gate XR of channel 1 in Fig. 5. The Walsh function $\text{wal}(1, 2^k\theta)$ required for driving the multiplier is produced by the binary counter BC. This counter is the stage $2^k$ of a parallel triggered multiple-stage counter. The trigger from a clock passes through the AND-gate AND$2^k$, if the stages $2^k + 1$, $2^k + 2$, ... are in the proper position. In turn, the output voltages $\text{wal}(1, 2^k\theta)$ are available to set the AND-gates of the stages $2^k - 1$, $2^k - 2$.

The analog sequency demultiplexer matching the multiplexer of Fig. 11 is shown in Fig. 12. It is identical with the first two channels of the demultiplexer in Fig. 6, except that the generation of the function $\text{wal}(1, 2^k\theta)$ is

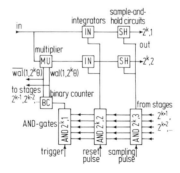

FIG. 12. Analog sequency demultiplexer ASD for repeated demultiplexing of digital signals.

shown. Also shown is a simple method of feeding sampling and reset pulses to the sample-and-hold circuits and the integrators. A pair of such pulses is generated at the highest required rate by a timing circuit. The gates AND$2^k$, 1 to AND$2^k$, 3 are driven like the AND-gate in Fig. 11. Hence, only every second, fourth, eighth, ... sampling or reset pulse passes through.

Figure 13 shows an example of the enormous flexibility one can achieve in a multiplex system by using the basic multiplexer of Fig. 5 and the analog

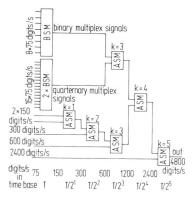

FIG. 13. Example of a sequency multiplex system that accepts digits at the rates $75 \times 2^n$ digits/sec. The digits fed to the units BSM must be binary; the others do not have to be binary.

multiplexer of Fig. 11. Following the standard of $75 \times 2^n$ digits/sec for digital communication channels, we start out with a basic multiplexer BSM that accepts 8 channels with 75 binary digits/sec each. A typical source for this data rate is a teletypewriter. The output of this multiplexer consists of binary multiplex signals.

Two multiplexers BSM being connected as shown in Fig. 9a accept 16 channels with 75 binary digits/sec each. The output is, however, a quarternary multiplex signal.

The total data flow from the 8 and 16 input terminal multiplexers is $(8 \times 75 + 8 \times 75) = 1200$ digits/sec. The two output signals can thus be combined by one analog sequency multiplexer ASM accepting 600 digits at each input terminal. If the time base $T = 1/75$ sec is used for the multiplexers BSM one has to use the time base $T/8 = T/2^3$ for the multiplexer ASM combining their outputs. Hence, $k = 3$ applies for the circuit of Fig. 11.

Figure 13 shows further two input terminals accepting 150 digits per second each, one terminal accepting 300 digits/sec, one for 600 digits/sec, and one for 2400 digits/sec. Since analog multiplexers ASM are used, the digits do not have to be binary. All the blocks denoted ASM have the circuit shown in Fig. 11. The difference is only in timing, expressed by the different values of $k$. The trigger for the stage ASM with $k = 5$ would always pass through the AND-gate in Fig. 11. All the inhibit terminals would be grounded. For the stage ASM with $k = 4$, one inhibit terminal of the AND-gate would be connected to the output of the binary counter of the stage ASM with $k = 5$. Only every second trigger pulse would pass. The outputs of both stages ASM with $k = 5$ and $k = 4$ would be connected to the inhibit terminals of the AND-gates in the two stages ASM with $k = 2$ and $k = 1$.

*E. Matching Sequency Multiplex Signals to Frequency Multiplex Channels*

The signal coming out of the multiplex system in Fig. 13 is a step function with 4800 independent amplitudes per second. This signal has a Gaussian amplitude distribution but it is not frequency band limited. In principle, to make the signal frequency band limited, one has to sample the amplitude of the 4800 steps per second and produce Dirac-like pulses; feeding them through an idealized frequency low-pass filter with a cutoff frequency of $f_g = 2400$ Hz produces a pulse of the shape $(\sin \pi f_g t)/\pi f_g t$ for each sample, and having an amplitude equal to that of the sampled step. At the receiver, the amplitudes of the pulses $(\sin \pi f_g t)/\pi f_g t$ can be detected by sampling, and a hold circuit will produce the original step function.

Two refinements are needed before this principle can be implemented practically. First, the use of an idealized frequency low-pass filter and $(\sin \pi f_g t)/\pi f_g t$ pulses is unrealistic. It is known that so-called symmetric roll-off filters have to be used. The pulses created by these filters from Dirac-like input pulses can be separated by sampling like $(\sin \pi f_g t)/\pi f_g t$ pulses, but the requirement for mathematically exact synchronization is eliminated. Second, one usually does not want a frequency multiplex signal in the band $0 < f < 2400$ Hz but in the band, e.g., $600 < f < 3000$ Hz. Similarly, if sequency multiplexing is used to produce a signal requiring a group band for transmission, one does not want it in the band $0 < f < 48$ kHz but in the band $60 < f < 108$ kHz.

Figure 14 shows on top the required circuits. The sampler SA samples the incoming step function $S_r$ times per second. $S_r$ equals $75 \times 2^5 = 2400$ for a poor telephone channel and $75 \times 2^6 = 4800$ for a high quality telephone

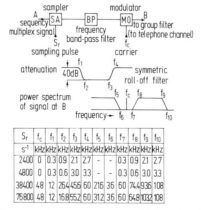

| $S_r$ | $f_c$ | $f_1$ | $f_2$ | $f_3$ | $f_4$ | $f_5$ | $f_6$ | $f_7$ | $f_8$ | $f_9$ | $f_{10}$ |
|---|---|---|---|---|---|---|---|---|---|---|---|
| s⁻¹ | kHz | kHz | kHz | kHz | kHz | kHz | kHz | kHz | kHz | kHz | kHz |
| 2400 | 0 | 0.3 | 0.9 | 2.1 | 2.7 | - | - | 0.3 | 0.9 | 2.1 | 2.7 |
| 4800 | 0 | 0.3 | 0.6 | 3.0 | 3.3 | - | - | 0.3 | 0.6 | 3.0 | 3.3 |
| 38400 | 48 | 12 | 26.4 | 45.6 | 60 | 21.6 | 36 | 60 | 74.4 | 93.6 | 108 |
| 76800 | 48 | 12 | 16.8 | 55.2 | 60 | 31.2 | 36 | 60 | 64.8 | 103.2 | 108 |

Fig. 14. Modem for matching a sequency multiplex signal to a telephone channel ($S_r = 2400, 4800$) or a telephone group channel ($S_r = 38400, 76800$).

channel. For a poor and a high quality group channel, the respective rates are $75 \times 2^9 = 38400$ and $75 \times 2^{10} = 76{,}800$. The sampler is followed by a frequency band-pass filter. The attenuation versus frequency of this filter is shown just below. There should be no attenuation in the band $f_2 < f < f_3$ and at least 40 dB attenuation for $f < f_1$ and $f > f_4$. The transition of the attenuation between $f_1$ and $f_2$ as well as between $f_3$ and $f_4$ should be that of a symmetric roll-off filter.

Values for the frequencies $f_1$ to $f_4$ describing the band-pass filter BP are shown for the sampling rates $S_r = 2400$, 3800, 38400, and 76800 in the table of Fig. 14.

For the sampling rates $S_r = 2400$ and 4800 the signal at the output terminal of the bandpass filter BP can be applied directly to a telephone channel. The modulator MO shown in the block diagram on top of Fig. 14 is not required.

For the sampling rates $S_r = 38{,}400$ and 76,000, one could in principle also do away with the modulator MO. The frequencies $f_1$ to $f_4$ of the band-pass filter BP would have to have the values shown for the frequencies $f_7$ to $f_{10}$ in the table of Fig. 14. For the sampling rate $S_r = 76{,}800$, one would obtain the passband from $f_8 = 64.8$ kHz to $f_9 = 103.2$ kHz. Good criteria for the difficulty of implementing the filter are the ratios $(f_8 - f_7)/(1/2) \times (f_8 + f_7) = 4.8/62.4 = 0.077$ for the lower band limit and the ratio $(f_{10} - f_9)/(1/2)(f_{10} + f_9) = 4.8/105.6 = 0.045$ for the upper limit. These ratios are too small for coil and capacitor type filters. Hence, the signal is first band limited at lower frequencies and then shifted by modulation of a carrier into the group band. For $S_r = 76{,}800$ one obtains from Fig. 14 the ratios $(f_2 - f_1)/(1/2)(f_1 + f_2) = 4.8/14.4 \doteq 0.33$ and

$$(f_4 - f_3)/(1/2)(f_4 + f_3) = 4.8/52.6 \doteq 0.091.$$

This filter is easier to implement. Amplitude modulation of a carrier with frequency $f_c = 48$ kHz in the modulator MO will shift the signal from the band $f_1 < f < f_4 = 12$ kHz $< f < 60$ kHz to the upper sideband $f_7 < f < f_{10} = 60$ kHz $< f < 108$ kHz; and to the lower sideband which is folded over in the band $0 < f < 12$ kHz. A group filter has the nominal passband 60 kHz $< f < 108$ kHz. The upper sideband fits right into it. The lower sideband starts 24 kHz below the nominal lower limit of the group filter and it will be rejected.

The carrier frequency of 48 kHz is a result of several compromises from the standpoint of simplest design for the band-pass filter BP as well as the group filter. The carrier frequency $f_c = 51$ kHz would be more advantageous. However, the frequency $f_c = 51$ kHz is neither available nor readily producible.

218    HENNING F. HARMUTH

At the receiving side the frequency band limited sequency multiplex signal has to be retransformed into a step function before it is fed into sequency demultiplexing equipment. The process is essentially the reverse process of the multiplexing side. The block circuit is shown in Fig. 15. Again, the demultiplexer DM is only required for the sampling rates $S_r = 38{,}400$ and 76,800.

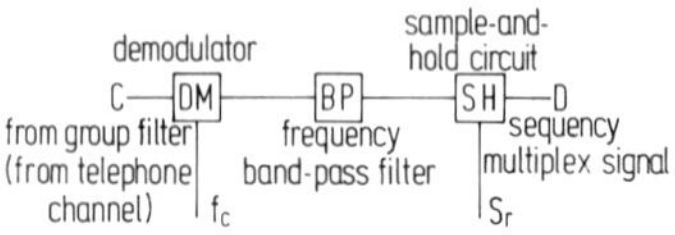

FIG. 15. Modem for matching a frequency band limited sequency multiplex signal to a sequency demultiplexer. The values of $f_c$ and $S_r$ as well as the characteristics of the frequency bandpass filter are shown in Fig. 14.

### III. FILTERS FOR SIGNALS WITH TWO SPACE VARIABLES

#### A. Principle of Spatial Filters

The Walsh–Fourier transform and the inverse transform of a space signal $F(x, y)$ in the interval $-1/2 < x < 1/2$, $-1/2 < y < 1/2$ are defined by

$$a(k, m) = \int_{-1/2}^{1/2} \int_{-1/2}^{1/2} F(x, y)\mathrm{wal}(k, x)\mathrm{wal}(m, y)\, dx\, dy,$$

$$F(x, y) = \sum_{k=0}^{\infty} \sum_{m=0}^{\infty} a(k, m)\mathrm{wal}(k, x)\mathrm{wal}(m, y).$$

A filtered signal $F_0(x, y)$ is obtained by multiplying the transform $a(k, m)$ with attenuation coefficients $K(k, m)$:

$$F_0(x, y) = \sum_{k=0}^{\infty} \sum_{m=0}^{\infty} K(k, m)a(k, m)\mathrm{wal}(k, x)\mathrm{wal}(m, y).$$

No shifts of $\mathrm{wal}(k, x)$ and $\mathrm{wal}(m, y)$ to $\mathrm{wal}[k, x - x(k)]$ and $\mathrm{wal}[m, y - y(m)]$ will be considered.

Figure 16 shows a circuit that generates the Walsh–Fourier transform $a(k,m)$ of a space signal $F(x, y)$. The signal is represented by $4 \times 4$ voltages $a$, $b, \cdots p$ in the $x$, $y$-plane as shown in the lower right corner.

Four printed circuit boards denoted by $\mathrm{wal}(0, x)$ to $\mathrm{wal}(3, x)$ are intersected by $4 \times 4$ wires; only seven of them are shown by dashed lines to avoid obscuring the picture. The voltages $a \cdots p$ are applied to these $4 \times 4$ wires.

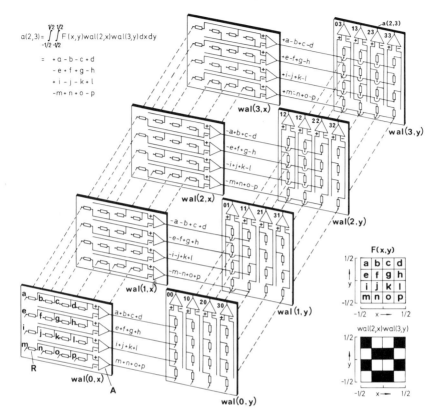

FIG. 16. Principle of a sequency filter for functions of two space variables $x$ and $y$. The circuit shown transforms $F(x, y)$ into $a(k, m)$. A similar circuit is required for the inverse transformation.

Resistors R feed them to the adding $(+)$ or subtracting $(-)$ inputs of summing amplifiers A. At the outputs of the board wal$(0, x)$ occur the four voltages $a + b + c + d, e + f + g + h, i + j + k + l$, and $m + n + o + p$; the topmost output of the board wal$(1, x)$ yields the voltage $-a - b + c + d$, etc. Note that the resistors of each board are connected to the adding or subtracting input of the summing amplifier according to the positive and negative sign of the respective Walsh function: $+ + + +$ represents wal$(0, x)$, $- - + +$ represents wal$(1, x)$, $- + + -$ represents wal$(2, x)$, and $+ - + -$ represents wal$(3, x)$.

A second set of four printed circuit cards is denoted by wal$(0, y)$ to wal$(3, y)$. These cards have the same circuitry as the cards wal$(0, x)$ to wal$(3, x)$ but are rotated $90°$. Again, only seven of the sixteen intersecting wires are shown by dashed lines. The output voltages of the cards wal$(0, x)$

to wal(3, x) are fed to these sixteen intersecting wires as shown. One may readily infer that the output voltage of the summing amplifier denoted by 00 is the sum

$$(a + b + c + d) + (e + f + g + h) + (i + j + k + l) + (m + n + o + p)$$

and thus represents the Walsh–Fourier coefficient $a(0, 0)$ except for an unimportant scaling factor. Generally, a summing amplifier denoted by $km$ yields the Walsh–Fourier coefficient $a(k, m)$. For instance, $km = 23$ yields

$$-(-a + b + c - d) + (-e + f + g - h)$$
$$- (-i + j + k - l) + (-m + n + o - p) = a(2, 3).$$

This sum is written in a square in the upper left corner of Fig. 16. The positive and negative signs of the voltages $a \cdots p$ correspond to the values $+1$ (=black) and $-1$ (=white) of the Walsh function wal(2, x)wal(3, y) shown in the lower right corner.

The "half filter" of Fig. 16 produces the Walsh–Fourier transform $a(k, m)$ of the signal $F(x, y)$. An essentially equal circuit will do the inverse transformation and reproduce $F(x, y)$ from the transform $a(k, m)$. A filtering effect is obtained by modifying the voltages $a(k, m)$. Before we can enter into a discussion of this filtering we must develop a suitable and simple representation of the processes to be studied. When investigating the filtering of time signals one may plot the signal as a function of time, its Fourier transform as a function of frequency, the attenuation and phase shift of the filter as a function of frequency, the Fourier transform of the filtered signal as a function of frequency, and finally the filtered signal as a function of time. All these plots can be done in a two-dimensional cartesian coordinate system.

The very same plots can be drawn for space signals $F(x, y)$ but one needs three-dimensional cartesian systems. Some simplification is required for their representation on two-dimensional paper. The first step is to replace the continuous function $F(x, y)$ by a sampled function $F(h, i)$. While $F(x, y)$ is defined for any value of $x$ and $y$ in an interval $-1/2 \le x < 1/2$, $-1/2 \le y < 1/2$, the function $F(h, i)$ is defined for nonnegative integer values of $h$ and $i$ only. Such a sample function is shown in Fig. 17. The Walsh–Fourier transform $a(k, m)$ of $F(h, i)$ is represented by Fig. 18.

Consider now the representation of the attenuation function $K(k, m)$ of the filter. Figure 19 shows such a function having the value $+1$ for $k + m \le 14$ and 0 for $k + m > 14$. It is shown in a further simplified form in Fig. 20. Instead of plotting the values of $K(k, m)$ we only indicate by black dots where these values should be written. The line $k + m = 14$ is then drawn. The attenuation function of Fig. 19 may be readily visualized by imagining the value $+1$ at all those dots for which $k + m$ is smaller than or equal to 14 and the value 0 for all other values of $k + m$.

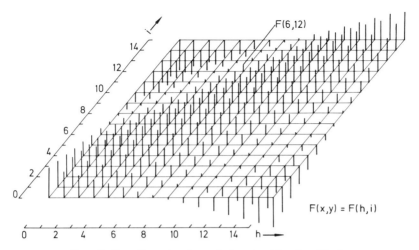

FIG. 17. Two-dimensional sample function $F(x, y) = F(h, i)$.

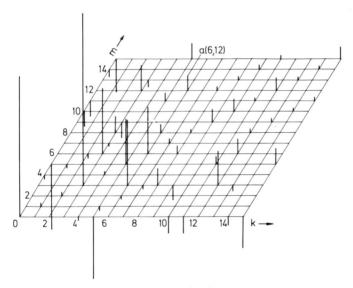

FIG. 18. Walsh–Fourier transform $a(k, m)$ of the function of Fig. 17.

A filter defined by the attenuation function $K(k, m) = 1$ for $k + m \leq 14$ and $K(k, m) = 0$ for $k + m > 14$ is referred to as a low-pass filter. The all inclusive name would be "two-dimensional spatial attenuation low-pass filter." One can build filters for signals with up to three space variables and the time variable, hence the term "two-dimensional spatial." The term

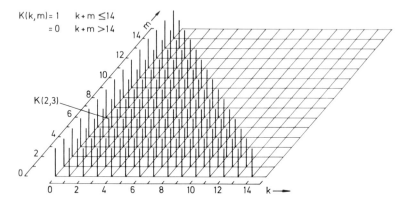

FIG. 19. Attenuation function $K(k, m)$ for a two-dimensional sequency low-pass filter.

"attenuation" is justified by the fact that one not only may have an attenuation function $K(k, m)$, but also a *resolution function* $R(k, m)$ which leads to *resolution low-pass filters*. Indeed, it is the resolution filter which is mainly of interest for TV picture transmission at the present time.

The dashed line $k^2 + m^2 = 14^2$ in Fig. 20 defines another low-pass filter: $K(k, m)$ equals 1 for $k^2 + m^2 \leq 14^2$ and 0 for $k^2 + m^2 > 14^2$. A third low-pass filter is defined by the condition $K(k, m) = 1$ for $k \leq 14$, $m \leq 14$ and $K(k, m) = 0$ for $k > 14$, $m > 14$.

The low-pass filters discussed so far let certain parts of the Walsh–Fourier transform $a(k, m)$ pass unchanged and suppressed others completely. The representation of Fig. 20 is, however, capable of handling more

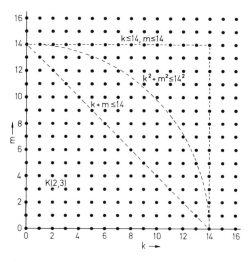

FIG. 20. The two-dimensional sequency domain with three band limits for low-pass filters.

complicated cases. Consider, for example, the attenuation function defined as follows:

$$K(k, m) = 1 \text{ for } k + m \leq 14,$$
$$K(k, m) = 1/2 \text{ for } 14 < k + m, k^2 + m^2 \leq 14^2,$$
$$K(k, m) = 1/4 \text{ for } 14^2 \leq k^2 + m^2, k \leq 14, m \leq 14,$$
$$K(k, m) = 0 \text{ for } 14 < k, 14 < m.$$

The part of the transform $a(k, m)$ between the limits $k = 0$, $m = 0$, and $k + m = 14$ in Fig. 20 will pass this filter unchanged; the part between the limits $k + m = 14$ and $k^2 + m^2 = 14^2$ will be attenuated to half its amplitude, etc.

Let us now turn from attenuation filters to resolution filters. The output voltages $a(k, m)$ of the 16 operational amplifiers in Fig. 16 denoted by 00 to 33 shall be fed to 16 analog/digital converters. Let some of the voltages be converted into binary numbers with many digits and the other voltages into binary numbers with a few digits only. The voltages $a(k, m)$ are thus represented with different resolution and the resolution is a function of $k$ and $m$.

Let us define a resolution function $R(k, m)$. The notation $R(k, m) = 5$ shall mean that the voltage $a(k, m)$ is represented by 5 binary digits. We can use Fig. 20 to define a resolution filter. For instance, a simple resolution low-pass filter is defined by:

$$R(k, m) = 5 \text{ for } k + m \leq 14,$$
$$R(k, m) = 0 \text{ for } k + m > 14.$$

A more sophisticated resolution low-pass filter is obtained by the following definition:

$$R(k, m) = 5 \text{ for } k + m \leq 14,$$
$$R(k, m) = 4 \text{ for } 14 < k + m, k^2 + m^2 \leq 14^2,$$
$$R(k, m) = 3 \text{ for } 14^2 < k^2 + m^2, k \leq 14, m \leq 14,$$
$$R(k, m) = 0 \text{ for } 14 < k, 14 < m.$$

In order to apply these results to TV pictures, let $i$ and $h$ in Fig. 17 assume the values 0 to 511 rather than 0 to 15. This corresponds to a TV picture with 512 lines and $512^2$ picture elements. $k$ and $m$ in Fig. 18 must also assume the values from 0 to 511. The implementation of a filter with $512^2$ inputs would be a formidable task if done according to Fig. 16. There are two methods to simplify such a filter. First, one can decompose the square of $512^2$ picture elements into smaller squares. For instance, a good choice is $32^2$ squares with $16^2$ picture elements each. A filter with $16^2$ inputs is then needed which transforms sequentially the $32^2$ square picture sections. Second, the circuit of Fig. 16 can be brought into a more practical form.

## B. Practical Implementation of Filters for Signals with Two Space Variables and Their Use for PCM Television Transmission

The two-dimensional Walsh–Fourier transform of a spatial sampled signal with $2^n \times 2^n$ samples,

$$a(k, m) = \sum_{h=0}^{2^n-1} \sum_{i=0}^{2^n-1} F(h, i)\mathrm{wal}(k, h/2^n)\mathrm{wal}(m, i/2^n),$$

can be performed in two steps:

$$a(k \mid i) = \sum_{h=0}^{2^n-1} F(h, i)\mathrm{wal}(k, h/2^n),$$

$$a(k, m) = \sum_{i=0}^{2^n-1} a(k \mid i)\mathrm{wal}(m, i/2^n).$$

Consider the case $2^n = 16$ and let the voltages $F(h, i)$ for a fixed value of $i$ and $h = 0 \cdots 15$ be fed to the input terminals $0, i \cdots 15, i$ of the circuit in

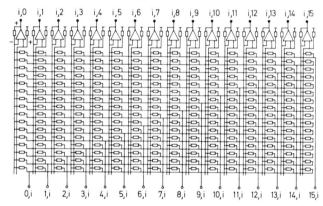

FIG. 21. Circuit for the Walsh–Fourier transform of a signal with one space variable. All resistors have the same value.

Fig.21. The voltages $-a(0 \mid i) \cdots -a(15 \mid i)$ are obtained at the output terminals $i, 0 \cdots i, 15$. For instance, the voltage $a(0 \mid i)$ at the output terminal $i, 0$ is the negative sum of all input voltages:

$$a(0 \mid i) = - \sum_{h=0}^{15} F(h, i) = - \sum_{h=0}^{15} F(h, i)\mathrm{wal}(0, h/16).$$

The voltage at output terminal $i, 1$ equals:

$$a(1 \mid i) = - \sum_{h=0}^{7} F(h, i) + \sum_{h=8}^{15} F(h, i) = - \sum_{h=0}^{15} F(h, i)\mathrm{wal}(1, h/16).$$

The second summation transforms $a(k \mid i)$ into $a(k, m)$. This process is the same as before except that the summation is now for $i = 0 \cdots 15$ rather than for $h = 0 \cdots 15$. Hence, cards with the circuit of Fig. 21 can again be used, but they must be positioned vertically as shown in Fig. 22.

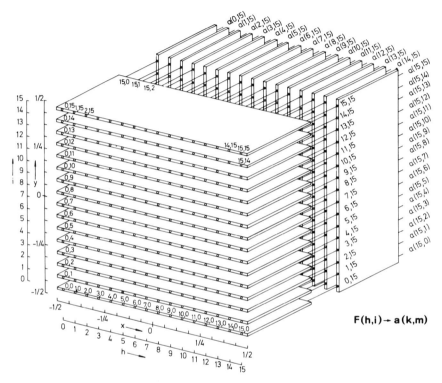

FIG. 22. Practical version of a sequency half-filter for functions of two space variables $x$ and $y$ or $h$ and $i$.

The circuit of Fig. 22 is much simpler than the one of Fig. 16 since all cards are equal and the connection between the horizontal and vertical cards can be made by a cable tree rather than by wires that have to be fiddled through the cards. Note that the operation of this filter does not depend on the use of Walsh functions. The circuit of Fig. 21 can be implemented, for example, for sinusoidal functions.

A major application of two-dimensional spatial filters is in television. Figure 23 shows the principle of television filtering. The image is scanned by an iconoscope. The resulting decomposition into block pulses must be undone by means of a matrix switch. In theory, the matrix switch should have one input terminal and, for example, $512 \times 512$ output terminals. This

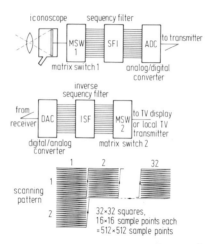

FIG. 23. Block diagram of a PCM television system using two-dimensional sequency filters with a usual block pulse scanner (iconoscope).

would lead to a very expensive filter with $512 \times 512$ input and output terminals. Experience from computer simulation has shown that little is to be gained by using a filter with more than $16 \times 16$ input terminals. Such a filter can be used if the TV image is not scanned line by line but in the checkerboard pattern shown in the bottom of Fig. 23. The $32 \times 32$ squares with $16 \times 16$ sample points each yield the total image with $512 \times 512$ sample points.

The sequency filter SFI at the top of Fig. 23 produces a Walsh, Fourier, or Karhunen–Loeve transform of the $16 \times 16$ voltage samples delivered by the matrix switch to the proper input terminal. An analog/digital converter produces PCM signals from the analog output signals of the filter. A filtering effect is obtained by converting only some of the voltages at the $16 \times 16$ output terminals into digital characters with equal information. The output voltages of certain terminals are converted into characters with 6 bit information, those at other terminals into characters with only 5, 4, ..., 0 bit information. The average information per sample is between 1/4 and 1/8 of the information per sample required without the transformation by the sequency filters. This translates into an increase in the number of possible TV channels by a factor 4 to 8.

The digital signals from the analog/digital converter ADC in Fig. 23 are transmitted. At the receiving side they are converted back to analog signals in the digital/analog converter DAC, the inverse transform is performed by the inverse sequency filter, and the matrix switch 2 changes the voltages from parallel to serial.

Let us see where equipment according to Fig. 23 would be economically feasible. It calls for large-scale PCM transmission with fairly expensive terminals. The emerging national satellite communication systems satisfy these requirements. They will carry television signals and they must be in PCM. Reception will not be done by the home TV receiver but by a local TV station, which permits the use of much more expensive equipment. At this time Canada has the first national satellite system. The United States, the Soviet Union, and India are working on such systems. The most advanced equipment, according to Fig. 23, exists in Japan (Shibata, 1969a, 1970a,b, 1971, 1972), but work has also been reported from West Germany (Kraus, 1972) and England (Clark and Walker, 1973).

The equipment in Fig. 23 is rather complicated but one may readily see that the primary cause of complication is the iconoscope which decomposes the image according to block pulses while the filter calls for Walsh, sine-cosine, or similarly more useful functions. If one had a scanning device based on these more complicated functions, one could do away with the matrix switch and the sequency filter. At the present we have no practical way to scan according to sine-cosine functions, but we can do so according to Walsh functions. The first such scanner has already been built in Japan (Inokuchi, 1972).

Figure 24 shows an example of the quality of PCM television images if a resolution filter is used and an average of 2.5 bits per sample are transmitted. Color images of comparable quality have been transmitted with 3.75 bits per sample. Unfortunately, color pictures cannot be reproduced here.

FIG. 24. Original analog TV signal (right) and PCM signal reduced to 2.5 bits per sample by a one-dimensional resolution filter with eight input terminals (left). (Courtesy K. Shibata of Kokusai Denshin Denwa Co., Tokyo.)

## C. Image Generation by Means of Two-Dimensional Filters

An object illuminated by waves produces scattered waves. The wavefront of these scattered waves at the object differs by a linear transformation from the wavefront received at some distance from the object. The inverse transformation recreates the wavefront at the object and thus yields an image of it. Electromagnetic waves are generally used to produce images, but one may use other waves, for instance acoustic waves.

There are presently three generally known principles for the generation of images by means of electromagnetic or acoustic waves. The classical method uses lenses. The echo principle, used in radar and sonar, is a second method. Holography is the third and youngest method for generating images. Holography requires a specific time variation of the wave. This time variation is always sinusoidal in practical cases, but there is no theoretical requirement that this must be so. Lens and echo principle do not require any particular time variation of the wave. The working of a lens can be discussed in terms of geometric optics while the echo principle can be explained with short pulses of any shape. Only the more refined uses of these two principles, such as color correction of lenses or the use of the Doppler effect, require a specification of the time variation of the wave.

A fourth method was provided by the two-dimensional electric filters. In their most general form, these filters do not require a specified time variation of the wave. However, a considerable simplification can be achieved by specifying a certain time variation. It will be assumed here that the time variation is sinusoidal, since the waves required for illumination of the object are readily available in sinusoidal form. Over the long run, one should not restrict oneself to sinusoidal waves, since different wave shapes provide different effects and some of these effects might be useful.

In order to define areas of application for image generation by two-dimensional electric filters, let us observe that these filters can be built at present without undue effort for transient times as low as 1 $\mu$sec or, in terms of sinusoidal functions, for frequencies up to 1 MHz. The filters are thus applicable in the whole range of useful acoustic waves. A typical application is the generation of images by sound waves of objects under water or even covered by mud on the sea bottom. Another application might be in observing from a helicopter the ground covered by the dense foliage of a jungle. Although this foliage is an almost perfect shield against observation by electromagnetic waves, it is readily penetrated by sound waves. In the field of geology one could produce images of structures deep underground by means of seismic waves. Electromagnetic waves have so far not yielded any application. The typical frequencies of radar are three orders of magnitude above the frequency range of two-dimensional filters; even over-the-horizon radar uses frequencies one order of magnitude too high. However, it is conceivable that the waves produced by so-called extra low frequency transmitters built for other purposes might be used to obtain images of geological structures many kilometers below ground.

The echo principle in the form of sonar is presently the most widely used one for generating images by means of sound waves. There are two known points of practical interest in which sonar imaging and sequency filter imaging differ. First, the obtainable signal-to-noise ratio is coupled with the resolution for sonar. There is no such coupling for sequency filters, and the signal-to-noise ratio can be increased arbitrarily by time filtering without reducing the resolution. Second, the theoretical limit of the resolution of sequency filter imaging is not the wavelength-to-aperture ratio but the signal-to-noise ratio. No relative velocity is required, as in synthetic aperture radar, to achieve this super-resolution.

For introduction to image generation by means of electrical filters, let us first discuss the principle of the lens. Figure 25 shows an optical or acoustic lens that receives parallel rays from three points P1, P2, and P3 at infinity. The lens combines the rays at the three points P1, P2, and P3 in the focal plane. The propagation time of a "ray" from a point at infinity to its image point in the focal plane is the same regardless of where the ray strikes the surface of the lens. The ray from point P1 at infinity requires more time to reach the point 1 on the surface of the lens than the points 2, 3, ..., 7. However, the propagation time from the surface of the lens to point P1 in the focal plane is shortest for the ray striking the lens at point 1 and becomes increasingly longer for rays striking at the points 2, 3, ..., 7. The lens together with the space between the lens and the focal plane act like many delay lines. There is a delay line for every point on the surface of the lens, and every angle of incidence. The heavy lines inside the lens in Fig. 25 show three "delay lines" originating at each one of the seven points on the surface of the lens.

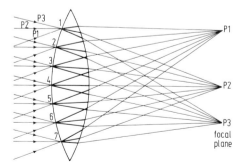

FIG. 25. Principle of the image generation by a lens.

Let us translate the action of the lens into electrical circuits. In order to remain practical, we will use sound waves rather than light waves for illustration. The seven points on the surface of the lens in Fig. 25 are replaced by seven microphones in Fig. 26 which transform the acoustic oscillations into electric oscillations. Let us assume that the electric oscillations travel through the delay circuits shown with the same velocity as the acoustic oscillations traveled before reaching the microphones; the other parts of the electric circuit produce negligible delays. The traveling time of a wave from point P1 at infinity, first as acoustic oscillation, then as electric oscillation, to the output terminal P1 of the summing amplifier, will thus be the same, regardless of which microphone received the sound wave.

The dashed lines in Fig. 25 are perpendicular to the line of propagation of the respective sound waves. One may readily see how these lines are used to determine the "length" of the delay circuits; the delay produced by these circuits is proportionate to their shown lengths.

Figure 26 shows seven microphones but only three points P1, P2, and P3 that are resolved. One will expect that an array of seven microphones can resolve seven points without ambiguity. This is shown in Fig. 27. There are again seven microphones, but delay circuits with taps are connected to the microphones instead of a multitude of delay circuits. The location of the taps is determined by lines perpendicular to the lines of propagation from the points P3, ..., P3. The resistors and summing amplifiers shown in Fig. 26 have been omitted in Fig. 27.

The image produced by the lens in Fig. 25 is side reversed since the sequence of points P1, P2, and P3 in the focal plane is the reversed sequence of the points at infinity. The same holds true for Fig. 26. No such reversal occurs in Fig. 27. It is quite obvious that the location of the output terminals

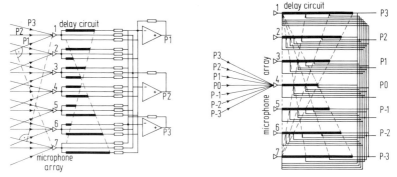

FIG. 26 (*left*). Electric equivalent of an acoustic lens according to Fig. 25, using microphones, electric delay circuits, and summing amplifiers.

FIG. 27 (*right*). Improved circuit of Fig. 26 yielding seven resolved points.

P1 to P3 in Fig. 26 or P–3 to P3 in Fig. 27 can be chosen at will. This possibility not only avoids side reversal, but also provides a means to eliminate the well known distortions of the simple lens for wide viewing angles.

The imaging process in Fig. 25 is explained for one dimension only but it is readily extended to two dimensions by rotating the cut of the lens around the line from point 4 on the lens surface to P2 in the focal plane. How does one accomplish the extension to two dimensions for Figs. 26 and 27? Let us assume the circuit of Fig. 27 is mounted on a printed circuit card, without the microphones, but with the resistors and summing amplifiers which were left out in Fig. 27. The transition to two dimensions is accomplished by stacking seven cards horizontally and seven cards vertically as shown in Fig. 28. The input terminals of the cards are represented by the terminals 1 to 7 in Fig. 27, the output terminals by the points P–3 to P3. The output terminals of the horizontally stacked cards in Fig. 28 are connected to the input terminals of the vertically stacked cards as indicated by the four dashed lines; the other connections are not shown, in order to avoid obscuring the picture.

A quadratic array of 7 × 7 microphones is connected to the input terminals of the horizontal stack in Fig. 28. The received acoustic waves are transformed into an image of electric oscillations at the output terminals of the vertical stack. The next task is to transform this electric image into a visible optical image. The oscillations must first be transformed into voltages proportionate to the power of the oscillations. This can be done by a quadratic rectifier. The resulting voltages can then be sampled and used to control the electron beam of a TV display. Any other two-dimensional optical display, in particular light emitting diodes, can also be used.

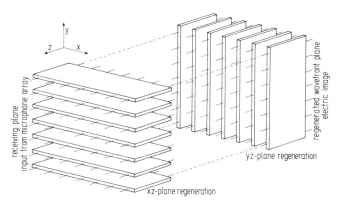

FIG. 28. Two-dimensional filter for spatial signals. The 7 × 7 voltages at the input terminals of the receiving plane are transformed instantaneously and linearly into 7 × 7 voltages at the output terminals of the regenerated wavefront plane.

An acoustic imaging system based on two-dimensional filters consists thus of a microphone array that converts acoustic signals into electric signals, a two-dimensional filter that transforms the electric signals, and a display that converts the transformed electric signals into visible optical images. The microphone array and the display are known items which will not be discussed any further.

One problem is immediately apparent from Fig. 27. The delay circuits are difficult to implement for acoustic waves. Hence, our objective will be to eliminate them.

### D. Image Generation without Delay Circuits

Two-dimensional filters for image generation can be built without delay circuits if one specifies the waveform. Since generators for sinusoidal waves in water or in air are readily available, we will use such waves.

To derive some geometric relations, consider the one-dimensional microphone array of Fig. 29. Nine microphones, or generally $2n + 1$ microphones, are spaced a distance $d$ apart. Let the directional characteristic of the microphones have the angle $\alpha$; ideally, a signal arriving within this angle will produce the same output voltage regardless of the actual angle of incidence

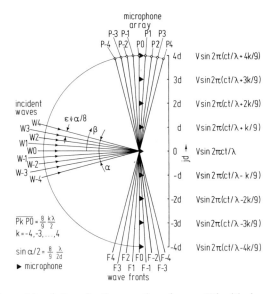

FIG. 29. Geometric relations for the reception of waves $Wk$ with sinusoidal time variation by a one-dimensional array of microphones. $\alpha$ is the reception angle of the microphones, $\lambda$ the wavelength of the sinusoidal wave used for illumination, and $d$ the distance between the microphones.

$\beta$, while a signal arriving from outside this angle will produce no output voltage. The angle $\alpha$ is subdivided into 8, or generally $2n$, angles for the nine incident waves denoted W–4 to W4 in Fig. 29. The respective wavefronts are denoted F–4 to F4. Using the sinusoidal wave $\sin 2\pi ct/\lambda$ for illumination, one obtains from Fig. 29 the relation between the reception angle $\alpha$ of the microphones, the wavelength $\lambda$, and the distance $d$ between the microphones:

$$\sin(\alpha/2) = (8/9)(\lambda/2d).$$

For $2n + 1$ instead of 9 microphones one obtains

$$\sin(\alpha/2) = [2n/(2n + 1)](\lambda/2d). \tag{1}$$

If $\alpha$ is small and the number of microphones large, one may use the simpler formula $\alpha \doteq \lambda/d$.

Figure 30 shows the geometric relations in the object plane. The reception angle $\alpha$ defines the observable width or the linear field of view $D$ in the object plane at a distance $L$:

$$D = 2L \operatorname{tg} \alpha/2.$$

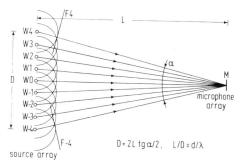

FIG. 30. Geometric relations for the object plane showing the emission of waves $Wk$ from an array of acoustic sources.

Let us return to Fig. 29. All waves $Wk$ produce the output voltage $V \sin 2\pi ct/\lambda$ at the microphone in the center of the array ($j = 0$). The same wave will arrive with a delay or advance $t' = (jd \sin \beta)/c$ at a microphone located at $jd$ if the angle of incidence is $\beta$. The output voltage $V \sin 2\pi[(ct/\lambda) + (jd \sin \beta)/\lambda]$ is produced. Using Eq. (1), one obtains

$$(jd \sin \beta)/\lambda = [j/(2n + 1)]n \sin \beta/\sin \alpha/2$$

Introduction of the resolution angle $\varepsilon$,

$$\sin \varepsilon = (1/n)\sin \alpha/2,$$

yields

$$(dj \sin \beta)/\lambda = [j/(2n + 1)]\sin \beta/\sin \varepsilon.$$

For $\sin \beta/\sin \varepsilon = k$ and $2n + 1 = 9$, one obtains the voltages $V \sin 2\pi \times (ct/\lambda + jk/9)$ written in Fig. 29. One could connect a delay line with several taps to each microphone as in Fig. 27. To obtain a circuit that does not require delay lines, we write $\sin 2\pi[(ct/\lambda) + jk/(2n + 1)]$ as follows:

$$\sin 2\pi[(ct/\lambda) + jk/(2n + 1)] = \sin 2\pi ct/\lambda \cdot \cos 2\pi jk/(2n + 1)$$
$$+ \cos 2\pi ct/\lambda \cdot \sin 2\pi jk/(2n + 1). \quad (2)$$

Let us substitute $\theta$ for $j/(2n + 1)$ and think of $\theta$ as a continuous variable. The purpose is to be able to use the continuous Fourier transform for explanation; later we will return to the discrete Fourier transform. Equation (2) is multiplied by $2 \sin 2\pi i\theta$, $2 \cos 2\pi i\theta$, or $2^{1/2}$ and integrated. The integration interval is $-1/2 < \theta < 1/2$ since $j/(2n + 1)$ runs from $-n/(2n + 1)$ to $+n/(2n + 1)$. The parameter $i$ assumes the values $1, 2, \ldots, n$. The angle of incidence $\beta$ shall only assume values such that $\sin \beta/\sin \varepsilon = k$ is an integer; $k = 0, \pm 1, \pm 2, \ldots$. One obtains the following three integrals:

$$2V \int_{-1/2}^{1/2} (\sin 2\pi ct/\lambda \cdot \cos 2\pi k\theta + \cos 2\pi ct/\lambda \cdot \sin 2\pi k\theta)\sin 2\pi i\theta \, d\theta$$
$$= \delta_{ik}(k/|k|)V \cos 2\pi ct/\lambda, \quad (3)$$

$$2V \int_{-1/2}^{1/2} (\sin 2\pi ct/\lambda \cdot \cos 2\pi k\theta + \cos 2\pi ct/\lambda \cdot \sin 2\pi k\theta)\cos 2\pi i\theta \, d\theta$$
$$= \delta_{ik} V \sin 2\pi ct/\lambda, \quad (4)$$

$$\sqrt{2} V \int_{-1/2}^{1/2} (\sin 2\pi ct/\lambda \cdot \cos 2\pi k\theta + \cos 2\pi ct/\lambda \cdot \sin 2\pi k\theta) \, d\theta$$
$$= \delta_{0k} V \sin 2\pi ct/\lambda. \quad (5)$$

Table I shows the values of the integrals for $k = -4, -3, \ldots, +4$. The rows show the Kronecker symbol according to Eqs. (3)–(5), and in addition "sin" for Eq. (3) and "cos" for Eq. (4). The wave W0 produces only one output voltage, while all the others produce two output voltages. Furthermore, the output voltages caused by waves W$k$ and W$-k$ are equal except for the sign change in the integrals denoted "sin." This difference of sign must be used to distinguish between waves coming from the upper and lower half of the object plane in Fig. 29.

TABLE I

VALUES OF THE INTEGRALS OF EQS. (3)–(5) FOR $k = -4, -3, \ldots, +4$

| $k \rightarrow$ | $-4$ | $-3$ | $-2$ | $-1$ | $0$ | $1$ | $2$ | $3$ | $4$ |
|---|---|---|---|---|---|---|---|---|---|
| | $W-4$ | $W-3$ | $W-$ | $W-1$ | $W0$ | $W1$ | $W2$ | $W3$ | $W4$ |
| $\delta_{0k}$ | 0 | 0 | 0 | 0 | $+V \sin 2\pi ct/\lambda$ | 0 | 0 | 0 | 0 |
| $\delta_{1k}$, sin | 0 | 0 | 0 | $-V \cos 2\pi ct/\lambda$ | 0 | $+V \cos 2\pi ct/\lambda$ | 0 | 0 | 0 |
| $\delta_{1k}$, cos | 0 | 0 | 0 | $+V \sin 2\pi ct/\lambda$ | 0 | $+V \sin 2\pi ct/\lambda$ | 0 | 0 | 0 |
| $\delta_{2k}$, sin | 0 | 0 | $-V \cos 2\pi ct/\lambda$ | 0 | 0 | 0 | $+V \cos 2\pi ct/\lambda$ | 0 | 0 |
| $\delta_{2k}$, cos | 0 | 0 | $+V \sin 2\pi ct/\lambda$ | 0 | 0 | 0 | $+V \sin 2\pi ct/\lambda$ | 0 | 0 |
| $\delta_{3k}$, sin | 0 | $-V \cos 2\pi ct/\lambda$ | 0 | 0 | 0 | 0 | 0 | $+V \cos 2\pi ct/\lambda$ | 0 |
| $\delta_{3k}$, cos | 0 | $+V \sin 2\pi ct/\lambda$ | 0 | 0 | 0 | 0 | 0 | $+V \sin 2\pi ct/\lambda$ | 0 |
| $\delta_{4k}$, sin | $-V \cos 2\pi ct/\lambda$ | 0 | 0 | 0 | 0 | 0 | 0 | 0 | $+V \cos 2\pi ct/\lambda$ |
| $\delta_{4k}$, cos | $+V \sin 2\pi ct/\lambda$ | 0 | 0 | 0 | 0 | 0 | 0 | 0 | $+V \sin 2\pi ct/\lambda$ |

Let us now turn from the continuous Fourier transform of Eqs. (3)–(5) to the discrete transform. Figure 31 shows the continuous functions for $\theta \geq 0$ used in these equations and the samples to be taken at $\theta = j/9 = 0$, $\pm 1/9$, $\pm 2/9$, $\pm 3/9$, $\pm 4/9$ for the discrete transform. A practical circuit for the discrete transformation is shown in Fig. 32. The microphones at the points $-4d$, $-3d$, ..., $4d$ in Fig. 29 are connected to the input terminals $-4d$, $-3d$,

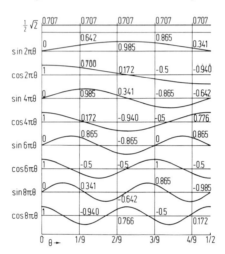

FIG. 31. Representation of the first nine functions of the Fourier series by sampled values.

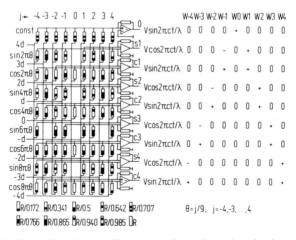

FIG. 32. Circuit for a discrete Fourier transformation using nine input samples at a time. The values of the resistors correspond to the sampled values in Fig. 31. The output voltages caused by the waves $W-4$ to $W4$ of Fig. 29 are shown on the right.

..., 4d of Fig. 32. The resistors and operational amplifiers perform multiplications and summations that represent the discrete transform corresponding to Eqs. (3)–(5). For instance, all voltages fed to the topmost operational amplifier are multiplied by 0.707 and summed. The factors 0.707 represent the constant shown on top of Fig. 31; the multiplication by this function and integration represents Eq. (5).

Figure 33 shows a circuit that can be connected to the output terminals of the circuit of Fig. 32 in order to distinguish between waves from the upper and lower half-plane in Fig. 29. The circuit transforms the cosinusoidal oscillations at the terminals sj by means of an integrator into sinusoidal

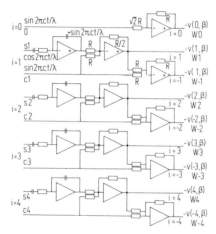

FIG. 33. Circuit to be connected to the output terminals of the circuit of Fig. 32 to distinguish between waves from the upper and lower half-plane in Fig. 29. This circuit works for a continuously variable angle of incidence $\beta$ but requires a fixed frequency of the wave.

oscillations. The sum of such a sinusoidal voltage from terminal sj with the one from terminal cj produces an output voltage at the terminal Wj, while the difference produces an output voltage at the terminal W-j. One may readily see from Fig. 29 that a wavefront Wj comes from the upper half-plane, and a wave W-j from the lower half-plane.

### E. Continuously Varying Angle of Incidence

It was assumed in the previous section that the incident waves $W-4$ to $W4$ in Fig. 29 came from certain discrete directions for which the ratio $\sin \beta/\sin \varepsilon = k$ was an integer. We will now permit this ratio to vary continuously. The integrals in Eqs. (3) to (5) then yield the following values:

$$2V \int_{-1/2}^{1/2} [\sin 2\pi ct/\lambda \cdot \cos 2\pi k\theta + \cos 2\pi ct/\lambda \cdot \sin 2\pi k\theta]\sin 2\pi i\theta \, d\theta$$

$$= V[\sin \pi(i - k)/\pi(i - k) - \sin \pi(i + k)/\pi(i + k)]\cos(2\pi ct/\lambda), \qquad (6)$$

$$2V \int_{-1/2}^{1/2} [\sin 2\pi ct/\lambda \cdot \cos 2\pi k\theta + \cos 2\pi ct/\lambda \cdot \sin 2\pi k\theta]\cos 2\pi i\theta \, d\theta$$

$$= V[\sin \pi(i - k)/\pi(i - k) + \sin \pi(i + k)/\pi(i + k)]\sin(2\pi ct/\lambda), \qquad (7)$$

$$2^{1/2}V \int_{-1/2}^{1/2} [\sin 2\pi ct/\lambda \cdot \cos 2\pi k\theta + \cos 2\pi ct/\lambda \cdot \sin 2\pi k\theta] \, d\theta$$

$$= 2^{1/2}V[(\sin \pi k)/\pi k]\sin(2\pi ct/\lambda). \qquad (8)$$

Let us take the indefinite integral of Eq. (6) and reverse the sign. The term $\cos 2\pi ct/\lambda$ is changed into $\sin 2\pi ct/\lambda$:

$$-(\lambda/2\pi c)V[\sin \pi(i - k)/\pi(i - k) - \sin \pi(i + k)/\pi(i + k)] \int \cos(2\pi ct/\lambda) \, dt$$

$$= V[\sin \pi(i - k)/\pi(i - k) - \sin \pi(i + k)/\pi(i + k)]\sin(2\pi ct/\lambda). \quad (9)$$

The process is performed by the integrators connected to the input terminals $si$ in Fig. 33. Adding Eqs. (7) and (9), multiplying by 1/2, and reversing the amplitude yields

$$-v(i, \beta) = -V[\sin \pi(i - k)/\pi(i - k)]\sin(2\pi ct/\lambda). \qquad (10)$$

This process is performed by the operational amplifiers with the resistor $R/2$ in the feedback loop in Fig. 33.

Adding Eqs. (7) and (10) and reversing the amplitude yields

$$-v(-i, \beta) = -V[\sin \pi(i + k)/\pi(i + k)]\sin(2\pi ct/\lambda). \qquad (11)$$

This process is performed by the operational amplifiers with the resistor $R$ in the feedback loop in Fig. 33.

The voltage defined by Eq. (8) must be multiplied by $1/2^{1/2}$ and amplitude reversed to conform to the voltages defined by Eqs. (10) and (11):

$$-v(0, \beta) = -V(\sin \pi k/\pi k)\sin(2\pi ct/\lambda). \qquad (12)$$

The variation of the voltages $v(0, \beta)$, $v(1, \beta)$, and $v(-1, \beta)$ as a function of the angle of incidence $\beta$ is shown in Fig. 34. These curves define in detail the resolving power of the microphone array and the filter. The first zero crossings of the curves have the distance $\varepsilon$ from their peaks; this gives the justification for the previously introduced term "resolution angle" for $\varepsilon$.

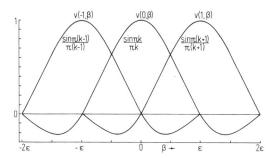

FIG. 34. Variation of the voltages $v(-1, \beta)$, $v(0, \beta)$, and $v(1, \beta)$ in Fig. 33 as functions of the angle of incidence $\beta$ of the received sound wave. The scale holds for small values of $\varepsilon$ and $\beta$; for larger values one must replace $\varepsilon$ and $\beta$ by $\sin \varepsilon$ and $\sin \beta$; $k = \sin \beta / \sin \varepsilon$.

One obtains for small values of $\varepsilon$:

$$\varepsilon \doteq [1/(2n + 1)]\lambda/d. \tag{13}$$

The length or aperture of an array of $2n + 1$ microphones equals $2nd$ according to Fig. 29. Substitution of $A = 2nd$ into Eq. (13) yields for large values of $n$:

$$\varepsilon \doteq \lambda/A. \tag{14}$$

The resolution angle of a telescope using a circular lens is defined by

$$\varepsilon \doteq 1.22\lambda/A, \tag{15}$$

where $A$ is the diameter of the lens. This formula is based on the first zero crossing of a diffraction pattern of a point source of light and is thus directly comparable to Eq. (14). It appears that Eq. (14) yields a better resolution. However, if we use a square array of microphones for two-dimensional reception as shown in Fig. 28, there is a question of what one should call aperture. If we use the side length $A = 2nd$ of the square array we obtain Eq. (14); if we use the diagonal $A = 2 \cdot 2^{1/2}nd$ we obtain $\varepsilon \doteq 1.41\lambda/A$, which is slightly larger than Eq. (15). A square array having the same area as a lens, $(2nd)^2 = A^2\pi/4$, yields $A = 4nd/\pi^{1/2}$ and one obtains $\varepsilon \doteq 1.13\lambda/A$ instead of Eq. (14). Hence, the resolution of a microphone array with a two-dimensional filter is for all practical purposes the same as that of a lens occupying the same area as the microphone array.

A major difference between imaging based on a lens or the radar principle on the one hand and the microphone array with two-dimensional electric filters on the other hand is clearly shown by Eqs. (10)–(12). The sinusoidal voltages $\sin 2\pi ct/\lambda$ can be filtered by resonant filters to yield a good signal-to-noise ratio. There is no theoretical limit for the bandwidth of these filters, while for the radar principle the resolution varies inversely to the bandwidth. An optical lens would also permit an arbitrarily narrow bandwidth, if we knew how to produce it, and if we had powerful coherent light sources for illumination.

### F. Two-Dimensional Filters without Delay Circuits

The transition from a one-dimensional filter according to Fig. 32 to a two-dimensional filter is not as straightforward as in the case of the filter of Fig. 27 which uses delay circuits. For an explanation refer to Fig. 35. A

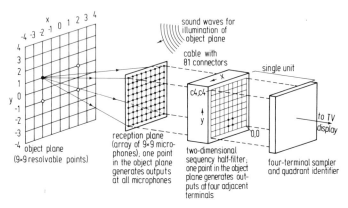

FIG. 35. Principle of obtaining an image by means of sound waves using a microphone array, a two-dimensional sequency filter without delay, and a television display.

source of sinusoidal sound waves for illumination—or *sonification*—of the object of interest is shown on top. The sound waves are scattered in the object plane. A total of $9 \times 9$ points can be resolved in the reception plane.

The output voltages of the microphone array are fed to a two-dimensional filter consisting of stacks of cards as shown in Fig. 28, but stacks of nine rather than seven cards each are required now. The individual cards all have the circuit shown by Fig. 32. This circuit poses a practical problem due to leakage currents caused by the finite amplification of the operational amplifiers. However, it is known how to overcome it by using a circuit derived from the fast Fourier transform.

The card in Fig. 32 produced output voltages at two terminals for a wave with angle of incidence $+\beta$ or $-\beta$. As a result, the two-dimensional filter in Fig. 35 produces output voltages at four adjacent terminals, indicated by black dots. Four points in the object plane having the same absolute value of the $x$- and $y$-coordinates produce output voltages at the same four terminals of the filter. The points must be resolved by using the polarity of the voltages.

Figure 36 shows the two-dimensional image transformation in more detail. The object plane is divided into four quadrants: $x < 0$, $y < 0$; $x < 0$, $y > 0$; $x > 0$, $y < 0$; $x > 0$; $y > 0$. Waves $W(\pm k, m)$ and $W(\pm k, -m)$ with the same absolute value for the variable $x = |\pm m|$ produce the output voltage $V \sin 2\pi(ct/\lambda + jk/9)$ at the terminal $cl$ of each one of the horizontally stacked cards. If the wave comes from the half-plane $x > 0$, the output voltage at the terminal $sl$ of all cards will be $V \cos 2\pi(ct/\lambda + jk/9)$, while the voltage $-V \cos 2\pi(ct/\lambda + jk/9)$ is produced for $x < 0$.

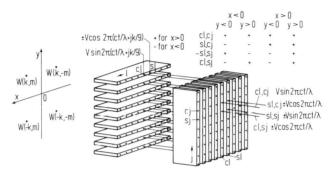

FIG. 36. Signs of the output voltages of a two-dimensional sequency filter caused by waves originating at the points $-k$, $-m$ to $+k$, $+m$ in the four quadrants of the object plane.

The output voltages from all terminals $cl$ of the horizontal stack of cards are fed to the card $cl$ of the vertical stack, and the output voltages from the terminals $sl$ to the card $sl$. The transformation of the voltages $V \sin 2\pi[ct/\lambda + jk/(2n + 1)]$ by card $cl$ is defined by Eqs. (3)–(5). The transformation of $V \cos 2\pi[ct/\lambda + jk/(2n + 1)]$ is similar:

$$\cos 2\pi[ct/\lambda + jk/(2n + 1)] = \cos 2\pi ct/\lambda \cdot \cos 2\pi jk/(2n + 1)$$
$$- \sin 2\pi ct/\lambda \cdot \sin 2\pi jk/(2n + 1)$$

$$2V \int_{-1/2}^{1/2} [\cos 2\pi ct/\lambda \cdot \cos 2\pi k\theta - \sin 2\pi ct/\lambda \cdot \sin 2\pi k\theta]\sin 2\pi i\theta \, d\theta$$

$$= -\delta_{ik}(k/|k|)V \sin 2\pi ct/\lambda$$

$$2V \int_{-1/2}^{1/2} [\cos 2\pi ct/\lambda \cdot \cos 2\pi k\theta - \sin 2\pi ct/\lambda \cdot \sin 2\pi k\theta]\cos 2\pi i\theta \, d\theta$$

$$= \delta_{ik} V \cos 2\pi ct/\lambda$$

$$\sqrt{2} V \int_{-1/2}^{1/2} [\cos 2\pi ct/\lambda \cdot \cos 2\pi k\theta - \sin 2\pi ct/\lambda \cdot \sin 2\pi k\theta] \, d\theta = \delta_{0k} V \cos 2\pi ct/\lambda$$

The signs of the output voltages at the terminals *cl, cj* to *sl, sj* in Fig. 36 follow from these relations for waves originating in the four quadrants of the object plane. These signs are shown in Fig. 36. They represent the first four Walsh functions and thus form an orthogonal set. The quadrant ambiguity may be resolved by transforming the sinusoidal voltages $V \sin 2\pi ct/\lambda$ at the terminals *cl, cj* and *sl, sj* into cosinusoidal voltages $V \cos 2\pi ct/\lambda$ by means of integrators. The resulting voltages are multiplied by $+1$ or $-1$, according to the signs in the table in Fig. 36, and summed. As a result of the orthogonality of Walsh functions, three of the four output voltages must be zero and the quadrant ambiguity is resolved.

A practical version of the circuits for acoustic imaging is shown by Fig.37. Voltages from an array of $16 \times 16$ hydrophones are fed to the input terminals of the two-dimensional sequency filter which performs a Fourier transform in about 10 $\mu$sec. The quadrant ambiguity resolver follows. The phase of the waves must be preserved up to the output terminals of this resolver. After the resolver one may insert time filters to obtain a better signal-to-noise ratio or Doppler resolution, without regard to phase shifts. One should observe that the arrangement of the printed circuit cards in Fig. 37 is only intended to make the interconnection of the cards more readily understandable. Also, the cards of the quadrant ambiguity resolver are ordinary cards and not blocks with two rows of input and output terminals as shown in the illustration.*

Table II shows some representative numerical values for acoustic imaging in water by means of the two-dimensional, delay-free filters discussed so far. Let us point out that essentially the same numerical values for size and resolution would hold for imaging by a lens.

The classical limit for the resolution angle $\varepsilon$ that can be achieved with a wavelength $\lambda$ and an aperture $A$ is defined by $\varepsilon \doteq \lambda/A$. Any method that yields a smaller value of $\varepsilon$ for the same values of $\lambda$ and $A$ is said to yield super-resolution. Synthetic aperture radar was the first method to achieve super-resolution by means of the Doppler effect. Two-dimensional sequency filters can yield super-resolution without using the Doppler effect. The reason is most easily explained by an analogon in filtering of time signals. A time signal having passed through a frequency low-pass filter of bandwidth $W$ is determined by $2W$ amplitude samples per second. These samples are usually spaced equally in time so that $n$ samples require the time $n/2W$. At the expense of signal-to-noise ratio, one can take the $n$ samples during the time $\delta n/2W$ and no samples during the remaining time $(1 - \delta)n/2W$, where

---

* The development of such equipment was supported by the Office of Naval Research under contract No. N00014-67-A-0377–009 at The Catholic University of America, Washington, D.C.

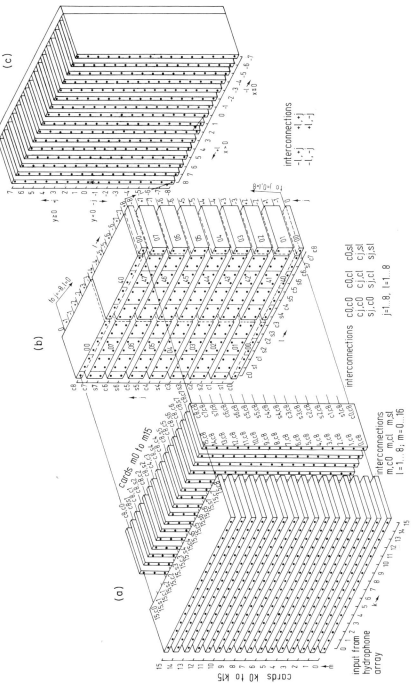

Fig. 37. Practical version of the electrical circuits for acoustic imaging showing (a) the two-dimensional sequency filter ($2 \times 16$ cards $k0$ to $k15$ and $m0$ to $m15$) performing a Fourier transform, (b) the quadrant ambiguity resolver (16 cards $q00$ to $q48$ shown partly divided), and (c) a light emitting diode display (16 cards $dl$, $l = -7$ to $+8$) with the associated driving circuits. This illustration shows how the various printed circuit cards are interconnected. The actual cards are mounted in the usual way and interconnected by cable trees.

## TABLE II

REPRESENTATIVE FIGURES FOR ACOUSTIC IMAGING IN WATER BY MEANS OF TWO-DIMENSIONAL SEQUENCY FILTERS

| Frequency of illuminating sinusoidal wave | 10 kHz | | 100 kHz | | 1 MHz | |
|---|---|---|---|---|---|---|
| Wavelength λ | 15 cm | | 1.5 cm | | 1.5 mm | |
| Microphone array | 31 × 31 | 255 × 255 | 31 × 31 | 255 × 255 | 31 × 31 | 255 × 255 |
| Number of microphones | 961 | 65,025 | 961 | 65,025 | 961 | 65,025 |
| **α = 30°** | | | | | | |
| $d = \lambda/2 \sin(\alpha/2)$ | 29 cm | | 2.9 cm | | 2.9 mm | |
| Size of microphone array | 8.7 × 8.7 m² | 74 × 74 m² | 0.87 × 0.87 m² | 7.4 × 7.4 m² | 8.7 × 8.7 cm² | 74 × 74 cm² |
| $D = 2L\,\mathrm{tg}(\alpha/2)$; $D \times D$ field of view for $L = 200$ m | 106 × 106 m² (all) | | | | | |
| Distance of resolvable points at $L = 200$ m | 3.54 m | 41.4 cm | 3.54 m | 41.5 cm | 3.54 m | 41.5 cm |
| Field of view for $L = 20$ m | 10.6 × 10.6 m² (all) | | | | | |
| Distance of resolvable points at $L = 20$ m | 35 cm | — | 35 cm | 4.2 cm | 35 cm | 4.2 cm |
| **α = 10°** | | | | | | |
| $d = \lambda/2 \sin(\alpha/2)$ | 86 cm | | 8.6 cm | | 8.6 mm | |
| Size of microphone array | 26 × 26 m² | 206 × 206 m² | 2.6 × 2.6 m² | 20.6 × 20.6 m² | 26 × 26 cm² | 206 × 206 cm² |
| $D = 2L\,\mathrm{tg}(\alpha/2)$; $D \times D$ field of view for $L = 200$ m | 35 × 35 m² (all) | | | | | |
| Distance of resolvable points at $L = 200$ m | 1.17 m | — | 1.17 m | 13.8 cm | 1.17 m | 13.8 cm |
| Field of view for $L = 20$ m | 3.5 × 3.5 m² (all) | | | | | |
| Distance of resolvable points at $L = 20$ m | — | | 11.7 cm | 1.38 cm | 11.7 cm | 1.38 cm |

$\delta < 1$. This concentration of samples is usually of little interest for time signals. The same principle applied to space signals permits a reduction of the aperture $A$ to $\delta A$, while the area of a two-dimensional microphone array, used to receive an acoustic wavefront, is reduced by $\delta^2$. The numbers given for "size of microphone array" in Table II clearly show why super-resolution is important in acoustic imaging.

Mathematically, super-resolution by means of two-dimensional filters requires the orthogonalization of a linearly independent set of functions. The methods for doing this are well known. Two practical difficulties are encountered. As pointed out before, the signal-to-noise ratio is degraded. The second difficulty is the extreme accuracy of components required.

Why can two-dimensional filters achieve super-resolution, at least theoretically, while lenses and parabolic mirrors cannot do it? The reason is the greater generality of linear transformations that is possible. Lenses and parabolic mirrors can only produce transformations by means of delays. Super-resolution would require that in addition a certain attenuation can be produced for each point on the surface of the lens or the mirror and each angle of incidence. If a glass lens is replaced by glass fibers having various lengths to produce proper delay, and various grayness to produce in addition proper attenuation, one could achieve super-resolution.

## IV. Nonsinusoidal Electromagnetic Waves

### A. Decomposition of Light into Walsh Waves

It is frequently believed that an electromagnetic wave must have an electric and magnetic field strength that varies with time like a sinusoidal function. Actually the generation of sinusoidal waves is a considerable technological feat. Heinrich Hertz never succeeded in producing anything close to a sinusoidal wave. His experiments with wave propagation were done with what we would call colored noise today. The generation of reasonably sinusoidal waves was a vexing problem for some twenty years following Hertz's experiments, and it was not solved satisfactorily until the invention of the electronic tube. Anybody working with fast switching circuits knows that the problem is not how to produce nonsinusoidal waves but now *not* to produce them. Indeed we cannot switch an electric lamp on or off without generating nonsinusoidal waves.

Visible light is a form of electromagnetic waves for which the sinusoidal time variation is often believed to have been proved by interference experiments. This is not so. Let us investigate in some detail what interference experiments prove.

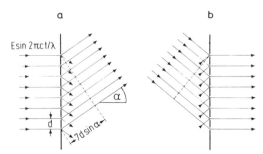

$\lambda = d \sin\alpha$ first maximum; $\lambda = 2d \sin\alpha$ first minimum

FIG. 38. Linear, time invariant diffraction grating with fixed light source and movable detector (a), and with fixed detector and movable light source (b).

Figure 38a shows a diffraction grating with eight transparent slots at a distance $d$. A sinusoidal wave with wavelength $\lambda$ coming from the left produces eight spherical waves on the right side of the grating. The waves add up for the angle $\alpha = 0$, they cancel to yield a first minimum for $\sin \alpha = \lambda/2d$, they add again to yield the first maximum for $\sin \alpha = \lambda/d$, and so forth.

If a sum of sinusoidal waves $\sum E_i \sin(2\pi ct/\lambda_i + \beta_i)$ is received from the left one obtains minima and maxima for each wave. This means that the incident light signal is decomposed into sinusoidal functions. Since we know from Fourier analysis that almost any signal can be decomposed approximately into sinusoidal functions, the pattern of minima and maxima produced by the diffraction grating will prove to us that this device has the necessary features to actually perform such a decomposition. In other words, the diffraction pattern proves that the diffraction grating is a linear, time invariant device. One will suspect that a time variable diffraction grating will decompose light into some other system of functions.

Figure 38b shows a modification of the diffraction grating of Fig. 38a. The detector observes the light emitted vertically to the grating while the light source is moved to provide angles of incidence from $\alpha = 0$ to $\alpha = 90°$. In Fig. 38a, on the other hand, the detector has to be moved while the light source is fixed.

Figure 39 shows how the diffraction grating of Fig. 38b has to be changed to decompose light into Walsh waves. The Walsh functions have the parameters time base $T$ and sequency $\phi$ which are significant for our purpose, while only the frequency or wavelength was important for sinusoidal functions. The time base is determined by choosing the angle of incidence $\alpha$. If the diffraction grating has eight slots with distance $d$ one has to choose $\alpha$ according to the formula

$$cT = 7d \sin \alpha,$$

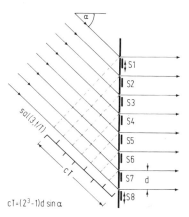

$cT = (2^3 - 1)d \sin \alpha$

FIG. 39. Linear, time variable diffraction grating with fixed detector. The time variation is provided by on–off shutters S1 to S8.

where $c$ is the velocity of light. In the general case of $2^n$ slots, one has to substitute $2^n - 1$ for 7.

The Walsh functions $sal(i, t/T)$ and $cal(i, t/T)$ with time base $T$ but various normalized sequencies $i$ are separated by making the transparent slots of the diffraction grating time variable. In the case of the Walsh functions, this time variation is provided by on–off shutters S1 to S8 that are either open or closed. For other systems of functions one would generally need a more complicated form of time variation. Let us ignore for the moment how such shutters could be implemented and let us find the rule for their operation.

Figure 40 shows the operation of the shutters S1 to S8 to permit a Walsh wave $sal(4, t/T)$ to pass. The lines 1 to 8 show the function represented by samples with time shifts $0, T/8, 2T/8, \ldots, 7T/8$. These time shifts correspond to the arrival of such a wave at the eight transparent slots in Fig. 39. The shutters S1 to S8 are open (white) or closed (black) as shown. One may see that there are always four positive samples of the Walsh wave that are superimposed by the properly opened shutters while the negative samples are blocked. One may readily see that the opening of the shutter S1 coincides with the four positive samples of $sal(4, t/T)$ in Fig. 40a and with those of $sal(3, t/T)$ in Fig. 40b. The opening times of shutters S2 to S8 are obtained from a cylical shift of the opening times of the shutter S1.

Figure 41a shows the shutters operated for passage of $sal(3, t/T)$ but the function $sal(4, t/T)$ is applied. Two positive and two negative samples are passed through the diffraction grating at any time and their sum yields zero. Figure 41b shows the shutters operated for passage of $sal(4, t/T)$ but $sal(3, t/T)$ is applied. Again the waves cancel at all times.

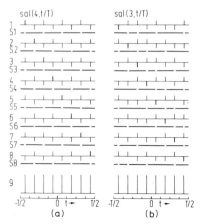

FIG. 40. Operation of the shutters S1 to S8 for passage of the waves sal(4, $t/T$) and sal(3, $t/T$); black: shutter closed, white: shutter open. The amplitude samples shown by a heavy line illustrate the delay between the eight waves.

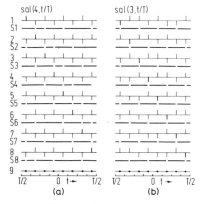

FIG. 41. (a) Shutters S1 to S8 operating to pass sal(3, $t/T$) block sal(4, $t/T$). (b) Shutters operating to pass sal(4, $t/T$) block sal(3, $t/T$).

Figure 42 shows the operation of the shutters S1 to S8 for separation of the first eight Walsh functions. The functions shown in heavy line are passed, all the others cancel by interference. Note that the constant function wal(0, $\theta$) = wal(0, $t/T$) is always passed in addition to the desired function. A short reflection of Fig. 38 shows that the same holds true for a time invariant diffraction grating—but there is no light wave with frequency or sequency zero.

To obtain an estimate for the operating times of the shutters, let us observe that the frequency of visible light lies between $4 \times 10^{14}$ and $8 \times 10^{14}$ Hz. This calls for switching times on the order of $10^{-16}$ sec to

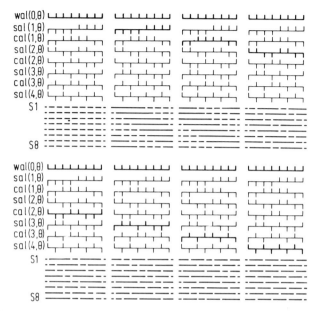

FIG. 42. Operation of the shutters S1 to S8 in Fig. 39 for separation of the first eight Walsh functions.

permit the use of the sampled amplitude representation of Walsh functions as shown in Figs. 40–42. One cannot rule out that this will eventually be possible, since light travels about 300 atomic diameters during $10^{-16}$ sec, but presently known effects for the implementation of shutters, such as the Kerr effect, are some 5 to 6 orders of magnitude slower. An obvious advantage of time variable spectrometers would be that they would not only show the relative power of different spectral lines but also uncover a relationship between the times of emission.

## B. Radiation of Electromagnetic Walsh Waves in the Radio Range

We have seen that light can be decomposed into the functions of a complete orthogonal system other than that of the sinusoidal functions, and that additional information can be obtained by doing so, but we have at present no way of building these more general spectrometers.* Such a technical barrier does not exist for waves in the radio range. The use of nonsinusoidal waves in this range has been handicapped by other reasons. It is almost

---

*Optical *Walsh–Hadamard spectrometers* have been developed by Briotta, Decker, Despain, Harwit, King, Phillips, Soroko, Vanassee, and others, but these spectrometers use grilles based on Walsh or Hadamard functions to obtain a higher sensitivity and/or a better signal-to-noise ratio. The diffraction grating is time invariant and decomposes light into sinusoidal functions.

generally believed that essentially any useful function in communications can be represented by a superposition of sinusoidal functions. The mathematical tools for doing this are Fourier series and Fourier transform. We know that Fourier series and transform converge in the sense of a vanishing mean-square-error only, and not in the much more restrictive sense of uniform convergence. Hence, there are solutions of the wave equation or of Maxwell's equations that cannot be represented by a superposition of sinusoidal functions. Are those solutions of any practical interest? To answer this question, let us look at either the sine or the cosine functions in Fig. 1, and let them represent the output voltage of a radio receiving antenna. The only important parts of the functions are their zero crossings. If we know their location, we know the frequency. The magnitude of the amplitude is of no interest as long as the signal-to-noise ratio is large enough; the automatic or manual gain control in our receivers will then provide the desired output amplitude. The sign of the amplitude is of no interest since a radio carrier $-\sin \omega t$ is as good as a carrier $+\sin \omega t$. This means that we do not care whether a function changes from positive to negative or from negative to positive at a particular zero crossing.

Let us now apply these results to the Walsh functions of Fig. 1 and let us consider either the functions $\text{cal}(i, \theta)$ or $\text{sal}(i, \theta)$. Again, the only interesting part of these functions is the location of the zero crossings. The magnitude of the amplitudes is of minor interest for the same reason as for sinusoidal functions. One may also verify that a periodically continued function $\text{cal}(i, \theta)$ or $\text{sal}(i, \theta)$ differs only by a time shift from the functions $-\text{cal}(i, \theta)$ or $-\text{sal}(i, \theta)$. Hence, the sign of the amplitude is no more important than for sinusoidal functions.

Let the Walsh functions be approximated by a Fourier series or a Fourier transform. The approximation converges uniformly everywhere except at the zero crossings, where the Gibbs phenomenon makes it divergent. Hence, we have uniform convergence everywhere except where we need it.

There are generalizations of the Fourier series and transform that do converge at the jumps of the Walsh functions, but this is of no practical help. No engineer would want to produce currents with the time variation of a Walsh function by a superposition of frequency-, phase-, and amplitude-stable sinusoidal currents. Indeed it is common practice to produce stable sinusoidal oscillations by a superposition of square waves $\text{sal}(i, \theta)$ and $\text{cal}(i, \theta)$ with $i = 1, 2, 4, 8, \ldots$ and not the other way around. The practical way to produce Walsh-shaped currents is by means of a switch, for instance in the form of a transistor or such lesser known semiconductor devices as the Trapatt diode.

Having shown that the use of more general than sinusoidal functions is theoretically and practically nontrivial, the engineer is faced with the task of showing that something useful can be done with nonsinusoidal waves. So far the investigation has been restricted essentially to Walsh waves. Five basic differences between Walsh and sinusoidal functions have been found that might be turned into useful applications:

1. The technology for the implementation of equipment is different.

2. The differentiation of a sinusoidal function yields the same function except for a changed amplitude and phase, while the differentiation of a Walsh function yields a differently shaped function.

3. The sum of several sinusoidal functions with arbitrary amplitudes and phases but equal frequency yields a sinusoidal function with the same frequency. This is the basis for interference effects. Walsh functions are summed differently and their interference effects are thus different.

4. The Doppler effect can transform a sinusoidal function into another for any velocity ratio $v/c$ while a ratio $v/c = 3/5$ or more is required to transform a Walsh function into another one of the same system.

5. A reversal of the amplitude of a sine function has the same effect as a shift by half a period. This is not so for functions that are not *polarity symmetric*.

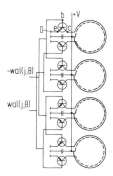

FIG. 43. Array of four Hertzian magnetic dipoles.

Let us look at examples of these effects. Figure 43 shows a radiator for Walsh waves implemented by four Hertzian magnetic dipoles. Currents will flow clockwise in the loops shown by dashed lines if the function wal($j$, $\theta$) has a positive value. A negative value of wal($j$, $\theta$) will cause currents to flow counterclockwise in the loops shown by solid lines. The diameter of the loops is small compared with $cT$, where $T$ is the time base of the Walsh function wal($j$, $\theta$) = wal($j$, $t/T$).

The time variation of the electric and magnetic field strengths produced by a Hertzian magnetic dipole are shown in Fig. 44. The first line shows the idealized current $i(t) = I$ cal(3, $t/T$) flowing in the dipole. $\mathbf{E}(1/r^2, t)$ is a component of the electric field strength declining in proportion to $1/r^2$, which is the near zone component. The far zone component declining in proportion to $1/r$ is represented by $\mathbf{E}(1/r, t)$. The magnetic field strength consists of the three components $\mathbf{H}(1/r^3, t)$, $\mathbf{H}(1/r^2, t)$, and $\mathbf{H}(1/r, t)$, which decline in proportion to $1/r^3$, $1/r^2$, and $1/r$.

The time variation of the far zone components $\mathbf{E}(1/r, t)$ and $\mathbf{H}(1/r, t)$ is the first derivative of the dipole current $i(t)$. The components $\mathbf{E}(1/r^2, t)$ and $\mathbf{H}(1/r^2, t)$ vary like the current $i(t)$, and the component $\mathbf{H}(1/r^3, t)$ varies like the integral of the current $i(t)$. These relations between the time variation of the dipole current $i(t)$ and the components of the electric and magnetic field strengths hold true for any current $i(t)$. If $i(t)$ varies like a sinusoidal function its derivative and its integral will vary like $i(t)$ except for phase shifts. Hence, the near and far zone components of sinusoidal waves are hard to separate while those of Walsh waves can be much more readily separated due to their different shape. If one can separate $\mathbf{E}(1/r, t)$ and $\mathbf{E}(1/r^2, t)$, one may compare their power and derive the distance of the receiver. One interesting aspect of this *passive* distance measuring effect is that one might be able to obtain a signal-to-noise power ratio that decreases like $1/r^3$, while the *active* distance measurement by the radar principle yields a signal-to-noise power ratio that decreases like $1/r^4$. However, the radar principle permits a much higher accuracy of distance measurement.

Figure 45 shows the idealized antenna current $I\,\mathrm{cal}(3, t/T)$ replaced by a more realistic current $i(t)$ with a switching time $\Delta T$ from $+I$ to $-I$ or from $-I$ to $+I$. The first derivative consists of rectangular pulses of duration $\Delta T$ and magnitude $2I/\Delta T$. When $\Delta T$ approaches zero one obtains the Dirac pulses of Fig. 44. The energy of a rectangular pulse is proportionate to $(2I/\Delta T)^2\,\Delta T = 4I^2/\Delta T$. Hence, the average power of the electric and magnetic field strength in the far zone can not only be increased by a larger amplitude $I$ of the antenna current but also by a shorter switching time $\Delta T$. This indicates that a small antenna can radiate a large average power.

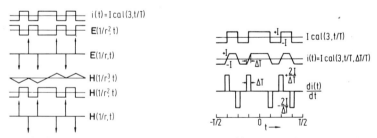

FIG. 44 (*left*). Time variation of the antenna current $i(t)$ in a Hertzian magnetic dipole and the time variations of the produced electric and magnetic field strengths declining in proportion to $1/r$, $1/r^2$, and $1/r^3$.

FIG. 45 (*right*). Walsh shaped antenna current $i(t)$ with a finite rise time $\Delta T$ and its first derivative $di(t)/dt$.

Sinusoidal waves in free space are generally dipole waves. The main reason seems to be that quadrupole and higher order multipole waves are radiated with less power by a radiator of fixed size. This is not so for Walsh waves.

Figure 46 shows on the left vertically and horizontally polarized dipoles denoted "dipole 21" and "dipole 22." Furthermore, there are three quadrupoles denoted 41 to 43. The right side of Fig. 46 shows two-dimensional

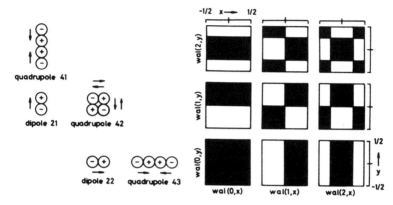

FIG. 46. Classification of dipoles, quadrupoles, and multipoles by two-dimensional Walsh functions wal($k$, $x$)wal($m$, $y$); $k$, $m$ = 0, 1, 2.

Walsh functions wal($k$, $x$)wal($m$, $y$) with $k$, $m$ = 0, 1, 2. The black areas represent value $+1$, the white areas the value $-1$. One may readily see that the positive and negative signs of the dipoles on the left correspond to the positive and negative signs of the two-dimensional Walsh functions. This correspondence greatly simplifies a discussion of radiation modes. For instance, it is obvious from Fig. 46 that there is no "unipole" corresponding to the all black function wal(0, $x$)wal(0, $y$). Unipole radiation does not exist for electromagnetic waves due to the preservation of charge but it is the major mode of radiation of acoustic waves.

According to Fig. 46, a quadrupole 41 consists of two dipoles fed with currents flowing in the opposite directions. Two such electrical Hertzian dipoles are shown in detail in Fig. 47. The electric and magnetic field strengths in the far zone, declining in proportion to $1/r$, vary with time like the second derivative of the current flowing in the dipoles. The resulting time variation is shown in Fig. 48. On top is the nominal dipole current $I$cal(3,$t/T$). The second line shows the more realistic current with finite transient time between $+I$ and $-I$; furthermore, the transients are rounded while those in Fig. 45 had sharp breaks. As a result, the first derivative shown in line 3 consists of trapezoidal rather than rectangular pulses of

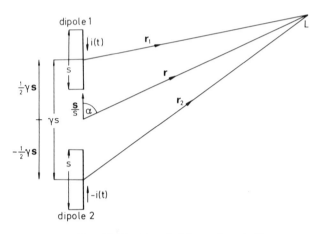

FIG. 47. Hertzian electric quadrupole 41.

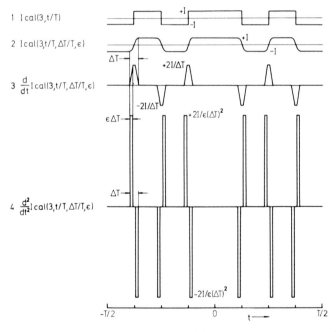

FIG. 48. Time plots for quadrupole radiation. (1) nominal antenna current; (2) realistic antenna current; (3) first derivative of the antenna current; (4) second derivative of the antenna current or time variation of the electric and magnetic field strength in the far zone.

width $\Delta T$. The second derivative consists of pairs of rectangular pulses of width $\varepsilon \Delta T$. The energy of a rectangular pulse is proportionate to

$$[2I/\varepsilon(\Delta T)^2]^2 \varepsilon \, \Delta T = 4I^2/\varepsilon(\Delta T)^3.$$

Hence, the average power of the electric and magnetic field strength in the far zone increases faster with decreasing switching time than for dipole radiation. This means that it is theoretically possible to radiate more power in the quadrupole mode than in the dipole mode for an antenna of given size if one succeeds in decreasing the switching time sufficiently. The different time variation of the electric and magnetic field strengths of dipole and quadrupole radiation indicates that interference effects will be different. Generally speaking, interference effects of quadrupole radiation yield a better resolution than interference effects of dipole radiation, and this holds even more so for higher order multipole radiation.

## C. Practical Radiators and Receivers

The four dipoles in Fig. 43 are arranged in one dimension. Two- and three-dimensional antennas are of great practical interest, but they are difficult to implement for sinusoidal waves. For instance, the radiator in the focal point of a parabolic dish should be in three dimensions as close to a point as possible. This requirement is best met by a three-dimensional radiator with the shape of a sphere.

The radiator of Fig. 43 can be transformed into a three-dimensional radiator by arranging the loops according to Fig. 49. There are several variations of this circuit. One may drive currents clockwise and counter-clockwise by using a pair of *pnp* and *npn* transistors as drivers. All eight

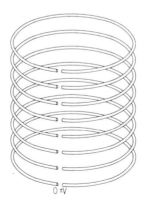

FIG. 49. Three-dimensional magnetic dipole array.

rather than only four loops in Fig. 49 may then carry currents at any one time. The radiated power is thus increased by a factor four.

A radiator according to Fig. 49 is useful as long as the switching times between positive and negative values of the current $i(t)$ in Fig. 45 are not much below 1 nsec. For shorter switching times one may use double helix antennas as shown in Figs. 50 and 51. Let a short positive current pulse be applied to the terminal $\pm V$ in Fig. 50. A magnetic field directed upward is produced when the pulse travels around the first loop. The second loop turns in the opposite direction and the current pulse produces a magnetic field directed downward when traveling around that loop. If a negative current pulse is fed to the terminal $\pm V$ at that time, one obtains the sum of the two magnetic fields. The process is repeated for every loop. This antenna can radiate the so-called Rademacher waves $\text{wal}(2^n, t/T)$, which are particularly simple Walsh waves.

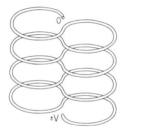
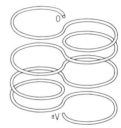

FIG. 50 (*left*). Simplest double helix antenna for Walsh waves with very short switching times.
FIG 51 (*right*). More complicated double helix antenna for Walsh waves.

A double helix antenna for more complicated Walsh waves is shown in Fig. 51. A short positive current pulse at the terminal $\pm V$ produces a magnetic field that points upward during the time interval $0 < t < T/8$; the magnetic field points downward in the following time interval $T/8 < t < 3T/8$, which is twice as long; in the time interval $3T/8 < t < T/2$ it points upward again. A summation of field strengths is achieved if a negative current pulse is applied to the terminal $\pm V$ at the time $T/2$. The flow of the current in the sense of a right- and left-hand screw can be represented by the signs $+--+-++-$. This is the Walsh function $\text{wal}(5, t/T)$ in the interval $0 < t < T$. The principle of this double helix antenna may readily be extended to other Walsh waves.

Sinusoidal waves have to be discriminated at the receiver by their frequency only. Walsh waves have the two parameters time base and sequency which are not changed during transmission. A circuit which resonates with a Walsh function, having a certain time base and sequency, must be time variable. It is difficult to build such circuits for the low signal-to-noise ratios encountered at the input terminal of a radio receiver. However, one can build time invariant circuits that resonate with any signal of proper time base. Such a prefiltered signal has a much higher signal-to-noise ratio. It may be fed to a time variable circuit that resonates with Walsh functions of proper time base and sequency.

Figure 52 shows the principle of a selective receiver for a periodic wave with arbitrary time variation. The delay line DL produces a delay equal to the period $T$ of the wave. The period of a Walsh wave $\mathrm{wal}(j, t/T)$ is in general equal to its time base but certain Walsh waves, like the Rademacher waves $\mathrm{wal}(2^n, t/T)$, have shorter periods. A shorter delay line can be used for their reception.

Let a periodic signal with amplitude $A$ arrive at the input "$a$" of the summing amplifier in Fig. 52, and let us assume that the attenuation of delay

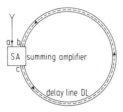

FIG. 52. Principle of a receiver for Walsh waves with certain time base $T$ and arbitrary sequency $i/T$.

line and amplifier is exactly zero. If the signal has the period $T$ equal to the delay time of the delay line DL, the delayed first period will be added to the newly arriving second period of the signal. A signal with amplitude $2A$ will then circulate in the delay line DL. Generally, the circulating signal will have the amplitude $nA$ after $n$ periods have been received. Signals with a wrong period or no periodicity will add more slowly. For instance, the mean-square-deviation of the distribution of the amplitudes of thermal noise will only increase in proportion to the square root of $n$. This difference in summation provides a filtering effect.

Let there be some attenuation in the circuit so that an input amplitude $A$ becomes $qA$ after one circulation through the delay line, where $q < 1$. After $n$ periods, one obtains the amplitude

$$A(1 + q + q^2 + \cdots + q^n) = A(1 - q^n)/(1 - q).$$

For most practical purposes one may assume that $n$ is so large that $q^n$ can be neglected and the sum becomes $A/(1 - q)$. For $q = 0.9$ one obtains the sum

$10A$, while $q = 0.99$ yields the sum $100A$. The second case provides a much better selectivity than the first, but the stability of the circuit must be very good to prevent $q = 0.99$ from becoming $q > 1$, which would make the circuit oscillate.

This is in perfect analogy to our experience with selective circuits for sinusoidal waves; the higher the selectivity the better must be the stability and vice versa. The analogy can be carried a good deal further. The selectivity of a receiver for sinusoidal waves can be increased by using several tuned circuits, the output of one always feeding the input of the next. The practical way of doing this for Walsh waves will be discussed later. The theory is as follows: The amplitude $A$ of the input signal is summed to $A/(1 - q)$ by the first delay line feedback circuit. This signal with amplitude $A/(1 - q)$ is summed to $[A/(1 - q)]/(1 - q) = A/(1 - q)^2$ in a second delay line feedback circuit. Generally, $m$ circuits will produce the amplitude $A/(1 - q)^m$ of the signal with correct period. By using delay lines with slightly different delay in each one of the staggered circuits, one may produce the effect of a narrow bandpass filter rather than the effect of a resonant filter.

The principle of the circuit of Fig. 52 is also known for sinusoidal waves. Since the circuit resonates with any wave of proper period, it is less selective than a circuit that resonates only with waves of proper period and sinusoidal time variation. This drawback for sinusoidal waves is an advantage for Walsh waves since one wants to receive the wave zone components $\mathbf{E}(1/r, t)$, $\mathbf{H}(1/r, t)$ as well as the near zone components $\mathbf{E}(1/r^2, t)$, $\mathbf{H}(1/r^2, t)$ in Fig. 44.

A practical implementation* of the principle of Fig. 52 is shown in Fig. 53. The input signal is fed to one input terminal of a wide band hybrid coupler, passes through a wide band amplifier, a second hybrid coupler, a

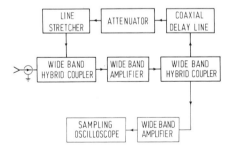

FIG. 53. Block diagram of a practical receiver for electromagnetic Walsh waves.

* The development of such equipment was supported by the Office of Naval Research under contract No. N00014-67-A-0377-0019 at The Catholic University of America, Washington, D.C. Similar equipment was developed by Fralick in 1972 at Stanford Research Institute, Menlo Park, California, under contract for the Air Force Office of Scientific Research.

coaxial cable used as delay line, an attenuator, and a line stretcher which is essentially a coaxial cable of variable length. Although all these components, except the coaxial cable, were designed for sinusoidal waves, they worked well in experimental equipment. The frequency bandwidth of available components is 1 GHz and more, and the phase distortions appear insignificant even though they are generally not specified. The attenuation of a coaxial cable increases with the square root of the frequency, but this turns out to be of little consequence since the attenuation of the cable used was fractions of a decibel at 1 GHz.

The second wide band hybrid coupler in Fig. 53 is used to feed the signal circulating in the feedback loop for display to an oscilloscope. In a receiver with several feedback loops one would connect the output terminal of this hybrid coupler to the input terminal of the next feedback loop.

Figure 54 shows on top the voltage $\text{wal}(j, \theta) = \text{wal}(1, t/T)$ with $T = 500$ nsec applied to the bases of the transistors in Fig. 43. The pulses below were received by a small loop at a short distance. It appears that this voltage represents the wave zone component $\mathbf{H}(1/r, t)$ of the magnetic field

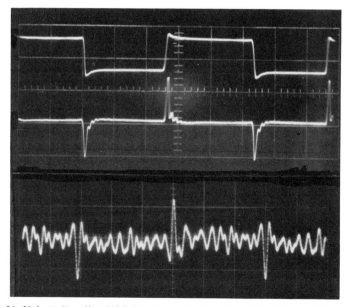

FIG. 54. Voltage $V \text{ wal}(1, t/T)$ fed to the transistors in Fig. 43 driving a radiator as shown in Fig. 49 (first line); output voltage of a magnetic dipole used as receiver (second line); the period $T$ is 500 nsec. Time variation of the electric field strength displayed by the sampling oscilloscope in Fig. 53, caused by a current $I \text{ wal}(1, t/T)$ fed to a Hertzian electric dipole as radiator; the period $T$ is about 50 nsec (third line).

in accordance with Fig. 44. The interpretation of this voltage is actually more complicated, since the current in the loops in Fig. 43 is not proportional to the input voltage.

The trace at the bottom of Fig. 54 shows pulses displayed on the sampling oscilloscope of Fig. 53. The radiated pulses again had the shape of the function wal$(1, t/T)$ but the period $T$ was about 50 nsec and the switching time about 1 nsec. The many small pulses occurring in this oscillogram are caused by reflections in the transmitting antenna and in the receiver, and also by reflections from the walls and the ceiling of the laboratory room. Such reflections would not be visible for sinusoidal waves due to the peculiar summation theorem of sinusoidal functions. We will see in the next section how this apparently undesirable effect can be turned into a useful effect.

### D. Applications of Electromagnetic Walsh Waves

Figure 55 shows an example of an interference effect in the reflection of a sinusoidal radar pulse from two pointlike targets B1 and B2. The reflected pulses are shown in lines $a$ and $b$. Line $c$ shows their sum, which is the signal received by the radar. Except for the bulges of duration $2(d_2 - d_1)/c$ at the beginning and end of the signal, there is nothing that indicates that this signal was caused by two targets rather than by a single larger target. A typical radar pulse will contain about 1000 carrier cycles; hence the relative energy of the bulges at the end is very small.

Let a Walsh wave in dipole mode be reflected by the two targets. The far zone components of the electric and magnetic field strengths consist of narrow rectangular pulses. The pulses reflected from B1 are shown in line $d$ and those reflected from B2 are shown in line $f$. The difference between a

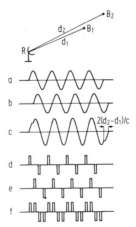

FIG. 55. Example of an interference effect of sinusoidal and Walsh shaped waves.

reflection from two small targets or one larger target is no longer represented by low energy effects at the beginning and end of a radar pulse.
Indeed, even a periodic Walsh wave would distinguish between one and two
targets.

Let us go one step further. A reflected and a scattered sinusoidal wave
differ in amplitude only. A small mirror returns the same wave as a large
scatterer from an incident wave. This is a unique feature of sinusoidal waves.

Figure 56 shows on top an idealized Walsh wave consisting of Dirac
pulses. A two-dimensional radar reflector is shown below. Regardless of
where the wave strikes the reflector, it requires the time $2R/c$ to travel from
the reference plane via the reflector back to the reference plane. This causes
the reflected wave to have the same time variation as the incident wave.

Consider now a spherical scatterer as shown in the bottom of Fig. 56. A
metallic sphere returns a complicated mixture of a reflected and a scattered

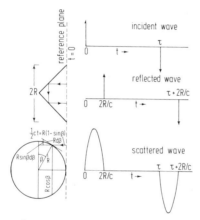

FIG. 56. Difference between a reflected and a scattered idealized Walsh wave.

wave. Only the scattered part is of interest here. The time variation of the
scattered wave is shown in the bottom of Fig. 56. The essential feature of this
scattered wave is the broadening of the original Dirac pulses into pulses of
duration $2R/c$. A sphere of three feet in diameter yields $2R/c = 3$ nsec.
Hence, this sphere can be distinguished from a reflector if the radiated wave
has a switching time $\Delta T$ of less than 3 nsec.

As an exploitation of the Doppler effect of Walsh waves, let us consider
the Doppler resolution of the relative velocity of a target in radar.

The solid line in Fig. 57 shows the well-known relation between relative
velocity $v/c$ of a target and Doppler resolution. A sinusoidal wave with $m$
periods produces at the receiver output a voltage as a function of $v/c$, which

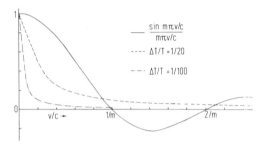

FIG. 57. Doppler resolution of the relative velocity $v/c$ by a sinusoidal wave (solid line) and by Walsh waves (dashed lines) with $m$ periods.

varies like this curve. The voltage is 1 for $v/c = 0$, decreases with increasing values of $v/c$, first slowly and then faster, becomes zero for $v/c = 1/m$, and has sidelobes for larger values of $v/c$. This curve has two drawbacks: (i) The sidelobes may simulate targets with velocities $v/c < 1/m$. (ii) The slow decrease of the curve for small values of $v/c$ limits the velocity resolution for a fixed number of periods, $m$.

Let us consider the Doppler resolution of Walsh waves. Figure 58a shows a Walsh shaped antenna current with linear transitions of duration $\Delta T$ between the amplitudes $+1$ and $-1$. The time variation of the electric field strength in the far zone is proportional to the first derivative of the current. It is shown in Fig. 58b. The pulse sequence of Fig. 58b stretched by a Doppler shift is shown in Fig. 58c. It can be shown that the output voltage of a receiver for Walsh waves depends on the ratio $\Delta T/T = $ (switching time)/(period of the wave). The two curves with $\Delta T/T = 1/10$ and $\Delta T/T = 1/50$ in Fig. 57 show the output voltage as a function of $v/c$ for $m$ periods of the Walsh wave. The curves decrease for small values of $v/c$ much faster than for the sinusoidal wave, and the "sidelobes" are significantly reduced.

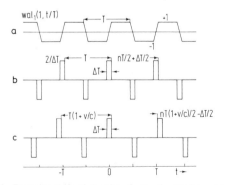

FIG. 58. Doppler shift of the time base of a Walsh wave.

The Doppler resolution with $m$ periods of a wave is mathematically equivalent to the angular resolution of an antenna array having $m$ elements. Hence, Fig. 57 also indicates an improved angular resolution by means of Walsh waves.

Let us look at the Doppler effect for a system of Walsh or sine functions rather than just one function. Figure 59a shows three Walsh and three sine functions as observed in a system of reference that has no relative velocity to the system of reference in which they are generated. Figure 59b shows the observed functions if the system in which they are generated moves away

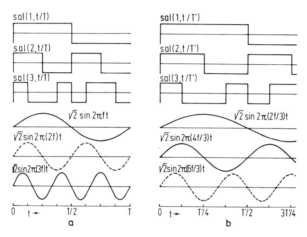

FIG. 59. Doppler effect for sinusoidal and Walsh waves: (a) functions generated and observed in the same system of reference; (b) functions observed in a system moving away with velocity $-v/c = 5/13$.

with velocity $-v/c = 5/13$. An observer cannot tell whether the observed sine function $\sin 2\pi(6f/3)t$ was produced by a transmitter without relative velocity radiating the function $\sin 2\pi(2f)t$ or one with relative velocity $-v/c = 5/13$ radiating the function $\sin 2\pi(3f)t$. In the case of the Walsh function, there is no difficulty telling that sal(3, $t/T'$) in Fig. 59b was caused by sal(3, $t/T$) in Fig. 59a and not by sal(2, $t/T'$). The same applies to the first or second derivatives of the Walsh functions.

If transmitter and receiver move away from each other, the relative velocity must be at least $-v/c = 3/5$, if they approach other it must be at least $+v/c = 3/5$ before a Walsh function with known time base $T$ is transformed into another. There are no such minimum velocities for sinusoidal functions.

## References

1. H. F. Harmuth, "Transmission of Information by Orthogonal Functions," 2nd Ed. Springer-Verlag, Berlin and New York, 1972.
2. "Proceedings of Applications of Walsh Functions (Washington, D.C.)." Nat. Tech. Inform. Serv., Springfield, Virginia 22151. AD 707,431 (1970); AD 727,000 (1971); AD 744,650 (1972); AD 763,000 (1973).
3. "Proceedings of Applications of Walsh Functions (Hatfield)." Dep. Elec. Eng. Phys., Hatfield Polytech., Hatfield, Hertfordshire, England, 1971 and 1973.
4. J. N. Bramhall, "An Annotated Bibliography on Walsh and Walsh Related Functions." Tech. Memo. TG 1198A. Johns Hopkins Univ. Appl. Phys. Lab., 8621 Georgia Ave., Silver Spring, Maryland 20910. [This bibliography is periodically updated and reissued.]
5. K. G. Beauchamp, "A Classified Bibliography for Walsh and Related Functions." Comput. Cent., Cranfield Inst. Technol., Cranfield, Bedford, England, 1972. [This bibliography is periodically updated and reissued.]

# Technology of Electron-Bombardment Ion Thrusters

## HAROLD R. KAUFMAN

*Colorado State University,
Fort Collins, Colorado*

## I. Introduction

The first electron-bombardment thruster, shown in Fig. 1, was operated in 1960 (*1, 2*). Electron bombardment was, of course, an accepted means of producing ions prior to 1960. The need in electric propulsion was for a high ion current, but with a low enough current density to permit the integration of the source with long-lived accelerators operating at the moderate exhaust velocities of interest for propulsion. The earlier bombardment sources either produced too low an ion current to be of interest, or too high a current density to be used with practical accelerator systems for moderate velocities (*3–11*). In addition, often inadequate information was available to evaluate these earlier sources for applications other than the intended ones.

The performance of early electron-bombardment thrusters showed substantial advantages over other ion thrusters of that period (*12*). Because of these advantages, a research program was initiated, first in the United States and later in other countries (particularly England and Germany).* This bombardment thruster work has obvious applications to space propulsion.

Fig. 1. First electron-bombardment thruster. (Courtesy of NASA.)

---

* Russia is also developing bombardment thrusters, but the results of their program are not generally available in open literature.

But there are also the ground applications of producing a wide range of plasma environments, cleaning and machining by ion beam bombardment, and plating with the sputtered particles produced by the ion beam.

Instead of trying to address all related aspects of work on electron-bombardment thrusters, the author has chosen to emphasize the basic thruster technology. Further, the presentation has been made as general as possible to facilitate the use of this information in all potential applications. In making this choice, the author has omitted the other components of electric propulsion systems, as well as the system and background knowledge required for the successful application of electric propulsion in space. For these omissions the author recommends the excellent monographs by Stuhlinger (*13*), Au (*14*), and Brewer (*15*).

Bombardment thrusters have been operated for the most part with mercury propellant so, of necessity, work with that propellant will be emphasized. Cesium has also been used, so some information is available for that propellant. Other propellants of increasing interest (but little data) are xenon and argon. The use of xenon would avoid the potential pollution problems of mercury in the near-earth environment, while argon appears to be the best choice for most ground applications.

Unless stated otherwise, all quantities are in SI (rationalized mks) units.

## II. Ion Production

The function of the ion source in electric propulsion is to generate an ion current of moderate, but nearly uniform current density, and to do this with low losses of energy and neutrals. The production of ions in an electron-bombardment thruster takes place within the ion or discharge chamber volume—see Fig. 2. This ion production involves a variety of processes which are, as yet, only partially understood.

### A. Theory

The simplest level of theory is a description of the conditions within an operating bombardment ion source. The numerical values given are typical of mercury propellant.

#### 1. Typical Operation

Electrons are emitted by the ion chamber cathode, which is maintained 30–40 V negative of the anode for mercury propellant. The emitted electrons are constrained by the ends of the chamber which are at cathode potential and by a magnetic field which is usually approximately parallel to the axis of

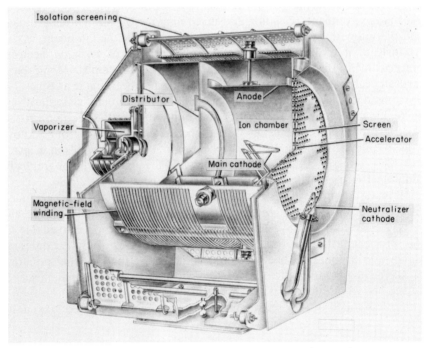

FIG. 2. Cutaway sketch of bombardment thruster. (SERT I.)

the ion chamber. The magnetic field is required for efficient ionization of the low density ($\sim 10^{18}/m^3$) propellant atoms. The electron (or ion density) is about an order of magnitude lower than the neutral density. Most of the electrons, usually well over 90%, are in a low energy Maxwellian distribution with a temperature of $\sim 5$ eV. The magnetic field is such that the small fraction of high velocity or primary electrons have a cyclotron radius of 0.1–0.5 of the ion chamber radius.

The path length for excitation or ionization of propellant is less than a meter for 30–40 eV primary electrons, while these same electrons would have a path length of more than 100 m for coulomb collisions (90° deflections) with other electrons. This disparity in path length leads to the separate population of primary electrons. The temperature of the low energy Maxwellian distribution is governed by the rapid decrease in the mercury excitation cross section below about 6 eV.

The Debye shielding length in the ion chamber plasma is less than 0.1 mm, so the extent of the plasma is large relative to this parameter. The potential variation in the plasma is usually of the order of a volt or so, while the sheath at the anode is $\sim 5$ V. Although the neutral density is at least several times the ion density, the escape rate of ions through the accelerator system is normally much greater than the neutral escape rate. This apparent discrepancy is caused by the fact that neutrals move at velocities consistent with the several hundred degrees centigrade wall temperature and the ions move (at least near the boundaries of the ion chamber) at 2–3 eV. This large ion velocity is consistent with Bohm's criterion for a stable sheath (*16*).

A more complete description of any of these processes or parameters can be found in any appropriate book on the physical processes—such as that by Jahn (*17*). Having looked briefly at conditions within the ion chamber, consideration of some of the gross effects of parametric changes is in order.

## 2. Dimensional Effects

Reader found that ion chambers with magnetic fields roughly parallel to the axes tend to perform in a similar manner if the magnetic field strength is varied inversely with ion chamber diameter (*18*). Such a scaling relationship makes the cyclotron radius of primary electrons a constant fraction of the ion chamber diameter. The radial diffusion of Maxwellian electrons can also be made similar if volume ion production rate is constant and if the ion chamber length is also a constant. This similarity, however, does not necessarily mean constant discharge loss per beam ion. If the ion chamber diameter is much less than its length, the probability of an ion being extracted into the ion beam (instead of recombining on the wall) is reduced substantially and the discharge loss per beam ion will rise accordingly.

Approximate similarity for a much wider range of conditions can be obtained by making the parameter $j_b l_i \sigma_i W$ a constant (*19*), where $j_b$ is ion beam current density, $l_i$ is ion chamber length, $\sigma_i$ is ionization cross section of the propellant, and $W$ is its atomic or molecular weight. This similarity parameter will still not apply to excessively large length-to-diameter ratios, but it permits comparison over a much wider range of operating variables.

Consider next the effect of decreasing the ion chamber diameter. If constant ion chamber length is maintained, the discharge losses per beam ion will rise at the small diameter, as mentioned earlier. If the ion chamber length is now much longer than the diameter, the discharge losses can be reduced by shortening the chamber. The maximum utilization of propellant will also drop (more neutrals escape) because the probability of ionizing the propellant will drop for the shorter chamber. But the ions that were formed in the upstream portion of the long chamber had little chance of being extracted by the accelerator system. Thus the drop in propellant utilization due to shortening the ion chamber will probably be less significant than the decrease in discharge losses. This trade-off between wall losses and ionization probability results in optimized designs of different diameters having some compromise between constant ion chamber length and a constant length-to-diameter ratio. Strongly divergent magnetic field designs modify this discussion (as will be shown) only to the extent of redefining ion chamber length.

One might expect the discharge losses to scale with the first ionization potential, but this does not appear to be the case. A correlation by Kaufman (19) indicated no clear trend with the first ionization potential. Ionization data with very high energy electrons shows some drop with ionization potential, but at a much less than linear relationship (20). This result perhaps indicates a larger fraction of incident electron energy goes into excitation when substances with a low ionization potential are bombarded by electrons. Regardless of the reason, though, it appears more correct to use constant losses per ion for all materials than to scale losses with the first ionization potential.

## 3. Comprehensive Theories

The initial theory of Kaufman (2) made two significant contributions. First, the separation (for mercury propellant) of the electron population into two groups was deduced from cross sections for the various electron interactions. Second, calculations of radial potential difference in the ion chamber indicated moderate magnetic fields (well under $10^{-2}$ Tesla or 100 G for a 10 cm diam source) were required to prevent the radial inflow of ions and the associated poor use of total accelerator system area. The Bohm criterion for a stable sheath was recognized, but it was not used to predict that ion densities could be much less than neutral densities while, at the same time, ions were being extracted from the chamber at a much greater rate than neutrals were escaping.

Following the initial simplified approach of Kaufman, a number of attempts were made towards a more comprehensive theory (19, 21–27). An excellent survey of theoretical attempts through 1969 was made by Milder (28). He concluded the usefulness of the attempts just mentioned was limited because of either simplifying assumptions or difficulty of solution. Milder also concluded that the most successful approach at that time (and to the present time in the author's opinion) was a semiempirical approach by Masek (29). This work of Masek has since been published in a journal (30).

The theory of Masek is semiempirical in that it requires detailed Langmuir probe surveys of the ion chamber. The basic assumption made in interpreting the Langmuir probe data is that the electron population consists of a fraction of monoenergetic primaries superimposed upon a Maxwellian distribution at a lower energy. This assumption permits a simple graphical analysis of probe data that is described by Strickfaden and Geiler

(31).* The validity of this electron distribution assumption has occasionally been questioned. Recent measurements by Martin (32) in argon indicate an actual distribution surprisingly close to that assumed.

Using Langmuir probe data obtained by him, Masek made a numerical evaluation of ion production by primary and Maxwellian electrons, energy losses of electrons in inelastic collisions, escape rate of electrons to the anode, and escape rate of ions to all ion chamber boundaries. Masek found that, although primaries were less than 10% of the electron population, about half the ions were produced by primaries and about half by Maxwellians. The total ion production he obtained was consistent with measured ion density and minimum Bohm velocity towards all ion chamber boundaries.† The collisional losses (atoms and ions) made up about 90% of discharge losses, with the remainder represented by electron energy transport to the anode. The minimum discharge energy for total ion production was about 40 eV/ion. Masek estimated a maximum of about half the ions could be extracted with the best use of conventional design techniques, which would give 80 eV/ion based on extracted ions. Masek guessed that the use of unconventional ion chamber designs might increase the fraction extracted so that discharge losses might eventually reach 60 eV/ion.

The theory of Masek is believed to be quite general and to apply in at least a qualitative manner to most propellants. The numerical results with cesium propellant, though, are quite different. Cesium has a first ionization potential of only 3.9 eV (compared with 10.4 eV for mercury) and is typically used with a 6–15 V discharge. Inasmuch as the coulomb cross section for emitted electrons varies inversely as the square of electron energy, the emitted electrons are rapidly randomized into a single 1–2 eV Maxwellian distribution (33). This type of discharge would be expected any time a very low discharge voltage is used and is normally efficient only with propellants that have a low ionization potential.

## 4. *Primary Electron Region*

Following the comprehensive theory of Masek, Kaufman made a study of propellant utilization (34). Comprehensive probe data, such as that given by Masek, can be used to show primary electrons are concentrated in only a

---

* The procedure of Strickfaden and Geiler is a specific application of a more general procedure described earlier by G. Medicus.

† Masek uses a Bohm criterion that is modified for the presence of primary electrons. In this modification the usual minimum ion energy of Bohm, based on Maxwellian electron temperature, is multiplied by the density ratio $n_i/n_m$. (Here $n_i$ is ion density and $n_m$ is Maxwellian electron density.) Equation 3 in Masek (30) is written in a misleading form that implies the reciprocal of this ratio, but the correct form is derived in Masek (29).

portion of the ion chamber volume. This portion is the volume enclosed by
the accelerator system and magnetic field lines that intersect the anode. The
reason for this restricted volume is that the small anode sheath does not
reflect energetic electrons that reach it. The energetic electrons have long
collision path lengths, so that they tend to escape to the anode if they are on
a magnetic field line that intersects the anode.

Because the bulk of ion production takes place within the primary electron region (both from primary and Maxwellian electrons), Kaufman
assumed that an analysis based on this region alone could be used to predict
performance. Using the tendency of ions to flow outwards in all directions
from the volume in which they are produced (35), together with the Bohm
criterion for ion loss rate, Kaufman found the neutral density tended toward
a constant as the primary electron density increased towards a maximum.
The loss of neutrals would thus be nearly constant at maximum utilization,
regardless of propellant flow rate. The level of this constant is proportional
to $A_p/V_p$, where $A_p$ is the area of the boundary around the primary electron
region and $V_p$ is the volume of that region. The ratio $V_p/A_p$ is a characteristic
length for the primary electron region, so that the neutral density required
for a given ion beam current density will increase as the size of a thruster is
decreased. Although it was not discussed by Kaufman, it should be noted
that the ratio of volume ion production costs found by Masek to the actual
costs per beam ion is about equal to the ratio of screen open area to $A_p$.

## B. Experimental Performance

Typical performance of an ion chamber at constant discharge voltage
and constant propellant flow rate is shown in Fig. 3. The discharge loss in
electron volts per ion is obtained by dividing discharge power by ion beam
current. Unless stated otherwise, these losses are per beam ion, and not for

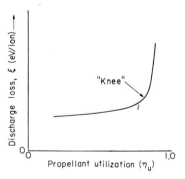

Fig. 3. Typical ion chamber performance for constant discharge voltage and propellant
flow rate.

total volume production of ions. Optimum operation is normally near the "knee" of the curve. If operation is allowed to shift to the left of this knee, the loss of un-ionized propellant increases sharply with only a small compensatory decrease in discharge losses per ion. If operation shifts upwards from the knee, the discharge losses increase sharply with only small savings in propellant. Throttled operation is therefore normally achieved by reducing both propellant flow and beam current in such a manner as to remain near the knee. The knee utilization tends to vary in a manner similar to the maximum utilization obtainable at high discharge losses when propellant flow rate is changed. Because maximum utilization of a configuration is at nearly constant neutral loss rate, the knee operation **also** tends to be at nearly constant loss rate. Throttled operation therefore implies decreased utilization.

While the need to save propellant is obvious in space propulsion, neutral losses are also a problem in ground applications. This is because increased neutral losses will normally increase facility pumping requirements as well as charge-exchange erosion.

## 1. *Magnetic Field Strength and Shape*

The magnetic field was investigated in an early study by Reader (*36*). He found the variation of discharge losses with field strength shown in Fig. 4. The losses first dropped rapidly with increasing magnetic field strength, then very slowly. The decreasing efficiency of the magnetic field at high values was attributed later to anomalous electron diffusion as a result of ion chamber noise measurements made by Cohen (*23*) and Pawlik *et al.* (*37*).

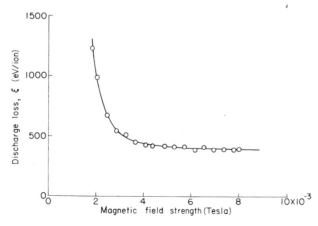

FIG. 4. Variation of discharge losses with magnetic field strength. From Reader (*36*).

Regardless of the reason, the effect of little or no improvement above some value of magnetic field has been found to be quite general.

Reader (36) also investigated the effect of magnetic field shape in his early study. Discharge losses were reduced to about half when a nearly uniform axial field was replaced by one that decreased in strength toward the accelerator system. For the moderately divergent design used by Reader, the axial field strength decreased from a maximum at the distributor to about 60% of that value at the screen, as indicated in Fig. 5. Typical performance obtained by Reader with a 10 cm (anode diameter) source operating at a beam current of 0.125 A was 80% utilization and 500 eV/ion. A 50% open screen was used in this investigation.

The field shape arrived at by Reader (Fig. 5) remained in general use until the SERT II development. In an extensive test program of over 100 configurations, Bechtel (38) developed the strongly divergent field shape of Fig.6, which gave much improved performance. With the aid of a 71% open screen grid in his 15 cm thruster, Bechtel obtained about 85% utilization at 200 eV/ion with a 0.25 A ion beam. The improvement in performance with a strongly divergent field design was felt to be due at least in part to primary electrons being able to cover the ion chamber cross section by following field lines, instead of working their way across field lines by collisions.

More recent magnetic field configurations have been reported by Knauer et al. (39), Moore (40), and Ramsey (41). The radial field design of Knauer et al. is shown in Fig. 7 and might be considered a further evolution from Figs.5 and 6. The 15 cm thruster operated by Knauer et al. obtained losses as low as 180 eV/ion at 90% utilization and a beam current of 0.265 A. In addition, a very flat beam profile was obtained. The multipole design of Moore and Ramsey is shown in Fig. 8 and might be considered an adapta-

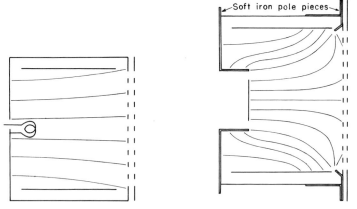

Fɪɢ. 5 (*left*). Moderately divergent magnetic field shape.
Fɪɢ. 6 (*right*). Strongly divergent magnetic field shape. (SERT II.)

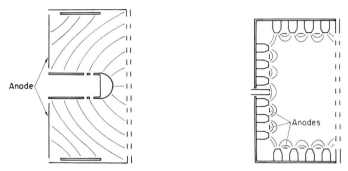

FIG. 7 (*left*). Radial magnetic field shape.
FIG. 8 (*right*). Multipole magnetic field shape.

tion of the "picket fence" concept of plasma containment in fusion research. The 12 cm multipole thruster operated by Ramsey on mercury propellant gave losses of 155–160 eV/ion at a utilization of 90% and a beam current of 0.325 A. The ion beam profile was not described by Ramsey, but it would be expected to be flatter than obtained with a SERT II type of chamber.

## 2. Ion Chamber Shape

Reader (*36*) also investigated different ion chamber configurations in his early study. The effect of ion chamber length for a 10 cm diam chamber is shown in Fig. 9. The optimum for these operating conditions appears to be 8 or 9 cm in length. The effect shown in Fig. 9 is quite general for moderately diverging magnetic fields, although the value of the optimum will change somewhat with the various operating parameters. The tendency for length to change little with changing diameter (as described in Section II,A,2) has long been known to knowledgeable workers in the field. For example, in 1964 a 50 cm diameter thruster (also with a moderately diverging field) had an ion

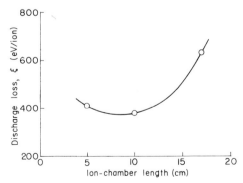

FIG. 9. Variation of discharge losses with ion chamber length for a 10 cm source. Data from Reader (*36*).

chamber length of only 13 cm (42), less than twice the optimum for the 10 cm thruster shown in Fig. 9. Despite the short length-to-diameter ratio, this thruster performed well—less than 400 eV/ion at 90% utilization and a 2.5 A ion beam. A 58% open screen was used.

Reader (36) also showed that for a given ion chamber length, with both a uniform and a moderately diverging field, the anode length did not significantly affect discharge losses. The downstream ends of both anodes were in about the same location, so that electrons had to diffuse to about the same magnetic field line to reach the anode. These results of Reader indicate that the part that intercepts the innermost field line is the most important part of the anode.* The only adverse effects of a short anode were harder starting and greater instability at low voltages.

After the introduction of strongly divergent magnetic field shapes, Kaufman investigated the effects of ion chamber dimensions with a fixed field of this shape (34). A variety of lengths and diameters were tried, but, as indicated in Fig. 10, there was little or no effect on performance. In fact, the

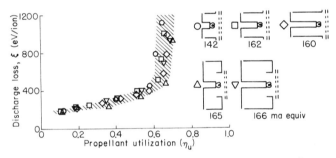

Fig. 10. Performance with different ion chamber shapes for a strongly divergent magnetic field. Primary electron region is constant for ion chamber shapes used. Propellant flow rate is indicated under each sketch. From Kaufman (34).

spread of data would probably be less than that shown if the propellant flow rates had been exactly the same. Using the concept of the primary electron region, the ion chamber changes shown in Fig. 10 did not affect performance because they did not affect the primary electron region.

The effect of chamber diameter on the primary electron region shape [from Kaufman and Cohen (43)] is shown in Fig. 11 for strongly divergent field designs. All three chambers shown were developed in extensive cut-and-try testing using similar approaches. The shapes shown therefore represent a substantial degree of optimization. The characteristic lengths of these chambers, $V_p/A_p$, are also shown. These lengths have a range of about 3.6 : 1 for a

---

* Several investigators have also shown since this early investigation that the majority of electron current also goes to the part of the anode that intercepts the innermost field line.

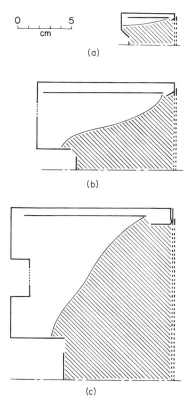

FIG. 11. Primary electron regions for thrusters with strongly divergent magnetic-field shapes. (a) Anode diam, 5 cm; $V_p/A_p$, 0.7 cm. (b) Anode diam, 15 cm; $V_p/A_p$, 1.4 cm. (c) Anode diam, 30 cm; $V_p/A_p$, 2.5 cm. From Kaufman and Cohen (43).

diameter range of 6 : 1. The smaller chamber has a clearly decreased length (relative to the larger chambers) to avoid excessive wall losses. Even so, the discharge losses of this smaller chamber are substantially higher than the others, as indicated in the knee performance data with mercury shown in Table I. The screens used on the three thrusters in Table I were all 65–70% open. Additional recent performance data with strongly divergent fields and mercury propellant can be obtained from Hudson and Banks (44) for an 8 cm thruster and from Rawlin (45) for 30 cm thrusters.

Performance data with cesium and moderately divergent fields has been given by Sohl et al. (46) for a 12.7 cm thruster and by Sohl et al. (47) for thruster diameters of 0.9, 2.5, 5.3, and 12.7 cm. The multipole design of Moore (40) gave a discharge loss of ~90 eV/ion at 95% utilization and 0.60 A ion beam current. This performance appears to be the best obtained to date in any ion chamber. An interesting anode configuration with cesium

TABLE I

Typical Knee Performance with Strongly Divergent Magnetic Fields[a]

| Anode diam (cm) | $J_b$ (A) | $\eta_u$ | $\xi$ (eV/ion) |
|---|---|---|---|
| 5 | 0.037 | 0.75 | 550–600 |
| 10 | 0.25 | 0.85 | 250–300 |
| 15 | 1.9 | 0.95 | 250–300 |

[a] From Kaufman and Cohen (43).

was used by Sohl et al. (48) in which longitudinal slots permitted alternate segments to function as the anode in alternate half-cycles of applied discharge voltage. The discharge chamber thus served as the rectifier for discharge power.

## 3. Propellants Other Than Mercury

Reader (49) obtained 10 cm ion chamber performance data with argon, krypton, and nitrogen in a moderately divergent magnetic field. Data with a 15 cm SERT II ion chamber were obtained with argon by Schertler (50) and xenon, krypton, argon, neon, nitrogen, helium, and carbon dioxide by Byers and Reader (51, 52). The performance with xenon was quite similar to mercury in both discharge loss and utilization. The lighter atomic weight of argon resulted in substantial reductions in utilization.

An interesting attempt was also made to use heavy molecule propellants. A straight experimental approach by Byers et al. (53) indicated excessive fragmentation.* A theoretical study by Dugan (54) indicated large polyring aromatics, such as anthracene, might be stable during the ionizing bombardment of electrons. But an experimental test of anthracene by Milder (55) indicated the same fragmentation that was found earlier by Byers et al. The consideration of heavy molecule propellants apparently stopped with these three investigations.

The cathode position should also be mentioned as a significant aspect of ion chamber shape, although data are less extensive than for major parameters such as length and diameter. A few qualitative conclusions can be drawn, though, from studies such as those by Reader (36) and Bechtel (38).

* If the reader has been examining the references, he knows by now that the devices discussed herein have been referred to in the literature as rockets, engines, motors, thrustors, and thrusters by various people at various times. The author is inclined toward the suggestion by P. D. Reader that since the device propels a spacecraft, we should scrap all the other names and call it a propeller.

Both high utilization and a flat ion beam profile require primary electrons to be injected so that they can reach (without collisions and without intercepting the anode) as much as possible of the ion chamber cross section near the accelerator. [This principle was apparently stated first by Poeschel *et al.* (56).] These remarks are intended, of course, for operation at a sufficiently high discharge voltage that separation of the electron population into primary and Maxwellian groups occurs. This would not be the case with cesium propellant.

### 4. *Propellant Injection*

Reader (57) investigated several modes of propellant introduction in a thruster with a moderately divergent magnetic field. The extremes in performance were obtained for the conventional straight-through feed and reverse feed with both modes illustrated in Fig. 12. At a 35 V discharge, where

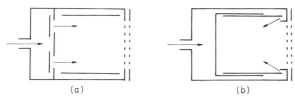

(a)                               (b)

FIG. 12. Propellant injection modes. (a) Conventional feed. (b) Reverse feed.

double-ionization effects should be small, Reader found about a 100 eV/ion, or 20% advantage for reverse feed. Various side feed modes fell at intermediate levels of improvement. The next extensive study of propellant injection was by Bechtel (38) with a strongly divergent magnetic field. Bechtel found as much as a 15% improvement for reverse feed, but only with a nonoptimum position of the cathode. When the best position of the cathode was used, no significant difference was found for conventional and reverse feeds. Bechtel did, however, find increased losses when the propellant was introduced near the cathode, particularly when the flow was directed toward the cathode.

Comparison of the data by Reader and Bechtel indicates that magnetic field shape and propellant introduction are not independent in their effects. Normal propellant injection with a uniform field results in the concentration of the ionization in the upstream end of the chamber because the neutral density is greatest in that location. Going to increasing degrees of divergence for the magnetic field offsets this neutral distribution by concentrating the primary electrons and the higher energy Maxwellian electrons at the downstream end of the chamber. The reverse feed of propellant also tends to shift the ionization toward the accelerator system, and thus produces some of the same effects as the divergent field. Apparently the effect of reverse feed is not significant in a further shift of ionization towards the accelerator system if a strongly divergent field is used properly.

The poorer performance found by Bechtel for injection near the cathode has an important implication for hollow cathodes. Ion chamber performance at the same discharge voltage would be expected to be best for minimum permissible flow through the cathode. For the 5 cm thruster this minimum flow is near the total propellant flow through the thruster; hence no division of flow is used. It is still possible to make some improvement. Holes in the cathode pole piece of the 5 cm thruster are covered with a fine mesh screen (58). The screen appears as a solid surface to charged particles, but permits neutrals to escape. The effective escape area for neutrals is thus more than just the opening by which the electrons leave the pole piece, and the neutral density is thereby decreased where the electrons are injected into the ion chamber.

## 5. Interaction with Accelerator System

Reader (36) found that increasing the accelerator system electric field decreased the discharge losses, so that some degree of interaction between ion chamber and accelerator system was recognized quite early. The interaction observed by Reader was put in a more general form by Byers (59) by plotting the discharge loss against the fraction of Child's law current carried by the accelerator system.

More aspects of the interaction were found by Kerslake (60) and Kerslake and Pawlik (61). Increasing the screen open area or decreasing the screen thickness were both found to decrease discharge losses, as indicated by the plot of four sets of data in Fig. 13a. These data were taken primarily for accelerator performance, which perhaps explains the greater than normal scatter in discharge losses. The four configurations used all had the same diameter for screen and accelerator holes and the same accelerator thickness. Some difference in interelectrode gap could not be avoided, but the effect of this variation should be small when plotted against normalized perveance. Normalized perveance is a conventional perveance per hole, $J/V_t^{3/2}$, multiplied by $l_e^2/d_s^2$. The effective acceleration distance $l_e$ is defined in Section III,B, while $d_s$ is the screen hole diameter. This form of perveance is shown in section III,B to be a way of correcting for space-charge effects due to differences in interelectrode gap. For comparison, a normalized perveance of $3.03 \times 10^{-9}$ corresponds to Child's law current density with mercury ions.

In addition to the screen open-area fraction ($F_s$) and relative thickness ($t_s/d_s$) effects, the data show the general trend found by Reader and Byers;

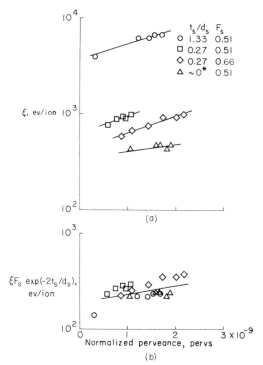

FIG. 13. Effect of relative screen thickness and open area on discharge losses. (a) Discharge loss. (b) Normalized discharge loss. *Screen hole angle $\alpha_s$ is 57 deg. The ratio of screen thickness to hole diameter is $t_s/d_s$, while the open area fraction is $F_s$. Data from Kerslake (60) and Kerslake and Pawlik (61).

that is, lower discharge losses for a lower fraction of Child's law current. This decrease in losses is presumably due to motion of the sheath in the upstream direction with a decrease in current, so that recombination on the screen grid would decrease.

An understanding of ion losses on the screen would be shown by correcting the data of Fig. 13a to essentially a single group of data for all screen configurations. This line of attack is used in a simplified manner in the following analysis. For a simple approximation of the ion losses through a screen hole (upstream of the plasma sheath), assume first that, at any plane normal to the hole axis, the ions recombine on a unit area of side wall at the same rate as they pass in the axial direction. Using the Bohm velocity for both the axial motion and the arrival rate at the side wall, the recombination losses lead to an exponential fall-off in plasma density with distance $x$ through the hole.

$$n_i/n_i' = \exp(-4x/d_s). \tag{1}$$

(The initial ion density upstream of the accelerator is $n_i'$, while $n_i$ is the ion density at location $x$.) We know that the ions actually have a Bohm velocity in the downstream direction at the entrance to the screen hole and therefore will not be deflected into the side wall of the hole as rapidly as they pass downstream. From the differences in losses shown in Fig. 13a for different screen thicknesses, the actual fall-off is about half as rapid as that given in Eq. (1). The ion density ratio across a screen grid of thickness $t_s$ will therefore be about

$$n_i/n_i' = \exp(-2t_s/d_s). \tag{2}$$

To normalize the data of Fig. 13a to a screen of zero thickness and 100% open area, the data are multiplied by the screen open-area fraction $F_s$ and the exponential term of Eq. (2). The result is shown in Fig. 13b.

The most noticeable thing about Fig. 13b is that the 15 : 1 spread in loss of Fig. 13a has been reduced to a spread of less than 2 : 1. Any remaining scatter can easily be attributed to differences in ion chamber operating conditions. It would be a mistake, however, to treat this method of normalizing screen effects as anything more than a first-order approximation.

It has for a long time been suspected that a sufficiently large screen open-area fraction, together with a small enough screen thickness and a low enough perveance per hole, could result in the near total extraction of all ions that arrive at the plane of the screen grid [see Free and Mickelsen (62)]. Recent data obtained by Rawlin (45) with a variety of accelerator systems indicate that this total extraction can occur with screen open-area fractions of about 0.7. With this condition the plasma sheath is believed to move upstream of the screen, permitting ions that would otherwise recombine on the screen to be deflected into extraction trajectories. Such an effect is clearly not included in the simple normalization used for Fig. 13b.

One might suspect that, although the ions are transmitted only partially through a thick screen, the neutrals might be held back to an even greater degree. As shown in Fig. 14, this is not true. The transmission coefficient for

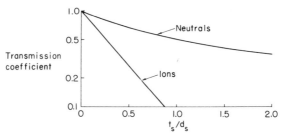

FIG. 14. Transmission of ions [Eq. (2)] and neutrals [Clausing (63)] as a function of the thickness-to-hole diameter ratio for the screen grid.

ions is taken from Eq. (2). The transmission coefficient for neutrals is taken from Clausing (63). It is clear that neutrals escape preferentially over ions for all screen thicknesses. This is more true if one considers the increase in neutral loss that results from recombination of ions within the screen hole. Thus, even if discharge losses are ignored, it is still desirable to use thin screen grids.

The effect is just the opposite for the accelerator grid. Assuming variables are not changed to the point of causing direct interception, the accelerator grid should be as thick as possible and have holes as small in diameter as possible. The beneficial effect of these accelerator changes on utilization was also shown by Rawlin (45).

## 6. Double Ionization

Normal operation of a bombardment ion thruster is expected to produce a beam consisting almost entirely of singly ionized propellant atoms. This type of operation is desirable because the presence of multiply ionized atoms would result in a spread of exhaust velocity, which would give an efficiency loss analogous to that caused by random thermal motion in a rocket exhaust. Singly ionized atoms are normally also desired in ground applications of this type of ion source.

Inasmuch as singly ionized atoms are desired, operating conditions are usually such that only singly and doubly ionized atoms need be considered. The first study of doubly ionized mercury propellant was made by Milder (64), using a mass spectrometer with a fixed magnetic field. The result of this study was felt to be fairly clear and unambiguous. Most of the ionization was believed, at that time, to result from primary electron collisions. Milder obtained a fraction of doubly ionized atoms that was only slightly greater than the ratio indicated by one-step ionizations at the discharge voltage energy. This difference could be explained by the plasma being 5 to 10 V positive of the anode, so that the actual primary electron energy was greater by that amount. The contribution of the two-step ionization process was felt to be small because of the large (Bohm) velocity with which an ion leaves the chamber after being formed.

This straightforward picture then became confused. Masek (30) found only about half of the ions were produced by the high energy primaries, and the low energy Maxwellians could produce essentially no double ionizations in a one-step process. Then too, measurements in thrusters larger than the 10 cm version originally used by Milder indicated higher double ionization fractions than Milder originally obtained (65, 66). This was particularly disturbing because these later thrusters used hollow cathodes, and the primary electrons emitted by a hollow cathode usually lose several electron volts before leaving the pole-piece plasma to enter the main ion chamber.

As a result of these difficulties, the problem of double ionization has been reexamined. The best agreement with experiment at present has been that of Wilbur (67). Wilbur calculated the cross section for the second ionization of singly ionized mercury using the method of Gryziński (68). This calculation required a summation for the single remaining $6_s$ electron and the 10 $5_d$ electrons, as well as an estimate of ionization potential for $5_d$ electrons in a singly ionized atom (estimated at 24 eV). Wilbur calculated cross sections of about 0, 0.7, 1.6, 2.2, and 2.6 × $10^{-20} m^2$ for this process at 20, 30, 40, 50, and 60 eV. This calculated cross section, together with the ion chamber description of Masek, gave the desired agreement with experiment. These calculations of Wilbur supported the earlier suggestion of Pawlik, Goldstein, Fitzgerald, and Adams that the two-step process was actually more important than the single-step process for producing doubly ionized atoms. Further, the two-step process explained the observed increase in the double ionization fraction with ion chamber size.

## 7. Durability and Reliability

Durability is generally not a problem for ion chamber parts other than the cathode (which is treated separately). Perhaps the most serious process for ion chamber parts is the possibility of selective erosion, with the subsequent flaking of the accumulated eroded material elsewhere in the ion chamber. This erosion has been most serious when parts protrude into dense portions of the ion chamber plasma. These flakes have occasionally caused shorting problems in the accelerator system, as described by Nakanishi (69). A recent detailed study on erosion and deposition in the ion chamber used by Nakanishi has been made by Power (70). It should be emphasized, though, that problems of the type described by Nakanishi and Power are the exception rather than the rule.

The effects of propellants other than mercury can be estimated from low voltage sputtering information. The voltage of the discharge and the fraction of double ionization are probably the most important parameters.

## C. Performance Correlations

In the absence of many detailed theoretical solutions, such as exist for accelerator systems, correlations of experimental performance are the best tools for approaching some new ion chamber design. The reader should keep in mind that there are many considerations in undertaking a new ion chamber design—such as operating at a discharge voltage where the ionization cross section is significant, avoiding direct access to the anode for primary electrons, local saturation of permeable pole pieces when used, baffle optimization for a hollow cathode when used, and proper electrical contact when refractory metal cathodes are used. The correlations presented herein are intended only as an aid to selection of overall dimensions and major operating parameters, and not as a substitute for these detailed considerations.

## 1. Residence Time

The residence time correlation of propellant utilization is shown in Fig. 15 for ion chambers with moderately divergent magnetic fields and a variety of propellants. The correlating parameter is $j_+ l_i Q_i W$. The data for

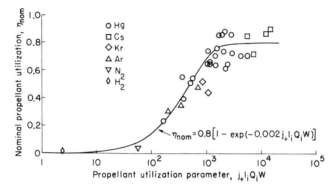

FIG. 15. Residence time correlation of nominal utilization, with the latter defined as the utilization obtained with discharge losses twice the minimum value. This correlation is for moderately divergent magnetic fields. From Kaufman (19).

this correlation were collected from a number of publications given in the original reference (19). Nominal propellant utilization was defined as that value where the discharge losses reached twice the minimum recorded value (at the same neutral flow rate). This arbitrary definition was a means of defining a knee in data that typically exhibited a broad poorly defined knee. The form of the curve shown in Fig. 15 was derived using a simplified representation of the plasma and the resultant probability of a neutral being ionized before it escapes through the accelerator system. Two constants were adjusted in this derived equation to give approximately a $\pm 20\%$ data fit.

A correlation of minimum discharge loss was also included in Kaufman (19), but it showed no significant effect other than a sharp rise below a value of about 10 for the parameter $j_0 l_i Q_i W (\bar{r}_c/r_i)^2$. Above this value the discharge losses were about 500 eV/ion $\pm 50\%$.

In the residence time correlation the total neutral current density $j_0$ is in equivalent $A/m^2$, where the equivalent ampere is the current that would result if each atom or molecule carried one electronic charge. Both the ion beam current density $j_+$ and the neutral current density $j_0$ are based on the entire ion chamber cross section. The ion chamber length $l_i$, ion chamber radius $r_i$, and cyclotron radius $\bar{r}_c$ are in the usual SI units. The cyclotron radius $\bar{r}_c$ is for the minimum magnetic field strength at the screen and an assumed Maxwellian energy of half the first ionization potential $\phi_i$. The first ionization cross section $Q_i$ is the maximum value and measured in units of $\pi a_0^2$ (one $\pi a_0^2$ unit equals $0.88 \times 10^{-20}$ m$^2$), while the atomic weight $W$ is in amu. The values for $\phi_i$ and $Q_i$ used in this correlation are given in Table II.

TABLE II

PROPERTIES USED IN RESIDENCE–TIME CORRELATION[a]

| Propellant | H | H$_2$ | N | N$_2$ | Ar | Kr | Cs | Hg |
|---|---|---|---|---|---|---|---|---|
| $\phi_i$(eV) | 13.6 | 15.6 | 14.5 | 15.5 | 15.8 | 14.0 | 3.9 | 10.4 |
| $Q_i(\pi a_0^2)$ | 0.8 | 1.1 | 1.5 | 2.8 | 4.0 | 5.5 | 13.0 | 6.5 |

[a] From Kaufman (19).

This correlation is easy to use because it ignores detailed differences such as specific magnetic field shape and specific accelerator structure. Much of the scatter in the correlations no doubt arises from these detailed differences. To give some guide in design selection, the magnetic field strength at the screen was typically 0.4–0.8 of the maximum (usually located near the upstream end of the chamber). Most accelerator systems used 50% open screen and accelerator grids.

## 2. Primary Electron Region

The neutral loss correlation for four different thrusters and four propellants is shown in Fig. 16, with the neutral loss parameter defined as $\dot{N}_0 m_i \sigma(V_p/A_p)/A_0$. The sources of the data presented are given in the original reference (43). This correlation is based on a simplified analysis of the primary electron region, treating that region as the determinant of thruster performance. The thrusters used in this investigation had high performance, so that the performance knees (see Fig. 3) tended to be well defined. There was, therefore, no need for an arbitrary definition similar to that used in the preceding correlation.

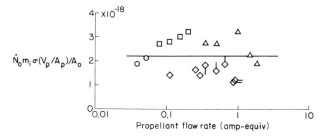

FIG. 16. Neutral loss correlation for "knee" of discharge loss curve for strongly divergent magnetic fields. Anode diameter (cm): ○, 5; □, 10; ◇, 15; △, 30. Propellant: ◇, mercury; ◇, xenon; ◇-, krypton; ◇, argon. From Kaufman and Cohen (43).

Thruster diameters of 5–30 cm were used in this correlation, including the three whose primary electron regions are indicated in Fig. 11. Only the 15 cm thruster was used with other than mercury propellant. Discharge losses at the knee tended to be between 250 and 300 eV/ion, except for the 5 cm thruster with losses of about 550–600 eV/ion.

The neutral loss rate through the accelerator system, $\dot{N}_0$, is defined for knee operation and given in equivalent amps. The utilization is simply

$$\eta_u = 1 - \dot{N}_0/J_0, \tag{3}$$

where $J_0$ is the total propellant flow rate into the ion chamber in equivalent amps. The ion mass $m_i$ is given in amu. The ionization cross sections $\sigma$ used in this correlation are given in Table III. The volume of the primary electron region $V_p$, the area of the boundary surrounding that region $A_p$, and the effective escape area $A_0$ are all given in SI units. The effective escape area is the sharp edged orifice area that will give the same restriction to free molecular flow as the accelerator system.

TABLE III

IONIZATION CROSS SECTIONS USED IN PRIMARY-ELECTRON-REGION CORRELATION[a]

| Propellant | Ar | Kr | Xe | Hg |
|---|---|---|---|---|
| $\sigma(m^2)$ | $3 \times 10^{-20}$ | $4.5 \times 10^{-20}$ | $5 \times 10^{-20}$ | $5 \times 10^{-20}$ |

[a] From Kaufman and Cohen (43).

This correlation gives neutral losses to $\pm 50\%$, which should result in closer estimates of utilization in the range of interest than the preceding correlation. For this more accurate result, additional information is required. The magnetic field shape can be obtained from existing hardware by the customary technique of iron filings on cardboard, fixed with a lacquer spray. (Designing to a given shape is far more difficult.) The calculation of $A_0$ is fairly straightforward. Clausing factors are used for accelerator system apertures. In the case of a screen with a high open-area fraction followed by a thick, more opaque accelerator, the accelerator will probably be the only significant flow restriction.

The earlier data using moderately divergent fields can also be included in a correlation such as that shown in Fig. 16. This was done in several cases, with the result that neutral loss parameters were 2–3 times as large as for the strongly divergent fields shown in Fig. 16. One can conclude that deep primary electron regions (typical of moderately divergent designs) are not efficient, especially when coupled with the poor distribution (without collisions) of primary electrons over ion chamber area (also typical of those designs).

### 3. Multipole Designs

The preceding approach using the primary electron region is readily adaptable to multipole ion chamber designs. The primary electron region in a multipole design is the entire ion chamber, so that $V_p$ and $A_p$ can be obtained from the hardware dimensions (smoothing across pole pieces and anodes). If the multipole design reduces ion losses to walls, as Moore (40) and Ramsey (41) believe, an appropriate reduction in $A_p$ would be required. If one ignores this possible reduction, the theory for the primary electron region approach gives a conservative answer.

Using no correction for reduced ion loss to the walls, the 95% utilization point of Ramsey corresponded to a neutral loss parameter of about $1 \times 10^{-18}$, at the low end of the range shown in Fig. 16. For discharge loss, if the extraction over the beam area $A_b$ is assumed to be nearly total due to the high open-area fraction of the screen, the discharge loss can be estimated by multiplying the 40 eV/ion value of Masek (30) by $A_p/A_b$. This gives $\sim 180$ eV/ion which is quite close to the $\sim 165$ eV/ion given by Ramsey for 95% utilization.

Bias experiments for the engine shell clearly indicated a low ion loss rate to the walls. But, on the other hand, the primary-electron-region approach and the volume discharge loss of Masek are both consistent with wall losses in the multipole design being treated the same as any other thruster. The solution to the apparent disagreement is not presently evident. Even if the wall losses of ions are higher than believed by Moore and Ramsey, the multipole design has a substantial advantage in the shape of the primary electron region. The ratio $V_p/A_p$ is the characteristic length for this region, and a large value is desired for low neutral losses. The value of $V_p/A_p$ for the design used by Ramsey is over 1.6 cm. Thus, with only a moderate ion chamber depth, a value of $V_p/A_p$ was obtained for a 12 cm thruster that was greater than the value of 1.4 cm for the thoroughly developed SERT-II thruster. In other words, the multipole approach appears to offer far greater flexibility for designing to specific end performance.

## III. Electrostatic Acceleration

The function of the accelerator system is to act upon the ions generated within the ion chamber and produce an approximately parallel flow of ions in an exhaust beam of the desired velocity. This function is constrained by the interaction of the accelerator system with the ion chamber (partially discussed in Section II,B,5), the electric field limit of electrical breakdown, the cosine thrust loss due to ion beam divergence, and the current density limitations of space-charge flow and charge-exchange erosion. These constraints and their interactions are described in this section.

### A. General Considerations

Space-charge effects on accelerator system performance can be obtained by solving Poisson's equation

$$\nabla^2 V = -\rho/\varepsilon_0, \tag{4}$$

where $V$ is the electrostatic potential and $\rho$ is the charge density of the charged particle flow. Solution of Poisson's equation for the one-dimensional planar case with zero electric field at the plane of charged particle origin yields the space-charge limited current density

$$j = (4\varepsilon_0/9)(2q/m)^{1/2}(V^{3/2}/l^2), \tag{5}$$

where $q/m$ is the particle charge-to-mass ratio and $V$ is the potential at the distance $l$ from the plane of origin. Equation (5) was originally derived by Child (71) in 1911 and is often called Child's law. Equation (5) was also derived independently by Langmuir (72) in 1913 and Schottky (73) in 1914, and is therefore also known as the Child–Langmuir and Child–Langmuir–Schottky law.

Child's law is useful for estimating space-charge current limits for configurations that approximate the planar case. Child's law can also be used for nonplanar electrodes because only the numerical constant in Eq. (5) changes with electrode shape. Thus the effects of changing the particle charge-to-mass ratio, the accelerating potential difference, or the electrode size can be predicted for any given electrode configuration.

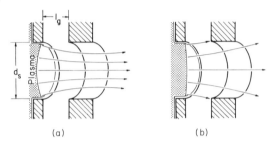

Fɪɢ. 17. Effect of plasma sheath shape on ion trajectories. (a) Normal operation. (b) Excessive ion production.

For an important example of the size effect, consider the accelerator system indicated in Fig. 17a. If all dimensions in this configuration are doubled, the shape is unchanged and Child's law can be used as a scaling relation. The increased value of $l^2$ in the denominator reduces the current density to one-fourth of its former value. But, because the area of the beam (proportional to $d^2$) is at the same time increased by a factor of four, the total beam current from this single aperture is unchanged by the doubling in size.

Because the current per hole is about constant at given operating conditions, the space-charge performance of an accelerator system is maximized by using many small apertures in closely spaced grids. The smallness of the apertures and the closeness of spacing are limited by electrical breakdown and charge-exchange erosion, as well as the more indefinite limits of ease of fabrication and mechanical stability. For the space-charge limit alone, though, the performance per unit area increases continuously as the holes and spacings become smaller.

The indicator of space-charge performance for an accelerator system is the perveance, $J/V^{3/2}$. The current $J$ can be the current through a single aperture, or it can be the total current through a large number of apertures. The potential difference $V$ is measured between whatever electrodes are felt to be appropriate. The units of perveance are pervs. Because there is no size effect on an accelerator system, a length dimension is not involved. A normalized perveance is used in this paper to compare performance on a per hole basis for different ratios of acceleration length to hole diameter. But the inclusion of a factor $l^2/d^2$ does not introduce a length dimension to normalized perveance. The perveances given in this paper are all for singly charged mercury ions. For protons, the mercury perveance should be multiplied by 14.11, while for electrons the factor would be 604.7.

The shape of the plasma sheath shown in Fig. 17a is typical for normal operation. As indicated in Fig. 17a, the ions leave the sheath in a normal direction and are thus directed away from the accelerator electrode. The sheath position shifts downstream (toward the accelerator) if the plasma density is increased. Sufficient increase in plasma density (and ion extraction) leads to a loss of focusing and an increase in accelerator impingement and erosion, as indicated in Fig. 17b. Ion production must, of course, be considered relative to accelerator system current capacity. The effect shown in Fig. 17b can also be obtained by reducing the accelerating potential difference at a constant ion beam current. Because of the movement of the plasma sheath, all accelerator solutions are space-charge limited. The limitation on maximum perveance is the maximum allowable impingement, rather than space-charge limited operation.

The preceding plasma sheath effects were described by Kaufman and Reader (1). These effects, though, are not unique to bombardment thrusters, but are encountered whenever ions are extracted from a plasma through apertures in electrodes. An earlier paper by Thoneman and Harrison (74), since pointed out to the author, described the same qualitative effects for a radio frequency proton source.

The region downstream of the accelerator must also be considered in accelerator system performance. The density of ions in this downstream region is generally sufficient for severe space-charge effects unless neutralized by electrons. The addition of the electrons to the ion beam results in *charge neutralization,* which is one of the two functions of the neutralizer. (The other function is *current neutralization* to avoid the building up of charge on the spacecraft.) The potential of the neutralized ion beam is near that of space and determines the final velocity of ions leaving the accelerator system. In practice the ion beam potential is made positive relative to the accelerator electrode, as indicated in Fig. 18. This sequence of potentials is called accel–decel and the net-to-total voltage ratio typically ranges from 0.5 to 0.8. The approximate upper limit of 0.8 is set by electron backstreaming through the accelerator system. The reason for the value of 0.8 instead of something nearer unity is that the onset of backstreaming is determined by

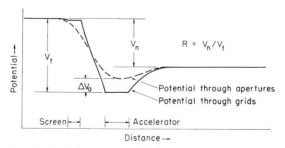

FIG. 18. Variation of potential through accelerator system.

the potentials of the "saddle points" in the grid apertures instead of just the accelerator potential. The approximate lower limit of 0.5 is set by increased erosion from charge exchange ions and increased loss in thrust due to ion beam spread. A general rule of charged particle optics is that acceleration focuses and deceleration defocuses.* A large amount of deceleration (low net-to-total ratio) thus results in a defocused beam and large off-axis cosine losses.

Closely spaced, mechanically simple, two-grid systems suit bombardment thruster requirements quite well. This is because high current densities are required at moderate potential differences, and poor beam collimation is usually a problem only if severe enough to give substantial cosine losses. Unfortunately, mechanical simplicity does not in this case correspond to analytical simplicity. The early bombardment-thruster accelerator systems were therefore the result of cut-and-try methods. Rubber sheet analogs were sometimes used, but gave rather poor quantitative results. Another shortcoming was that rubber sheet analogs are suited primarily to Laplace's equation (with no space charge), and are difficult to adapt to Poisson's equation.

### B. Space-Charge Flow Solutions

With the theoretical foundations for space-charge flow already in existence, solutions for the bombardment thruster waited on the development of practical methods suitable for this problem. The first reasonably convenient method utilized the electrolytic tank analog. This analog was used for solution of Laplace's equation by Pierce (75) and modified (by the use of current injection into the space-charge region) for Poisson's equation by Hollway (76). This method of solving Poisson's equation was developed in detail and used effectively by Brewer (77) and Van Duzer and Brewer (78). At this point the electrolytic tank analogue was suitable for use with problems in which charged particles originated at a well defined surface, such as the electron emitter in an electron gun or the ionizer in a contact thruster. But in an electron-bombardment thruster the ions effectively originate at a sheath, with the position of this sheath dependent on operating parameters of both the ion chamber and the accelerator system.

A solution of the plasma sheath position was first given by Hyman et al. (79). The sheath assumes a position such that everywhere on the sheath: (i) The potential is uniform at the plasma value. (ii) The electric field normal to the sheath is zero. (iii) The extracted ion current density is everywhere equal to the maximum (saturation) value.

---

* This is more rigorously stated in terms of positive and negative second derivatives of potential with distance.

The procedure described by Hyman *et al.* is essentially an electrolytic tank solution of the space-charge flow together with a trial and error determination of sheath location to meet the above criteria.

The development and widespread use of digital computers provided another convenient means for obtaining accelerator system solutions. A program for two-dimensional configurations (variables in only two cartesian coordinates) was published by Hamza and Richley (*80*), followed by a program for axisymmetric configurations by Hamza (*81*). A general program that could be used on either two-dimensional or axisymmetric configurations was then published by Bogart and Richley (*82*). The program of Bogart and Richley, together with the sheath conditions of Hyman *et al.*, provides at present the most efficient and widely used approach to space-charge flow problems in electron-bombardment thrusters.

A final point of space-charge flow theory concerns the plasma sheath immediately downstream of the accelerator system. The electrons and ions in the ion beam form a plasma of nearly uniform potential. The upstream end of this region is a plasma sheath that is the downstream boundary to the acceleration region. The location of this downstream boundary (or closure surface) is shown in Fig. 19. The ion trajectories are essentially straight lines

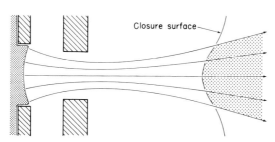

FIG. 19. Location of closure surface—the downstream boundary of the deceleration region.

after passing through this closure surface. An iterative procedure, similar to that used for the upstream plasma sheath, can be used to locate this closure surface. The criteria for location of the boundary are: (i) The boundary potential is uniform at the desired ion beam potential. (ii) The electric field normal to the boundary is zero.

A downstream boundary solution was first given by Pawlik *et al.* (*83*). Because the ions are moving rapidly near the downstream boundary, their motion is only slightly perturbed by changes in location of this boundary. The overall space-charge solution is therefore less sensitive to the accurate location of this plasma sheath than to the location of the sheath near the screen. As a result, simple flat planes are often used in place of more precisely defined surfaces.

A substantial number of space-charge flow solutions are available in the literature. Numerical values for solutions by Kramer and King (84), Lathem (85, 86), and Nudd and Amboss (87) are given in Table IV. All solutions except those of Kramer and King included simulation of the closure plane. The solutions of Kramer and King were included to show a comparison of conical and cylindrical screen holes obtained with the same method of solution. All conical screen holes have their smallest diameters facing the accelerator grid. All accelerator grids have cylindrical holes. The definitions of the different measurements are indicated in Fig. 20. It was necessary to calculate

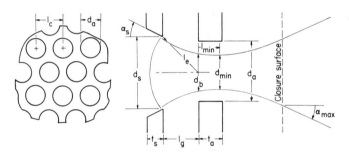

FIG. 20. Accelerator system measurements.

some of the parameters included in Table IV from trajectory plots given in cited references. In the case of Lathem's data, however, computer printouts were available for more precise values. Web thickness (the distance of closest approach between adjacent holes) ranged from $0.1d_s$ to $\infty$. The effect of this variable was found to be minor, so it was omitted from the tabulation.

To put the information of Table IV in a form suitable for the widest possible range of application, correlation (or normalization) parameters were used. Fundamental to the rest of the correlation was the selection of proper definitions for a normalized perveance of the form $(I/V^{3/2})(l^2/d^2)$. Fairly standard definitions have been used for most of the values. The current is that through a single hole, the voltage is the total screen–accelerator potential difference $V_t$, and the screen hole diameter was selected for the characteristic diameter.

For length, various combinations of separation gap $l_g$ and grid thicknesses $t_s$ and $t_a$ have been used in the past. Such definitions have a serious shortcoming that is illustrated in Fig. 21. When the spacing between the grids is large compared to the grid hole diameter, as shown in Fig. 21a, the separation $l_g$ is a fair approximation of the space-charge acceleration distance.

TABLE IV: Space-Charge Flow Solutions

| $\alpha_s$ (deg) | $t_s/d_s$ | $l_g/d_s$ | $t_a/d_s$ | $d_a/d_s$ | $R$, $(V_n/V_t)$ | $J/V_t^{3/2}$ (npervs/hole) | $d_b/d_s$ | $d_{min}/d_s$ | $l_{min}/d_s$ | $\alpha_{max}$ (deg) | $C_F$ | $R_{max}$ | Ref. |
|---|---|---|---|---|---|---|---|---|---|---|---|---|---|
| 0 | 0.25 | 0.50 | 0.50 | 1.00 | 0.80 | 4.62 | 0.75 | 0.75 | 0.05 | 16.0 | 0.974 | 0.869 | (85) |
| 0 | 0.25 | 0.50 | 0.50 | 1.00 | 0.80 | 3.28 | 0.62 | 0.62 | 0.10 | 11.0 | 0.989 | 0.872 | (85) |
| 0 | 0.25 | 0.50 | 0.50 | 1.00 | 0.80 | 1.90 | 0.43 | 0.38 | 0.40 | 4.7 | 0.998 | 0.882 | (85) |
| 0 | 0.25 | 0.50 | 0.50 | 0.665 | 0.80 | 3.28 | 0.54 | 0.53 | 0.15 | 9.4 | 0.992 | 0.937 | (85) |
| 0 | 0.25 | 0.50 | 0.50 | 0.665 | 0.80 | 2.83 | 0.54 | 0.52 | 0.15 | 8.9 | 0.994 | 0.945 | (85) |
| 0 | 0.25 | 0.50 | 0.50 | 0.665 | 0.80 | 1.82 | 0.38 | 0.31 | 0.70 | 3.3 | 0.999 | 0.935 | (85) |
| 0 | 0.25 | 1.00 | 0.50 | 0.665 | 0.80 | 1.18 | 0.43 | 0.43 | 0.20 | 5.2 | 0.998 | 0.964 | (85) |
| 0 | 0.25 | 0.50 | 0.95 | 0.665 | 0.80 | 3.28 | 0.54 | 0.52 | 0.10 | 10.9 | 0.989 | 0.966 | (85) |
| 0 | 0.25 | 0.50 | 0.25 | 0.665 | 0.80 | 3.26 | 0.54 | 0.53 | 0.15 | 7.8 | 0.995 | 0.892 | (85) |
| 0 | 0.20 | 0.253 | 0.20 | 1.00 | 0.75 | 4.10 | 0.60 | 0.59 | 0.18 | 13.2 | 0.987 | 0.760 | (86) |
| 0 | 0.20 | 0.253 | 0.20 | 1.00 | 0.75 | 2.20 | 0.45 | 0.36 | 0.50 | 5.2 | 0.998 | 0.786 | (86) |
| 0 | 0.20 | 0.253 | 0.20 | 1.00 | 0.75 | 1.11 | 0.32 | 0.17 | 0.77 | 7.7 | 0.997 | 0.811 | (86) |
| 0 | 0.20 | 0.505 | 0.20 | 1.00 | 0.75 | 3.85 | 0.70 | 0.70 | 0.03 | 14.1 | 0.982 | 0.808 | (86) |
| 0 | 0.20 | 0.505 | 0.20 | 1.00 | 0.75 | 1.19 | 0.31 | 0.25 | 0.56 | 7.0 | 0.997 | 0.829 | (86) |
| 0 | 0.25 | 0.50 | 0.50 | 1.00 | 0.60 | 4.61 | 0.75 | 0.75 | -0.05 | 18.5 | 0.963 | | (85) |
| 0 | 0.25 | 0.50 | 0.50 | 1.00 | 0.60 | 3.27 | 0.62 | 0.62 | 0.10 | 14.4 | 0.981 | | (85) |
| 0 | 0.25 | 0.50 | 0.50 | 1.00 | 0.60 | 1.89 | 0.43 | 0.38 | 0.35 | 7.3 | 0.996 | | (85) |
| 0 | 0.25 | 0.50 | 0.50 | 0.665 | 0.60 | 3.27 | 0.54 | 0.53 | 0.15 | 12.4 | 0.987 | | (85) |
| 0 | 0.25 | 0.50 | 0.50 | 0.665 | 0.60 | 1.82 | 0.38 | 0.31 | 0.62 | 5.9 | 0.998 | | (85) |
| 0 | 0.376 | 0.619 | 0.376 | 0.807 | 0.60 | 1.49 | 0.41 | 0.35 | 0.53 | 4.4 | 0.998 | | (86) |
| 0 | 0.376 | 0.619 | 0.376 | 0.807 | 0.60 | 0.60 | 0.17 | 0.07 | 0.53 | 17.0 | 0.990 | | (86) |
| 0 | 0.38 | 0.625 | 0.38 | 0.815 | 0.60 | 0.38 | 0.12 | 0.10 | 0.14 | 14.4 | 0.985 | | (86) |
| 0 | 0.38 | 0.625 | 0.76 | 0.815 | 0.60 | 0.38 | 0.12 | 0.10 | 0.14 | 14.6 | 0.985 | | (86) |
| 20 | 0.16 | 0.481 | 0.321 | 0.759 | 0.50 | 5.25 | 0.65 | 0.64 | 0.08 | 23 | 0.96 | | (87) |
| 20 | 0.16 | 0.481 | 0.321 | 0.759 | 0.50 | 2.98 | 0.51 | 0.49 | 0.19 | 14 | 0.98 | | (87) |
| 20 | 0.16 | 0.481 | 0.321 | 0.759 | 0.50 | 1.51 | 0.31 | 0.23 | 0.46 | 26 | 0.98 | | (87) |
| 20 | 0.16 | 0.374 | 0.321 | 0.759 | 0.50 | 5.89 | 0.59 | 0.57 | 0.17 | 20 | 0.965 | | (87) |
| 20 | 0.16 | 0.374 | 0.321 | 0.759 | 0.50 | 1.70 | 0.24 | 0.13 | 0.41 | 32 | 0.95 | | (87) |
| 0 | 0.15 | 0.48 | 0.15 | 0.75 | | 3.04 | 0.58 | | | | | | (84) |
| 45 | 0.15 | 0.48 | 0.15 | 0.75 | | 3.04 | 0.46 | | | | | | (84) |

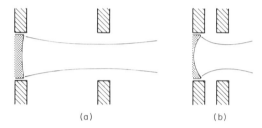

FIG. 21. Effect of interelectrode gap on shape of plasma sheath. (a) Large $l_g/d_s$. (b) Small $l_g/d_s$.

When that separation is relatively small, as shown in Fig. 21b, the separation distance is clearly too small. To account for the recession of the plasma sheath in normal operation, an effective length $l_e$ was defined.

$$l_e = (l_g^2 + d_s^2/4)^{1/2}. \tag{6}$$

This definition corresponds to the length from the center of the accelerator hole (upstream side of the accelerator) to the near edge of the screen hole, as indicated in Fig. 20.

### 1. Ion Beam Diameter

The diameter of the ion beam $d_b$ is shown for the data of Table IV in Fig. 22. There is no significant effect of different net-to-total voltage ratios, $R$, for the data shown. In fact, some data of Table IV are identical except for $R$, and have identical—or near identical—values of $d_b/d_s$. The most significant cause of data spread appears to be the shape of the screen holes, and separate curves are drawn for conical and cylindrical screen holes. It just happens that all of the Nudd and Amboss data are for conical holes and all the

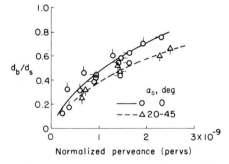

FIG. 22. Ion beam diameter at entrance to accelerator hole. References: O– (84); –O (85); Ó (86); Q (87). Data from Kramer and King (84), Lathem (85, 86), and Nudd and Amboss (87).

Lathem data are for cylindrical holes. Having shown no effect of conditions downstream of the accelerator on $d_b/d_s$ (no $R$ effect), the data of Kramer and King (84) can be used to show that the difference between the two sets of data is not the result of a systematic difference in the method of solution used by Lathem and by Nudd and Amboss. The two points of Kramer and King are at a normalized perveance of $1.46 \times 10^{-9}$ and are identical in dimensions and operating conditions, except that one has a conical screen hole and one a cylindrical screen hole. The spread between these two points matches quite closely the average spread between conical and cylindrical data.

Although Kramer and King were the first to show comparative space-charge flow solutions for the two types of screen holes, the possible advantage of conical holes had been indicated earlier by Kerslake (60) and Kerslake and Pawlik (61). Kerslake operated a thruster with conical holes that were larger on the downstream side. This was an attempt to approximate the first electrode of a Pierce-type accelerator system. This electrode showed some marginal value in reducing accelerator impingement current. Later, Kerslake and Pawlik showed that reversing the screen grid (large diameters upstream) reduced the impingement current even further. The advantage of conical screen holes has thus been clearly demonstrated by both experiment and theory but insufficient data exists to determine the optimum value of this angle. There is also the interaction with the ion chamber to be considered, in which the ion chamber losses decrease with increasing cone angle.

Figure 22 is useful for estimating the maximum current capacity that can be obtained without large direct impingement of the accelerator electrode. But some additional information is required on what happens after the beam enters the accelerator hole. The minimum diameter of the beam $b_{min}$ and the distance this minimum is downstream of the entrance to the accelerator hole $l_{min}$ are both shown in Fig. 23. The values of $d_{min}/d_s$ are primarily a function of normalized perveance and hole shape. This is also true for $l_{min}/l_e$ at normalized perveances above about $1 \times 10^{-9}$. Below this value, $R$ has a substantial effect. The reason that $R$ can have a noticeable effect on $l_{min}/l_e$, but not on $d_{min}/d_s$, in this range is simply that the diameter of the beam is small and the distance downstream generally large. Thus small angular changes and small diameter changes can correspond to a large change in the location of the minimum.

The normal perveance regimes for accelerator operation are indicated in Fig. 24. The data of Table IV and the figures of this section are limited to perveances less than that of direct interception, so that the data correspond to Figs. 24a and 24b. In general, data above $1 \times 10^{-9}$ normalized perveance correspond to Fig. 24b, while those below $1 \times 10^{-9}$ correspond to Fig. 24a.

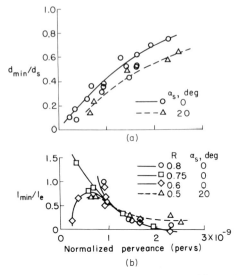

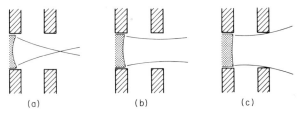

FIG. 23. Minimum ion beam diameter and its location. (a) Minimum beam diameter. (b) Distance from entrance to accelerator. Data from Lathem (85, 86) and Nudd and Amboss (87).

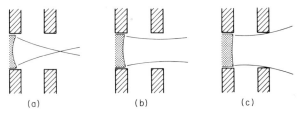

FIG. 24. Perveance regimes for accelerator system operation. (a) Low perveance, crossover. (b) Medium perveance, normal operation. (c) High perveance, direct impingement.

The effects of external fields are felt the most in the center of the accelerator aperture. One might therefore expect the condition of Fig. 24a to be most sensitive to variations in $R$ and, hence, expect the spread with $R$ below $1 \times 10^{-9}$ shown in Fig. 23b.

## 2. Beam Spread and Thrust Coefficient

A similar effect of $R$ is found at low perveances for maximum half-angle of beam spread $\alpha_{max}$ and thrust coefficient $C_f$. Both $\alpha_{max}$ and $C_f$ are determined at the closure surface downstream of the accelerator system. Now Figs. 22 and 23 are consistent with the view that all values of $l_e/d_s$ have similar behavior at the same normalized perveance, with lengths scaled in proportion to $l_e$. This scaling with $l_e$ can be stated in another form, that the trajectory angles relative to the axis are inversely proportional to $l_e/d_s$ for operation at the same normalized perveance. In line with this view, the normalized angle parameter is assumed to be the product of $\alpha_{max}$ and $l_e/d_s$. There is also an effect of $R$ on the angle of a trajectory after it leaves the accelerator hole. The decelerating electric field is primarily axial, so that deceleration results in a greater loss of axial velocity than transverse velocity. Assuming no effect of deceleration on transverse velocity, it can be shown that the off-axis angle of a trajectory will be increased by the factor $R^{-1/2}$ during deceleration. Including this last effect gives a complete normalized angle parameter of $\alpha_{max} R^{1/2} l_e/d_s$. ("Normalized" in this case means normalized to an $l_e/d_s$ of unity and an $R$ of unity.)

The plot of normalized angle data in Fig. 25 shows excellent correlation except at low normalized perveance—in the crossover region. The same approach was used for thrust coefficient $C_f$ by first converting $C_f$ into a mean off-axis angle by using $\cos^{-1} C_f$. This correlation is shown in Fig. 26 and is quite similar to the maximum angle correlation. It should be noticed that the data for an $R$ of 0.5 are also the data for the conical screen holes. The graphical techniques used to calculate $\alpha_{max}$ and $C_f$ values for these data

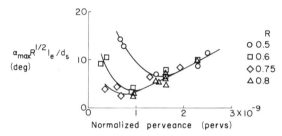

FIG. 25. Normalized beam spread. Data from Latham (85, 86) and Nudd and Amboss (87).

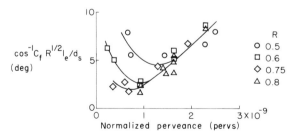

FIG. 26. Normalized thrust coefficient. Data from Latham (85, 86) and Nudd and Amboss (87).

are more uncertain than corresponding values obtained from computer print-outs. Thus, although Figs. 25 and 26 are drawn with only a single curve at high perveances, more accurate data may show reduced off-axis angles $(\alpha_{max})$ and reduced thrust losses (lower $\cos^{-1} C_f$) for conical screen holes.

## 3. Electron Backstreaming

Another correlation that was made of Table IV data concerns electron backstreaming. As discussed in connection with Fig. 18, backstreaming occurs when the minimum potential within the accelerator aperture is high enough to permit electrons from the ion beam to flow backwards through the accelerator system. The axial potential variations were available for all of Lathem's data. The minimum axial potential in the accelerator aperture approximates the minimum potential that the beam can reach without electron backstreaming, and can be used to calculate the maximum permissible value of $R$, or $R_{max}$. Only the data at $R$ values of 0.8 and 0.75 were used to calculate $R_{max}$, because lower $R$ values were further from the calculated $R_{max}$ values and, therefore, less accurate. The aperture effect of a circular hole in a thin electrode was used to obtain a generalization parameter. An idealized accelerator system is indicated in Fig. 27. For a spacing between

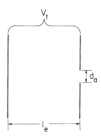

FIG. 27. Idealized accelerator system for accelerator aperture effect.

electrodes $l_e$ that is large compared to the aperture $d_a$, Spangenberg (88) gives the potential in the center of the accelerator hole relative to the accelerator potential as

$$|\Delta V_a| = (V_t/2\pi)d_a/l_e .  \qquad (7)$$

This ignores the electric field downstream of the accelerator, but that electric field is generally small compared to the electric field between screen and accelerator near backstreaming conditions. Inasmuch as

$$R_{max} = 1 - \Delta V_a/V_t ,  \qquad (8)$$

we can write

$$(1 - R_{max})l_e/d_a = 1/2\pi. \tag{9}$$

The expression $(1 - R_{max})l_e/d_a$ is thus the desired generalization parameter.

The aperture effect is exact for very large values of $l_e/d_a$. The effect of using moderate values of $l_e/d_a$, however, does not appear as serious as the effect of a finite accelerator thickness, which was not included in Eq. (7). The backstreaming data are plotted in Fig. 28 against normalized perveance.

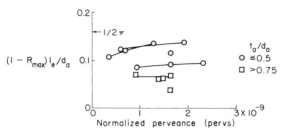

FIG. 28. Backstreaming parameter as a function of normalized perveance. Curves connect data that are identical except for ion current. Data from Lathem (*85, 86*).

The data all fall below the theoretical value of $1/2\pi$. More to the point, data with large $t_a/d_a$ fall below the data with small $t_a/d_a$. It should also be noted, though, that the effect of perveance is small over the range of interest, so that to the first approximation the perveance can be ignored. The same data are therefore plotted against accelerator thickness-to-diameter ratio $t_a/d_a$ in Fig. 29. The correlation obtained in Fig. 29 can be described adequately by the equation

$$(1 - R_{max})l_e/d_a = \exp(-t_a/d_a)/2\pi. \tag{10}$$

It appears that Eq. (10) could have been obtained entirely from aperture effects if the accelerator thickness effect had been included. Computer printouts of Laplace solutions were also available and indicate that this is not

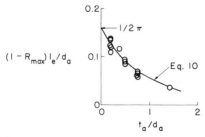

FIG. 29. Backstreaming parameter as a function of relative accelerator thickness. Data from Lathem (*85, 86*).

true, particularly for large $t_a/d_a$. While $R_{max}$ did not vary much with perveance for a given configuration, the difference between some space charge
and no space charge was often significant. Apparently the small beam
diameter at low perveance can result in a large change in potential in the
accelerator hole for the amount of space charge. At higher perveance a larger
amount of space charge is involved, but the larger beam diameter results in
only a small increase in maximum potential within the accelerator hole. The
percentage difference between the La Place and Poisson solutions increased
with $t_a/d_a$, as indicated earlier. For example, at the point with $t_a/d_a$ equal to
1.43, $(1 - R_{max})$ dropped from 0.034 with space charge to 0.003 without.

### C. Experimental Accelerator System Performance

When an accelerator system is operated with a thruster at a constant
beam current and a constant net-to-total voltage ratio, the accelerator current has the typical variation with total voltage shown in Fig. 30. Above a

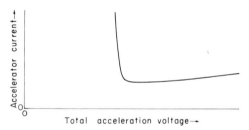

FIG. 30. Typical variation of accelerator current at constant values of beam current,
propellant utilization, and net-to-total voltage ratio $R$.

certain value of voltage the accelerator current is nearly constant, and results
from charge exchange and a small amount of direct impingement. Below this
voltage the accelerator current rises rapidly with decreasing voltage. The
gain in perveance below this "knee" is usually small compared with the
rapid increase in erosion, so such operation is usually avoided outside of
performance mapping. The exact definition of this knee is somewhat arbitrary, but usually corresponds to a rise in accelerator current of 20 to 50%
above the nearly constant level.

Operation at the knee means that the direct interception condition of
Fig. 24c is encountered somewhere in the accelerator system—usually near
the center. Surveys of ion beam density have been made close to the accelerator and they usually show a variation similar to one of the curves in Fig. 31.
The average perveance of an accelerator system operating at the accelerator
current knee thus depends strongly on the ratio of average-to-peak current

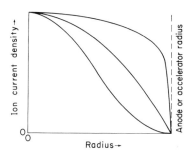

FIG. 31. Typical current density profiles for ions leaving accelerator systems.

density. Typical values of this current density ratio can be determined by comparing space-charge flow solutions with experimental values of perveance for operation at the accelerator current knee.

## 1. Perveance

The maximum perveance was evaluated for a large number of accelerator system configurations by Kerslake (60) and Kerslake and Pawlik (61), with most of this data obtained using the same 10 cm source. Kerslake and Pawlik recognized that thermal warping limited the validity of their small-gap data. The present rule of thumb is that the interelectrode gap should be a minimum of about 1/60 of the beam diameter if serious warping problems are to be avoided with flat grids. Data obtained with smaller gaps were therefore omitted from the experimental performance of Kerslake (60) and Kerslake and Pawlik (61) given in Table V.

Sufficient data are given in Table V to give a good indication of the range of normalized perveance that can be expected for $d_a = d_s$ and $R \geq 0.5$. These data are plotted in Fig. 32. From space-charge flow solutions for $d_a = d_s$, we

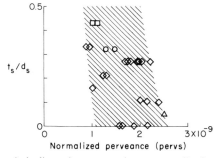

FIG. 32. Effect of screen hole dimensions on maximum normalized perveance for flat grids with equal screen and accelerator hole diameters. $d_s$ (mm): $\bigcirc$, 1.59; $\square$, 3.15; $\Diamond$, 4.76; $\triangle$, 9.52. Range of $R$ is from 0.5 to 0.8. Data from Kerslake (60) and Kerslake and Pawlik (61).

## TABLE V

Experimental Accelerator Performance (10 cm Thruster, $V_t$ = 2150–5830)

| $\alpha_s$ (deg) | $d_s$ (mm) | $F_s$ | $t_s/d_s$ | $l_g/d_s$ | $t_a/d_s$ | $d_a/d_s$ | $R$ $(V_n/V_t)$ | $J/V_t^{3/2}$ (npervs/hole) | Ref. |
|---|---|---|---|---|---|---|---|---|---|
| 0 | 1.59 | 0.51 | 0.321 | 1.245 | 0.642 | 1.00 | 0.8 | 0.82 | (61) |
| 0 | 1.59 | 0.51 | 0.321 | 1.245 | 0.642 | 1.00 | 0.6 | 0.72 | (61) |
| 0 | 1.59 | 0.51 | 0.321 | 1.245 | 0.642 | 1.00 | 0.4 | 0.49 | (61) |
| 0 | 3.15 | 0.22 | 0.162 | 0.686 | 0.162 | 1.762 | 0.8 | 2.76 | (61) |
| 0 | 3.15 | 0.50 | 0.403 | 0.322 | 0.403 | 1.00 | 0.8 | 2.93 | (60) |
| 0 | 3.15 | 0.50 | 0.403 | 0.322 | 0.403 | 1.00 | 0.6 | 3.15 | (60) |
| 0 | 3.15 | 0.50 | 0.403 | 0.322 | 0.403 | 1.00 | 0.4 | 3.02 | (60) |
| 0 | 4.76 | 0.51 | 0.309 | 0.784 | 0.267 | 0.334 | 0.8 | 1.15 | (61) |
| 0 | 4.76 | 0.51 | 0.309 | 0.784 | 0.267 | 0.334 | 0.6 | 1.10 | (61) |
| 0 | 4.76 | 0.51 | 0.309 | 0.784 | 0.267 | 0.334 | 0.4 | 0.94 | (61) |
| 0 | 4.76 | 0.51 | 0.107 | 0.384 | 0.107 | 1.00 | 0.72 | 6.03 | (60) |
| 0 | 4.76 | 0.51 | 0.107 | 0.384 | 0.107 | 1.00 | 0.6 | 5.08 | (60) |
| 0 | 4.76 | 0.51 | 0.107 | 0.795 | 0.107 | 1.00 | 0.72 | 2.44 | (60) |
| 0 | 4.76 | 0.51 | 0.160 | 0.368 | 0.160 | 1.00 | 0.72 | 2.64 | (60) |
| 0 | 4.76 | 0.51 | 0.160 | 0.368 | 0.160 | 1.00 | 0.6 | 2.68 | (60) |
| 0 | 4.76 | 0.51 | 0.160 | 0.368 | 0.160 | 1.00 | 0.4 | 2.42 | (60) |
| 0 | 4.76 | 0.51 | 0.214 | 0.410 | 0.214 | 1.00 | 0.72 | 3.14 | (60) |
| 0 | 4.76 | 0.51 | 0.214 | 0.410 | 0.214 | 1.00 | 0.6 | 2.99 | (60) |
| 0 | 4.76 | 0.51 | 0.267 | 0.492 | 0.267 | 1.00 | 0.8 | 3.61 | (61) |
| 0 | 4.76 | 0.51 | 0.267 | 0.492 | 0.267 | 1.00 | 0.6 | 3.69 | (61) |
| 0 | 4.76 | 0.51 | 0.267 | 0.492 | 0.267 | 1.00 | 0.4 | 3.38 | (61) |
| 0 | 4.76 | 0.51 | 0.267 | 0.655 | 0.267 | 1.00 | 0.8 | 2.87 | (61) |
| 0 | 4.76 | 0.66 | 0.273 | 0.576 | 0.273 | 1.00 | 0.8 | 3.14–3.83 | (60) |
| 0 | 4.76 | 0.66 | 0.273 | 0.576 | 0.273 | 1.00 | 0.6 | 2.91–3.52 | (60) |
| 0 | 4.76 | 0.66 | 0.273 | 0.576 | 0.273 | 1.00 | 0.5 | 3.43 | (60) |
| 0 | 4.76 | 0.66 | 0.273 | 0.576 | 0.273 | 1.00 | 0.4 | 2.62–3.52 | (60) |
| 0 | 4.76 | 0.66 | 0.273 | 0.576 | 0.273 | 1.00 | 0.3 | 1.94 | (60) |
| 0 | 4.76 | 0.51 | 0.326 | 0.624 | 0.326 | 1.00 | 0.8 | 2.25 | (61) |
| 0 | 4.76 | 0.51 | 0.326 | 0.624 | 0.326 | 1.00 | 0.6 | 2.13 | (61) |
| 0 | 4.76 | 0.51 | 0.326 | 0.624 | 0.326 | 1.00 | 0.4 | 1.70 | (61) |
| 0 | 4.76 | 0.51 | 0.326 | 0.773 | 0.326 | 1.00 | 0.8 | 1.52 | (61) |
| 0 | 4.76 | 0.51 | 0.326 | 0.773 | 0.326 | 1.00 | 0.6 | 1.36 | (61) |
| 0 | 4.76 | 0.51 | 1.334 | 0.655 | 0.326 | 1.00 | 0.8 | 1.44 | (61) |
| 0 | 4.76 | 0.51 | 1.334 | 0.655 | 0.326 | 1.00 | 0.6 | 1.47 | (61) |
| 0 | 4.76 | 0.51 | 1.334 | 0.655 | 0.326 | 1.00 | 0.4 | 1.35 | (61) |
| 45 | 4.76 | 0.51 | 0.267 | 0.538 | 0.267 | 1.00 | 0.8 | 3.18 | (61) |
| 45 | 4.76 | 0.51 | 0.267 | 0.538 | 0.267 | 1.00 | 0.6 | 2.89 | (61) |
| 45 | 4.76 | 0.51 | 0.267 | 0.538 | 0.267 | 1.00 | 0.4 | 3.00 | (61) |
| 45 | 4.76 | 0.51 | 0.267 | 0.790 | 0.267 | 1.00 | 0.8 | 1.70 | (61) |
| 45 | 4.76 | 0.51 | 0.267 | 0.790 | 0.267 | 1.00 | 0.6 | 2.61 | (61) |
| 57 | 4.76 | 0.51 | 0.235 | 0.410 | 0.267 | 1.00 | 0.9 | 5.21 | (61) |
| 57 | 4.76 | 0.51 | 0.235 | 0.410 | 0.267 | 1.00 | 0.6 | 4.45 | (61) |

*(continued)*

TABLE V—*continued*

| $\alpha_s$ (deg) | $d_s$ (mm) | $F_s$ | $t_s/d_s$ | $l_g/d_s$ | $t_a/d_s$ | $d_a/d_s$ | $R$ $(V_n/V_t)$ | $J/V_t^{3/2}$ (npervs/hole) | Ref. |
|---|---|---|---|---|---|---|---|---|---|
| 57 | 4.76 | 0.51 | 0.235 | 0.410 | 0.267 | 1.00 | 0.4 | 4.13 | (61) |
| 57 | 4.76 | 0.51 | 0.235 | 0.832 | 0.267 | 1.00 | 0.8 | 1.73 | (61) |
| 57 | 4.76 | 0.51 | 0.235 | 0.832 | 0.267 | 1.00 | 0.6 | 1.70 | (61) |
| 57 | 4.76 | 0.51 | 0.235 | 0.832 | 0.267 | 1.00 | 0.4 | 1.63 | (61) |
| 0 | 4.76 | 0.51 | 0.273 | 0.384 | 0.273 | 1.082 | 0.8 | 5.22–5.68 | (60) |
| 0 | 4.76 | 0.51 | 0.273 | 0.384 | 0.273 | 1.082 | 0.6 | 4.86–5.51 | (60) |
| 0 | 4.76 | 0.51 | 0.273 | 0.384 | 0.273 | 1.082 | 0.4 | 4.45–4.48 | (60) |
| 0 | 4.76 | 0.51 | 0.273 | 0.384 | 0.273 | 1.082 | 0.3 | 2.68 | (60) |
| 0 | 4.76 | 0.51 | 0.273 | 0.852 | 0.273 | 1.082 | 0.8 | 2.30 | (60) |
| 0 | 4.76 | 0.51 | 0.273 | 0.852 | 0.273 | 1.082 | 0.6 | 2.89 | (60) |
| 0 | 4.76 | 0.51 | 0.273 | 0.852 | 0.273 | 1.082 | 0.4 | 1.84 | (60) |
| 0 | 9.52 | 0.51 | 0.054 | 0.996 | 0.054 | 1.00 | 0.78 | 2.02 | (60) |

expect a normalized perveance of about $3 \times 10^{-9}$ before direct interception occurs. The range of perveance shown in Fig. 32 for very thin screens, therefore, indicates an average-to-maximum current density ratio from about 0.4 to 0.8. For a simple ion chamber configuration similar to that used by Kerslake, then, a ratio of 0.4 to 0.5 should be easy to obtain, a ratio of 0.6 might be expected with some development, while a ratio of 0.7 to 0.8 might take a great deal of development. More recent ion chamber designs such as either the radial field or multipole configurations should reach ratios of 0.7 to 0.8 with less difficulty.

The experimental range in Fig. 32 shifts slightly to lower values of perveance for thicker screen grids. This shift is not felt to indicate a change in average-to-maximum current density. It is also not an absolute size effect (to be discussed in connection with dished grid performance in Section III,D,3) because the limits are defined by one size of screen hole. Instead, it is felt to indicate an effect of screen thickness on the maximum perveance for a single hole. This effect of screen thickness has no counterpart in present space-charge flow solutions.

The method of solution in space-charge flow is to let the plasma sheath move to whatever location is required for a constant density plasma. There is usually some minimum approach distance used for originating trajectories near the side of the screen hole. This minimum distance is based on the qualitative knowledge that plasma density falls off near this surface. But, the method of solution is basically independent of screen thickness, and the screen can be assumed to extend any arbitrary distance back into the plasma. In the actual case, the fall-off in plasma density as the wall is approached would result in the plasma sheath moving upstream an amount roughly in proportion to the decrease. This means that the sheath near the edge of the screen hole directs ion trajectories more toward the accelerator as indicated in Fig. 33. If direct interception is to be avoided, there must be a corresponding decrease in current, and hence perveance, for that screen hole. This decrease in maximum perveance with a thick screen would, of course, also increase off-axis losses at a given perveance. The perveance and off-axis losses due to screen thickness, together with the previously described ion chamber losses, require that the screen be made as thin as possible consistent with structural and fabrication limitations.

The overall variation in ion current density also causes problems near the edge of the ion beam. The very low density at the edge of the ion beam usually results in crossover trajectories (Fig. 24a) with direct impingement. Whether the crossover condition is sufficient for direct impingement (and observed erosion) is not entirely clear. The normalized perveance at the edge

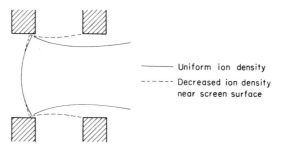

FIG. 33. Effect of decreased ion density near screen surface on shape of plasma sheath and resultant trajectories.

holes can be below the range covered by the solutions of Table IV, so that there may be greater crossover effects than indicated anywhere in the tabulation. A contributing factor, though, is the plasma density variation across the edge holes. This variation tends to focus trajectories more towards the outside of the overall ion beam. It appears that, in at least some cases, crossover trajectories alone cannot explain the magnitude of direct impingement near the edges of the ion beam.

Perveance matching is one approach to the drop-off of ion current density at the edge of the beam. This approach was first used on bombardment thrusters with cesium propellant by Sohl et al. (46). Perveance matching can be done by increasing just the gap between accelerators with increasing accelerator radius, or by increasing both the gap and the hole size. Changing the hole size in several steps does break up the regular pattern (usually hexagonal) of a single hole size. As shown by Sohl et al., though, a low blockage design can still be obtained.

It is probably not accidental that perveance matching was used first with cesium propellant. The electron emitting properties of thin cesium coatings cause special problems in accelerator design, particularly when coupled with regions of high impingement. The refractory materials one would like to use for accelerator systems emit electrons at some moderate temperature in the presence of cesium vapor. The theoretical choices for operation with such materials are two: (1) operate at a low enough accelerator temperature so that many monolayers of cesium are present and the thin film electron emitting phenomenon is thereby suppressed, (2) operate at a high enough accelerator temperature so that the cesium coverage of the electrode is negligible and the phenomenon again suppressed. From a practical viewpoint, only the former is acceptable, because the latter requires passing through the critical temperature region during every start-up. This means that a refractory accelerator is normally covered with cesium when operating with that propellant. Localized regions of high impingement can therefore cause localized heating and thinning of this coating, thereby leading to excessive electron emission. Sohl et al. (89) reported 230°C as the maximum safe temperature of a cesiated molybdenum accelerator in a bombardment thruster. An alternative approach that has been used with cesium propellant is to use an accelerator material that does not exhibit enhanced electron emission with a cesium coating—at least not to the degree shown by refractory metals. Sohl et al. (46) reported that Al, Cu, Mo, Ta, and Ti accelerators had been investigated, but only aluminum and copper did not show enhanced emission with cesium. Aluminum has been the preferred choice for cesium-bombardment thrusters because it offers the additional advantages of lightweight and low sputtering yield.

Both of these cesium-propellant approaches are less effective with larger thruster sizes. Refractory accelerators become more difficult to keep cool in the center of large beams, and aluminum has limited temperature capability and large thermal expansion. These difficulties probably explain why no bombardment thruster with cesium propellant has been used effectively in a size larger than about 12.7 cm.

## 2. Electron Backstreaming

Experimental measurements of maximum net-to-total voltage ratios (for backstreaming) tend to have substantial scatter. This scatter is probably due to varying degrees of ion beam neutralization. Electron backstreaming, when it occurs, is a function of the ion beam potential relative to the accelerator. The assumption usually made is that the ion beam potential is the same as the target potential, and the target potential used instead. The beam potential is typically 5 to 20 V positive relative to the target, but can climb to 100 V or more under conditions of poor or partial neutralization. The tendency, then, is to measure a maximum $R$ value that is higher than would be obtained with good neutralization. Kerslake (90) and Nakanishi *et al.* (91) included a number of experimental values for $R_{max}$, which were calculated in the manner just described. These data are shown in Fig. 34, with

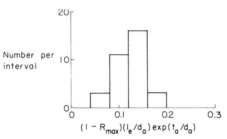

FIG. 34. Distribution of backstreaming parameter for experimental data with flat grids. Data from Kerslake (90) and Nakanishi *et al.* (91).

frequency plotted against aperture parameter $(1 - R_{max})(l_e/d_a)\exp(t_a/d_a)$. From Eq. (10), this parameter should be 0.1592 (that is, $1/2\pi$), or greater, to avoid backstreaming. Although the data spread over a wide range of aperture parameter, there is a sharp drop-off in frequency above the expected value. To be larger than all the experimental data shown in Fig. 34, an aperture parameter 20% larger than 0.1592 must be used. This is the recommended safety margin for design. This margin should, of course, be applied to the tolerance extremes that favor backstreaming—such as the shortest gap between electrodes, the thinnest accelerator grid thickness, and largest accelerator grid holes. The last two should also include wear effects of up to 10% for long duration applications.

### 3. *Electrical Breakdown*

Breakdown data in which a discharge was sufficient to trip overload protection on high voltage power supplies are plotted in Fig. 35. These data were obtained from Byers (59) and Kerslake (90). To extend the breakdown information to smaller gaps, nonbreakdown data were used from Rawlin (92). Little or no breakdown problems were encountered with these latter data, so they should be used to determine a lower limit for that region. The dished grids used by Rawlin were known to have a variation in gap across the grids due to unequal thermal expansion. Because only a mean gap

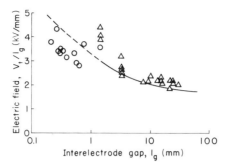

FIG. 35. Experimental breakdown data. $\triangle$, breakdown; $\bigcirc$, no breakdown. Data from Byers (59), Kerslake (90), and Rawlin (92).

distance was available, the data of Rawlin should be raised by some unknown amount. Additional breakdown data were available from earlier work by Kerslake (60), but these earlier data were obtained under poorer vacuum conditions and showed considerably more scatter.

The data shown in Fig. 35 were all obtained in the low $10^{-6}$ Torr range and are felt to be representative of mercury propellant. The fairing of the curve in Fig. 35 is felt to be conservative below 1–2 mm. Further experience may show it to be overly conservative in that range. As a design value, 0.8 to 0.85 of the faired electric field is recommended. For example, from Bechtel (38) the SERT II total voltage was 5 kV and the accelerator system gap was 2.50 ± 0.25 mm. This gives a maximum of 2.22 kV/mm at a gap of 2.25 mm, which is about 0.82 of the faired value at that gap.

The breakdown values of Fig. 35 are well below the voltages that can be sustained across nonoperating grids, even with propellant flow. The lower values during operation are probably due to the steady bombardment of the accelerator by charge-exchange ions. Breakdowns are usually frequent when a set of grids is operated for the first time. It is customary to advance the voltages up to design values over the time of an hour or so during initial operation. After this initial conditioning the voltages can be applied quite rapidly as long as the grids have been exposed to vacuum long enough to outgas and have not been allowed to become dirty.

During prolonged operation at high vacuum ($\leq 10^{-6}$ Torr) the mean interval between breakdowns was found to be about 5 hr by Kerslake *et al.* (93). Essentially the same rate was also reported by Kerslake *et al.* for the orbital SERT II test, which was in the far better vacuum environment of space. Pressures above the $10^{-6}$ Torr range increase the frequency of breakdown and decrease the allowable electric field. Because the primary application of thrusters has been in space, the effects of poorer vacuums have largely been ignored.

One might expect the breakdown rate to be zero under ideal vacuum conditions, but the SERT II flight experience is in disagreement with this conclusion. Brewer (15) describes the growth of protrusions or whiskers in regions of high electric fields and applies this process to electrical breakdown in ion thrusters. Given a whisker that is sufficiently long and sharp in an accelerator system gap, an electrical breakdown will follow. The critical question is how the whiskers continually form during prolonged operation. With a cesium coating on aluminum or copper (which is the case of most interest to Brewer), the thermal mechanism described by Charbonnier et al. (94) and Bennette et al. (95) seems likely. For a thin mercury coating (or no coating) on molybdenum, the normal operating voltages appear too low for this formation process. For the case of molybdenum, a better explanation of whisker growth would appear to be the preferential deposition of sputtered material in regions of high electric field due to induced dipole attraction. In any event, the formation of whiskers during prolonged operation appears very likely and is consistent with the rough texture observed on interelectrode surfaces after endurance tests. The low breakdown rate observed during long tests would be that required to burn away whiskers as rapidly as they are formed.

There is much less information on the breakdown limits of other propellants. From Sohl et al. (46), long duration operation with cesium is satisfactory at 2.4 kV/mm with a 1.75 mm gap. From Reader (49) and Schertler (50), it is clear that thrusters designed for mercury operated well at voltages normal for mercury with a variety of gaseous propellants. These results suggest that Fig. 35 be used for propellants other than mercury in the absence of any better information.

### D. High Perveance Accelerator Systems

The need in electric propulsion is to obtain dense high current ion beams at the moderate voltages associated with the desired exhaust velocities. This need results in the emphasis on high perveance designs. From the earlier discussion of Child's law, high perveance means closely spaced grids with many holes.

### 1. Composite Grids

Most accelerator systems use two grids separated by vacuum. Composite grids use a single metallic grid that is bonded on the ion chamber side of this grid to an insulator, as indicated in Fig. 36. The initial test of this one-piece accelerator system was made by Margosian (96) using a layer of alumina bonded to a molybdenum grid.* A fabrication technique for using high

---

* The combination of an insulator next to a metallic grid (not bonded) was tested earlier by H. W. Loeb in a thruster that used high frequency fields to ionize the propellant.

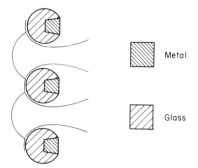

FIG. 36. Composite accelerator grid as used with high temperature glass.

temperature glass as the insulator was developed by Banks (97). Glass coated composite grids in the form developed by Banks exhibited higher overall perveance than any accelerator system available previously. This high perveance was due in part to the ability to use very short effective gaps relative to the beam diameter.

Despite their high perveance, composite grids were found to have several serious shortcomings. Deterioration towards failure was found to be highly nonlinear. A grid that had completed 80% of its life might look very much as it did originally. Associated with this nonlinearity was a large degree of variation in lifetime of individual units. Another problem that took considerable effort to solve was the interaction with backsputtered material from the facility [Bechtel *et al.* (98)]. It was found that a conductive coating on the part of the glass visible from the downstream direction was responsible for the very short effective acceleration lengths and very high perveances. Not only that, but the amount of backsputtered material had to be in a fairly narrow range to give both good performance and long life. This combination of problems led to most space-propulsion work on composite grids being terminated after the Bechtel *et al.* publication. The extremely high perveance of composite grids may still be useful in an application in which a several hundred hour life is acceptable.

## 2. Interelectrode Supports

As mentioned earlier, the interelectrode gap should not be smaller than about 1/60 of the beam diameter if warping problems are to be avoided. This limit could presumably be bypassed by using a number of insulating interelectrode supports within the beam diameter. Substantial thermal deflections would then be permissible, as long as the supports caused both grids to move together at a nearly fixed gap.

A short laboratory test of a high perveance accelerator system with one center support was made by Nakanishi *et al.* (*99*). Although this test seemed to give proof of concept, the interelectrode support used was adequate for only a short test. The design of interelectrode supports with the insulator sputter shielding required for long duration proved to be more difficult.* The major problem is finding room for the required structure. If the structure extends downstream of the accelerator, it must fit within the restricted cone shape of trajectories from adjacent apertures. If enough screen apertures are blocked to give adequate structure volume, a loss in ion chamber performance would also be involved. Byers (*100*) found another problem resulted from blocked apertures. The local charge-exchange erosion was focused into the center of the blocked area, in the manner of the pit formation in the triangles between holes elsewhere on the accelerator. The volume erosion rate went up with the number of holes blocked. It was 2 to 3 times as large for 7 holes as for 1, and about 8 times as large for the single case of 19 blocked holes. Byers found that insulated caps over the high erosion area greatly reduced the erosion, probably due to a positive floating potential of the cap. But the use of such an insulated cap by Collett (*101*) showed that erosion of the cap had been exchanged for erosion of a groove in the accelerator next to the cap—a problem at least as serious. The interelectrode support used by Collett extended several centimeters back into the ion chamber and was made of molybdenum. Although the upstream exposed surface of the support was at screen potential, it eroded fairly rapidly. This erosion was perhaps due to its higher temperature relative to the screen, inasmuch as the upstream structure was surrounded on all sides by the densest plasma in the ion chamber. (The screen showed the usual negligible erosion.)

It is perhaps still possible to build the required structure of an interelectrode support in a small enough volume to keep both ion chamber losses and erosion to a reasonable level. For space-propulsion purposes, however, the problems of interelectrode supports within the beam diameter are severe enough to make the dished grid approach far more attractive at present.

## 3. *Dished Grids*

The dished grid concept is simple: grids in the shape of spherical segments (see Fig. 37) should be far more stable to temperature gradients than flat grids. But the application of this concept as described by Rawlin *et al.* (*102*), took a large effort in a number of disciplines.

---

* For reliable long-term operation, it is necessary for the insulator shielding to be more than just line-of-sight. A very small fraction of sputtered particles is apparently reflected from the first impact, so that a more complicated, larger volume shield is required.

Fig. 37. Dished grid accelerator system. (Courtesy of Hughes Aircraft Company.)

One of the first difficulties to be solved was that of fabrication. Large variations in grid-to-grid spacing over the beam area were obtained with all forming processes except simultaneous hydroforming of both grids. With both molybdenum grids formed at the same time, any irregularities in the forming process tended to be duplicated in both grids. The gap could therefore be nearly constant despite these irregularities. A unique aspect of the hydroforming used was the firm clamping of the edges. No slippage was permitted, resulting in all grid deformation being tensile. (The usual approach with edge slippage resulted in circumferential compression and wrinkles near the outer edge. These wrinkles could not be removed from molybdenum sheets by stress relieving while clamped between dies.) Details of this grid fabrication process were included in the original preprint by Rawlin *et al.* (*103*), but were omitted in the journal version (*102*).

The holes were made by photochemical etching, which resulted in the cross sections indicated in Fig. 38. The 50–50 etch means that the etching

Fig. 38. Cross sections of etched grids. (a) 50–50 etch. (b) 90–10 etch.

was done equally from both sides of the sheet, while the 90–10 etch means that 90% of the etching was done from one side of the sheet. The many thousands of holes used in each dished grid made etching a far more convenient process than drilling. Originally the etching was done prior to dishing, but the photographic preparation for etching was found to withstand the dishing deformation, so that the dishing could be done on a uniform unperforated sheet.

The center of the screen grid in an electron-bombardment thruster is typically at 400–500°C during operation, while the center of the accelerator is 50–100°C cooler. The edges of the grids operate 100–300°C cooler than the centers, with the largest differences found in the accelerators of large thrusters. The radial temperature differences result in compressive and tensile stresses sufficient to cause some permanent warping in flat molybdenum grids, particularly if not stress relieved after fabrication. With sufficient dishing depth, the radial temperature differences result only in bending stresses that are well within the elastic capability of molybdenum. The dish depth increases as the center of the grid heats up relative to the edge support. The difference in screen and accelerator motion results in a differential change in spacing that is greatest at the center of the grids. For grids that are convex when viewed from the downstream direction, the spacing is decreased during operation. Dished grids for the 30 cm thruster were usually mounted in this manner, although the opposite orientation was useful for determining the magnitude of gap change during warmup. Grids with gaps of 0.33 and 0.86 mm were mounted so that the small gap opened up during operation (concave from downstream) and the large gap decreased. These two grids gave the same perveance at operating temperatures, indicating that half the difference in gap (about 0.26 mm) was the effective mean change in gap during warmup. This change is 30 to 40% of the cold gap (for convex from downstream), and is an appreciable factor in performance. This change in gap during warmup could be reduced by using a greater dishing depth, but off-axis losses would then become a greater problem.

The problem of off-axis losses is illustrated in Fig. 39. The two grids are fabricated with no misalignment when they are in contact, as indicated in Fig. 39a. When separated to the desired gap, the holes are no longer in alignment. The misalignment increases with radius and results in the trajectories being deflected away from the local normal to the dished grids—in the direction away from the thruster axis as indicated in Fig. 39b. The thrust loss with no misalignment of holes would be about 2% for the 2.54 cm dishing depth used by Rawlin et al. Depending on the gap used, the misalignment effect added another 2 to 6% thrust loss.

The misalignment effect can be eliminated by increasing the center-to-center spacing of the grid on the convex side. This can be done either by radially stretching that grid before hydroforming or by using a slightly larger pattern in the photochemical etching of that grid. The center-to-center spacing can also be changed enough to deflect the local trajectories toward the beam axis, instead of away from it, as was demonstrated by Danilowicz et al. (104). Preliminary tests by Danilowicz et al. indicated that the total thrust loss for a 30 cm thruster with dished grids could be reduced to 2 or 3% in this manner. The thrust coefficient measurements of Danilowicz et al. would

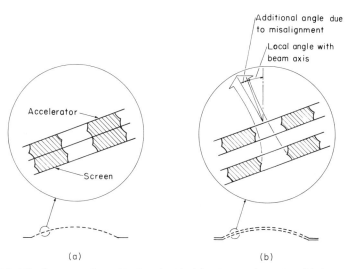

FIG. 39. Misalignment effect with dished grids. (a) Fabricated together. (b) Operated with interelectrode gap.

be interesting to compare with the space-charge flow values, but the dishing and corrective grid misalignment both introduce complications that make direct comparison difficult.

The data of Rawlin (*92*) and Rawlin *et al.* (*102*) for dished grids on a 30 cm thruster are given in Table VI. The maximum normalized perveance per hole for $d_a = d_s$, as shown in Fig. 40, is typically one-half to two-thirds that obtained with the flat grids in Fig. 32. The lower perveance per hole is more than made up by the larger number of grid holes that can be used effectively in a dished grid. But the lower perveance per hole still needs

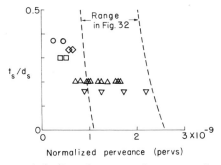

FIG. 40. Effect of screen hole dimensions on maximum normalized perveance for dished grids with equal screen and accelerator hole diameters. Range of $R$ is 0.57–0.75. $d$ (mm): $\bigcirc$, 1.02; $\square$, 1.27; $\diamondsuit$, 1.52; $\triangle$, 1.91; $\triangledown$, 2.41. Data from Rawlin (*92*) and Rawlin *et al.* (*102*).

TABLE VI

EXPERIMENTAL ACCELERATOR PERFORMANCE WITH DISHED GRIDS (30-cm THRUSTER)

| Etch | $d_s$ (mm) | $F_s$ | $t_s/d_s$ | $l_g/d_s$ | $t_a/d_s$ | $d_a/d_s$ | $R$ ($V_n/V_t$) | $J/V_t^{3/2}$ (npervs/hole) | $V_t^a$ | $R_{max}$ | Ref. |
|---|---|---|---|---|---|---|---|---|---|---|---|
| 50–50 | 1.02 | 0.51 | 0.373 | 0.304 | 0.373 | 1.00 | 0.69 | 0.84–1.28 | 900–1080 | 0.90 | (92) |
| 50–50 | 1.27 | 0.51 | 0.299 | 0.480 | 0.299 | 1.00 | 0.67 | 0.75–1.01 | 1300–1690 | 0.90 | (92) |
| 50–50 | 1.52 | 0.51 | 0.336 | 0.336 | 0.336 | 1.00 | 0.67 | 1.50–1.77 | 920–1680 | 0.90 | (92) |
| 50–50 | 1.91 | 0.67 | 0.199 | 0.110 | 0.267 | 0.796 | 0.67 | 2.72 | 790 | 0.82 | (92) |
| 50–50 | 1.91 | 0.67 | 0.199 | 0.225 | 0.267 | 0.796 | 0.67 | 2.10–3.36 | 940–1090 | 0.84 | (92) |
| 50–50 | 1.91 | 0.67 | 0.199 | 0.251 | 0.267 | 0.796 | 0.67 | 1.31–3.46 | 810–1240 | 0.85 | (92) |
| 50–50 | 1.91 | 0.67 | 0.199 | 0.351 | 0.267 | 0.796 | 0.67 | 2.64 | 1280 | 0.86 | (92) |
| 50–50 | 1.91 | 0.67 | 0.199 | 0.136 | 0.398 | 0.796 | 0.67 | 2.75–3.27 | 960–1110 | 0.80 | (92) |
| 50–50 | 1.91 | 0.67 | 0.199 | 0.199 | 0.398 | 0.796 | 0.67 | 2.10–2.91 | 940–1200 | 0.84 | (92) |
| 50–50 | 1.91 | 0.67 | 0.199 | 0.377 | 0.398 | 0.796 | 0.67 | 1.40–2.17 | 1060–1460 | 0.86 | (92) |
| 50–50 | 1.91 | 0.51 | 0.200 | 0.211 | 0.200 | 1.00 | 0.65–0.75 | 3.48–5.46 | 630–1090 | 0.72 | (102) |
| 50–50 | 1.91 | 0.51 | 0.200 | 0.316 | 0.200 | 1.00 | 0.65–0.75 | 2.74–4.41 | 740–1420 | 0.75 | (102) |
| 50–50 | 1.91 | 0.51 | 0.200 | 0.311 | 0.200 | 1.00 | 0.65–0.75 | 2.85–3.98 | 720–1160 | 0.89 | (102) |

| | | | | | | | | | | | |
|---|---|---|---|---|---|---|---|---|---|---|---|
| 50–50 | 1.91 | 0.51 | 0.200 | 0.589 | 0.200 | 1.00 | 0.65–0.75 | 1.98–2.58 | 920–1550 | 0.83 | (102) |
| 90–10 | 1.91 | 0.51 | 0.200 | 0.316 | 0.200 | 1.00 | 0.65–0.75 | 2.74–4.41 | 740–1420 | 0.81 | (102) |
| 50–50 | 1.91 | 0.51 | 0.199 | 0.162 | 0.199 | 1.00 | 0.65–0.75 | 4.48–5.92 | 680–1040 | 0.75 | (92) |
| 50–50 | 1.91 | 0.51 | 0.199 | 0.215 | 0.199 | 1.00 | 0.65–0.75 | 3.52–5.52 | 630–1090 | 0.75 | (92) |
| 50–50 | 1.91 | 0.51 | 0.199 | 0.293 | 0.199 | 1.00 | 0.65–0.75 | 2.77–4.13 | 740–1650 | 0.76 | (92) |
| 50–50 | 1.91 | 0.51 | 0.199 | 0.319 | 0.199 | 1.00 | 0.65–0.75 | 2.76–4.46 | 740–1420 | 0.76 | (92) |
| 50–50 | 1.91 | 0.51 | 0.199 | 0.387 | 0.199 | 1.00 | 0.65–0.75 | 2.40–3.10 | 1030–1380 | 0.78 | (92) |
| 50–50 | 1.91 | 0.67 | 0.199 | 0.215 | 0.199 | 1.00 | 0.65–0.75 | 2.73–3.77 | 790–1010 | 0.74 | (92) |
| 50–50 | 1.91 | 0.67 | 0.199 | 0.251 | 0.199 | 1.00 | 0.65–0.75 | 2.24–3.18 | 900–1130 | 0.76 | (92) |
| 50–50 | 2.16 | 0.67 | 0.176 | 0.356 | 0.352 | 0.801 | 0.69 | 2.04–3.09 | 1130–1360 | 0.82 | (92) |
| 50–50 | 2.41 | 0.67 | 0.158 | 0.137 | 0.158 | 1.00 | 0.57 | 6.39–8.10 | 610–1120 | 0.70 | (92) |
| 50–50 | 2.41 | 0.67 | 0.158 | 0.212 | 0.158 | 1.00 | 0.67 | 5.01–5.64 | 940–1380 | 0.71 | (92) |
| 50–50 | 2.41 | 0.67 | 0.158 | 0.320 | 0.158 | 1.00 | 0.67 | 2.56–3.72 | 1020–2830 | 0.78 | (92) |
| 50–50 | 2.41 | 0.67 | 0.158 | 0.212 | 0.315 | 0.801 | 0.67 | 3.80–4.31 | 1130–1260 | 0.80 | (92) |
| 50–50 | 2.41 | 0.67 | 0.158 | 0.245 | 0.315 | 0.801 | 0.67 | 3.48–3.84 | 1200–1360 | 0.82 | (92) |
| 50–50 | 2.41 | 0.67 | 0.158 | 0.286 | 0.315 | 0.801 | 0.69 | 2.38–3.32 | 1180–1500 | 0.83 | (92) |
| 50–50 | 2.41 | 0.67 | 0.158 | 0.415 | 0.315 | 0.801 | 0.67 | 1.82–2.96 | 1410 | 0.84 | (92) |

[a] Highest voltage usually goes with highest perveance.

explaining. The reader might suspect that a difference in $l_e/d_s$ is the reason for the difference in Figs. 32 and 40, inasmuch as the average value of $l_e/d_s$ is lower for Fig. 40. There is, however, even a trend of decreasing normalized perveance with decreasing size within the data of Fig. 40, where the ranges of $l_e/d_s$ were roughly the same for all sizes.

Kerslake and Pawlik (61) felt the ion chamber plasma sheath thickness might be significant compared to the size of the smaller holes used by them, and that this would explain an apparent absolute size effect. Kerslake and Pawlik did not attempt to separately evaluate the screen-grid thickness effect and this hypothesized absolute size effect. The data of Fig. 40 do show a clear ordering with size, even when the screen thickness effect from Fig. 32 is taken into account. The debye shielding distance is only about 5% of the smallest (1 mm) holes, but it might still have a defocusing effect analogous to that shown in Fig. 33 for ion density variation. There is also an alternative explanation that should be considered. Fabrication tolerances and misalignment effects (the latter not corrected in the grids of Table VI) are bigger fractions of the smaller holes; hence would lead to separation on an absolute size basis. Data from Table VI for $d_a = 0.8\ d_s$ show an average perveance about 0.6–0.7 times that for $d_a = d_s$.

For backstreaming, the frequency is plotted against aperture parameter for the data of Table VI in Fig. 41. The effect of better experimental

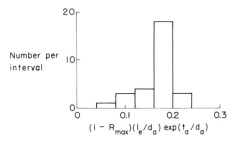

FIG. 41. Distribution of backstreaming parameter for experimental data with dished grids. Data from Rawlin (92) and Rawlin et al. (102).

technique (and less probability of a poorly neutralized beam) is apparently shown in the sharper peak in frequency when Fig. 41 is compared with Fig. 34. But the maximum value of aperture parameter to avoid all backstreaming has also increased about 20% over that of Fig. 34. This increase is perhaps due to a larger variation in interelectrode gap for dished grids, and the fact that backstreaming is largely determined by that part of the grid with the minimum gap.

The use of photochemical etching for dished grids made the use of noncircular screen holes relatively easy. Both Rawlin (92) and Danilowicz et al.

(*104*) report the use of hexagonal screen holes permitted the attainment of a screen open-area fraction equal to 0.77. Although this approach appears promising, there are still trade-offs that must be considered. A higher open-area fraction for the screen in general means better ion chamber performance, but Rawlin (*45*) has shown that this effect becomes negligible or disappears at a high enough open-area fraction. Also, the beam from a hexagonal screen hole has 6 sharp cusps on it when it passes through the accelerator,* so that a larger accelerator hole must be used to avoid interception at the same beam current.

To summarize, the present dished grid designs require sophisticated design and fabrication techniques, but offer a combination of long life and performance unequaled in any other accelerator approach. Current designs permit the use of an interelectrode gap of only 1/1000 of the beam diameter, or less. Demonstrated total perveances of designs with over 10,000 apertures in each grid exceed 80 $\mu$pervs. Such a level of performance appears adequate for the most demanding electric propulsion applications in space and may also be valuable in applications on earth.

### 4. Single Fine Mesh Screen

A very high perveance accelerator has been used by LeVagurèse and Pigache (*106*). A shorter description is also given by Pigache (*107*). The screen is of electroformed nickel, 10 $\mu$ thick, with holes in a square pattern on 100 $\mu$ centers and 50% transparency. Using argon as the propellant, 20 eV ions can be extracted with a current density of 3.8 mA/cm$^2$. The holes are smaller than the Debye shielding length so that essentially one-dimensional ion flow approaches the screen from the ion chamber side. The lifetime is given as about 50 hr when producing 20 eV ions. Although the lifetime would be expected to fall drastically for higher energy beams, this accelerator concept appears unmatched in the very low energy range.

### E. Thrust Vectoring

Two types of thrust vector control have been used within the accelerator system of a bombardment thruster; electrostatic and grid translation. This use has been limited at present to small attitude-control and stationkeeping sizes. The electrostatic approach appears too complicated for application to 30 cm thrusters, but the simpler grid translation approach appears feasible for this application.

---

* The effect of various screen hole shapes was investigated by Byers and Banks (*105*). Polygon screen holes appeared at the accelerator as segments of curves connected by sharper angles than the original polygon. By suitably distorting the screen hole, a square beam could be produced at the accelerator.

1. *Electrostatic*

Electrostatic thrust vectoring resembles closely the control of an electron beam in an oscilloscope cathode-ray tube. Instead of using separate deflection plates for the transverse electric field, this function is incorporated into the accelerator system of an ion thruster by segmenting the accelerator electrode. The deflection equation for a beam passing between parallel plates of length $t_a$ and separation $d_a$ is

$$\alpha_d = (K_f/2)(V_d/V_t)(t_a/d_a), \tag{11}$$

where $V_t$ is the total acceleration potential of the beam and $V_d$ is the deflection potential between the plates. (One accelerator segment is at $V_t - V_d/2$ while the opposite is at $V_a + V_d/2$.) This equation is often written with $\tan \alpha_d$ in place of $\alpha_d$. Most deflection systems are slightly nonlinear so that greater than expected deflection results at large angles, which would tend to offset the use of $\alpha_d$ in place of $\tan \alpha_d$. Further, experimental data agree well with the use of $\alpha_d$.

The factor $K_f$ is for the fringing effects in entering and leaving the deflection region. For $t_a \gg d_a$, this constant is unity. The practical range of interest is for $0.5 \leq t_a/d_a \leq 2$. For this range of two-dimensional configurations (infinite plate length normal to the beam direction) the factor $K_f$ is given within several percent by

$$K_f = 1 + d_a/t_a. \tag{12}$$

This constant is, however, strongly affected by accelerator configurations actually used. A value should therefore not be used from Eq. (12) without careful consideration of the configuration to which it is to be applied.

For a correlation or normalization parameter, all the variables of Eq. (11) can be put on one side.

$$\alpha_d(V_t/V_d)(d_a/t_a) = 28.6. \tag{13}$$

The constant $K_f$ has been assumed unity for reasons that will become evident, and the constant $1/2$ has been expressed in degrees rather than radians. For a broad multiaperture accelerator system, the transverse velocity would again be substantially independent of $R$. For such an accelerator system, the normalized deflection would therefore be $\alpha_d R^{1/2}(V_t/V_d)(d_a/t_a)$, with a nominal theoretical value of 28.6 deg.

Electrostatic deflection data from Collett *et al.* (*108*), Lathem (*109*), and King *et al.* (*110*) are given in Table VII. Inasmuch as deflection is close to linear with deflection voltage, no effect of deflection voltage magnitude is

TABLE VII

ELECTROSTATIC VECTORING DATA

| $d_s$ (mm) | $t_s/d_s$ | $l_g/d_s$ | $t_a/d_s$ | $d_a/d_s$ | $l_c/d_s$ | $R$ $(V_n/V_t)$ | $J/V_t^{3/2}$ (npervs/ hole) | $\alpha_d V_t/V_d$ (deg) | Ref. |
|---|---|---|---|---|---|---|---|---|---|
| 3.18 | 0.160 | 0.239 | 0.799 | 0.840 | 1.40 | 0.667 | 3.50 | 47 | (110) |
| 3.18 | 0.160 | 0.239 | 0.799 | 0.840 | 1.40 | 0.667 | 3.25 | 46 | (110) |
| 3.18 | 0.160 | 0.239 | 0.799 | 0.840 | 1.40 | 0.667 | 2.92 | 46 | (110) |
| 3.18 | 0.160 | 0.239 | 0.799 | 0.840 | 1.40 | 0.667 | 2.87 | 47 | (110) |
| 3.18 | 0.160 | 0.239 | 0.799 | 0.840 | 1.40 | 0.667 | 2.72 | 47 | (110) |
| 3.18 | 0.160 | 0.239 | 0.799 | 0.840 | 1.40 | 0.667 | 2.39 | 46 | (110) |
| 3.18 | 0.160 | 0.239 | 0.799 | 0.840 | 1.40 | 0.667 | 2.33 | 45 | (110) |
| 3.18 | 0.160 | 0.239 | 0.799 | 0.840 | 1.40 | 0.667 | 1.91 | 44 | (110) |
| 3.18 | 0.160 | 0.239 | 0.799 | 0.840 | 1.40 | 0.667 | 1.90 | 46 | (110) |
| 3.18 | 0.160 | 0.239 | 0.799 | 0.840 | 1.40 | 0.667 | 1.60 | 45 | (110) |
| 3.91 | 0.130 | 0.279 | 1.138 | 0.813 | 1.14 | 0.542 | 2.09 | 66 | (108) |
| 3.91 | 0.130 | 0.455 | 0.780 | 0.877 | 1.14 | 0.500 | 2.12 | 31 | (108) |
| 3.91 | 0.130 | 0.235 | 0.650 | 0.877 | 1.14 | 0.500 | 3.14 | 36 | (108) |
| 3.91 | 0.130 | 0.235 | 0.650 | 0.877 | 1.14 | 0.500 | 2.87 | 34 | (108) |
| 3.17[a] | 0.199 | 0.281 | 0.558 | 0.801[a] | 1.67 | 0.609 | 2.18[a] | 60[a] | (109) |

[a] Two-dimensional slots instead of holes. Perveance is per square of screen slot, instead of per hole.

shown. Normalized deflections for these data are plotted in Fig. 42. With two exceptions, the data fall in essentially one group. As was discussed by King *et al.* (*111*), one configuration had "L" shaped accelerator electrodes in which the ends of the "L" were wrapped around in such a manner as to

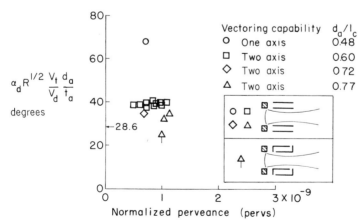

FIG. 42. Normalized deflection for electrostatic vectoring data. Data from Collett *et al.* (*108*), Lathem (*109*), and King *et al.* (*110*).

act counter to the main deflection direction (see insert in Fig. 42). This
configuration gave the low point which will not be considered further. The
high point is a two-dimensional design (only one-axis vectoring control).
The one configuration that was operated over a range of perveance showed a
small increase in normalized deflection with normalized perveance. This is
the direction that the effect would be expected to take from nonlinear electrostatic lens considerations.

There are several factors to consider when comparing the data in Fig. 42
with the nominal theoretical value of 28.6 deg. The biggest of these is the
effect of placing deflection systems for adjacent holes close together, which
would tend to offset most of the fringing effect of Eq. (12). That is, plates

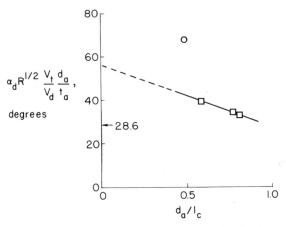

FIG. 43. Effect of adjacent apertures on normalized deflection for electrostatic vectoring
data. Vectoring capability: ○, one-axis; □, two-axis. Data from Collett et al. (*108*), Lathem
(*109*), and King et al. (*110*).

from adjacent apertures with opposite deflection potentials are much closer
together than $d_a$; hence the fringing field falls off much more rapidly than
assumed for Eq. (12). This effect is shown clearly in Fig. 43, where the average for each of the four values of $t_a/d_a$ in Table VII is plotted against $d_a/l_c$
(the ratio of electrode spacing to center-to-center aperture spacing). The
two-axis configurations show a trend that would extrapolate back to about
twice the theoretical value with no interference from adjacent apertures
($d_a/l_c = 0$), and thus be in approximate agreement with Eq. (12) at this
extrapolated point. There are two other effects that are in opposite directions for two-axis configurations: (i) the accelerator aperture effect which

results in the ions moving slower than would be expected from $V_t$, and hence give greater turning: (ii) the effect of using square apertures rather than two-dimensional ones, where the additional electrodes would reduce the average deflecting electric field about 20% from the two-dimensional case. These two offsetting effects are believed to be of the same order of magnitude.

The deflection of the one-axis configuration would, of course, have to be reduced about 20% to be comparable to the two-axis configurations. The aperture effect for two-dimensional configurations is also about twice that of the square apertures used here. These two effects are believed to explain the difference between the one-axis configuration and the trend of the two-axis configurations.

The trends of Figs. 42 and 43 should be useful for estimating the electrostatic-vectoring sensitivity of new designs. The maximum deflection angles are set by direct interception and can easily be as large as 10 to 15 deg with development. An approach for developing large deflection angle capability is described by Hudson et al. (112). This approach uses thin layers of copper plating to observe the effects of direct interception in short runs. In this manner the effect of different configuration changes on direct interception can be evaluated rapidly.

Some sacrifice in perveance relative to nonvectoring designs is apparently necessary to provide room for the beam motion within the accelerator aperture. Figure 42 shows a maximum normalized perveance of $1.1 \times 10^{-9}$ as compared to over $2 \times 10^{-9}$ for nonvectoring designs in Fig. 32.

Although electrostatic vectoring offers the absence of moving parts, it has some shortcomings that should be considered before making a commitment to this approach. The complexity of a two-axis electrostatic vectoring system, with notched interlocking deflecting plates as shown in Fig. 44, is not suitable for high perveance designs with many apertures. The two-axis designs of Table VII have 89 apertures, which is probably within a factor of 2 of the maximum practical number with the technology used. Another problem is electrical breakdown. The continual arcing that serves to maintain a reasonably smooth surface in a conventional accelerator system is not present for all the gaps between the interlocking plates. Occasional shorting between deflection plates was observed during a long duration test of this type of vector grid by Nakanishi (69). These shorts could be cleared by high current discharges, but perhaps the problem of shorts offsets the advantage of no moving parts. This particular test was a severe one inasmuch as similar accelerator-screen shorts also had to be burned away. Accelerator-screen

FIG. 44. Accelerator system with two-axis vectoring of 5 cm ion beam. (Courtesy of Hughes Aircraft Company.)

shorts are rare and were attributed in this test to excessive flaking of sputtered material in the particular ion chamber used. If a single axis of thrust deflection is sufficient, the one-axis electrostatic approach offers far less chance of shorting as well as a simple design that is suitable for ion thruster diameters up to 10 or 15 cm.


## 2. Grid Translation

The possibility of using grid translation to vector an ion beam was suggested by Lathem (113) in a study of electrode misalignment effects. The application of this concept to a bombardment thruster was carried out by Sohl and Fosnight (114). Additional theoretical solutions are available from Nudd and Amboss (87) as well as Lathem and Adam (115). Both the Lathem and the later Lathem and Adams solutions do not account for plasma sheath motion in the ion chamber, while the Nudd and Amboss solutions do. A more serious problem with the Lathem and the Lathem and Adam solutions is that the actual deflections were measured with two-dimensional techniques. The aperture effect is about twice as great for a two-dimensional slot as for a circular hole, so that values about twice the correct ones would be expected with this two-dimensional approach. These solutions were found to give about twice the experimental values of normalized deflection parameter. The far more rigorous analysis by Nudd and Amboss uses to advantage the techniques developed earlier by Amboss (*116*) for electron guns.

For a generalizing parameter, the lens effect of an aperture will be used. Consider a circular aperture of diameter $d_a$ with an electric field on one side of the aperture produced by a surface a distance $l_e$ from the aperture and at a potential $V_t$ relative to the aperture electrode. For $d_a \ll l_e$, Davisson and Calbick (*117*) found the circular aperture equivalent to a lens of focal length

$$f = -4l_e . \tag{14}$$

The negative sign indicates a defocusing effect. (A two-dimensional aperture has a focal length of $-2l_e$.) This lens effect of a circular aperture leads to an angular deflection of a beam when the aperture is displaced relative to the beam. If the displacement is $\delta_a$ the angular deflection is

$$\alpha_d = \delta_a/4l_e . \tag{15}$$

With multi-aperture accelerator systems, $R$ will again affect axial velocity more than transverse velocity. The normalized deflection parameter for grid translation is, therefore, $\alpha_d R^{1/2} l_e/\delta_a$ with a nominal value of 14.3 deg (1/4 rad).

Experimental data on thrust vectoring by grid translation from Sohl and Fosnight (*114*), Fosnight *et al.* (*118*), and Collett *et al.* (*108*) are shown in Table VIII. Also shown in Table VIII are the theoretical solutions by Lathem (*113*), Nudd and Amboss (*87*), and Lathem and Adam (*115*). The beam deflection is very close to linear with electrode displacement, so no effect of displacement magnitude is shown. The data of Table VIII are plotted in Fig. 45, with the exception of solutions by Lathem and Lathem and Adam. (These latter solutions were included in Table VIII for completeness and are felt to be more representative of two-dimensional grid translation.)

The theoretical solutions of Nudd and Amboss are in good agreement with the experimental solutions, which range from about 29 to 38 deg for the normalized deflection parameter. All solutions are about twice the nominal value, presumably because of (i) the lower velocity than indicated by $V_t$ (an

## TABLE VIII

### TRANSLATING GRID VECTORING DATA

| $\alpha_s$ (deg) | $d_s$ (mm) | $t_s/d_s$ | $l_g/d_s$ | $t_a/d_s$ | $d_a/d_s$ | $R$ $(V_n/V_t)$ | $J/V_t^{3/2}$ (npervs/ hole) | $\alpha_d d_s/\delta_a$ (deg) | Ref. |
|---|---|---|---|---|---|---|---|---|---|
| 20 | —[a] | 0.160 | 0.481 | 0.321 | 0.759 | 0.500 | 5.32[a] | 77[a] | (87) |
| 20 | —[a] | 0.160 | 0.481 | 0.321 | 0.759 | 0.500 | 1.54[a] | 66[a] | (87) |
| 0 | —[b] | 0.188 | 0.619 | 0.376 | 0.807 | 0.600 | 1.50[b] | 116[b] | (113) |
| 0 | —[b] | 0.188 | 0.619 | 0.376 | 0.807 | 0.600 | 1.50[b] | 110[b] | (113) |
| 0 | —[b] | 0.200 | 0.253 | 0.200 | 1.00 | 0.747 | 4.13[b] | 112[b] | (115) |
| 0 | —[b] | 0.200 | 0.253 | 0.200 | 1.00 | 0.747 | 2.20[b] | 117[b] | (115) |
| 0 | —[b] | 0.200 | 0.505 | 0.200 | 1.00 | 0.747 | 3.87[b] | 95[b] | (115) |
| 0 | —[b] | 0.200 | 0.505 | 0.200 | 1.00 | 0.747 | 2.20[b] | 90[b] | (115) |
| 0 | —[b] | 0.200 | 0.505 | 0.200 | 1.00 | 0.747 | 1.16[b] | 87[b] | (115) |
| 0 | 2.54 | N.A. | 0.620 | 1.250 | 0.860 | 0.769 | 1.19 | 46 | (114) |
| 0 | 2.54 | N.A. | 0.550 | 0.200[c] | 0.860 | 0.833 | 2.60 | 51 | (118) |
| 0 | 2.54 | N.A. | 0.550 | 0.200[c] | 0.860 | 0.800 | 2.44 | 52 | (118) |
| 0 | 2.54 | N.A. | 0.550 | 0.200[c] | 0.860 | 0.769 | 2.30 | 54 | (118) |
| 0 | 2.54 | N.A. | 0.550 | 0.200[c] | 0.860 | 0.741 | 2.18 | 54 | (118) |
| 0 | 2.54 | N.A. | 0.550 | 0.200[c] | 0.860 | 0.714 | 2.06 | 53 | (118) |
| 0 | 2.54 | N.A. | 0.550 | 0.200[c] | 0.860 | 0.690 | 1.96 | 52 | (118) |
| 0 | 2.54 | N.A. | 0.630 | 0.200[c] | 0.860 | 0.833 | 2.69 | 40 | (118) |
| 0 | 2.54 | N.A. | 0.630 | 0.200[c] | 0.860 | 0.800 | 2.53 | 42 | (118) |
| 0 | 2.54 | N.A. | 0.630 | 0.200[c] | 0.860 | 0.764 | 2.38 | 45 | (118) |
| 0 | 2.54 | N.A. | 0.630 | 0.200[c] | 0.860 | 0.741 | 2.25 | 47 | (118) |
| 0 | 2.54 | N.A. | 0.630 | 0.200[c] | 0.860 | 0.714 | 2.13 | 47 | (118) |
| 0 | 2.54 | N.A. | 0.630 | 0.200[c] | 0.860 | 0.690 | 2.02 | 47 | (118) |
| 0 | 2.4 | 0.262 | 0.479 | 0.529 | 1.00 | 0.500 | 1.29 | 77 | (108) |

[a] Theoretical, not experimental.

[b] Theoretical, with two-dimensional procedure used to evaluate deflection angles.

[c] Thickness of cylindrical portion of accelerator. Conical section is downstream of cylindrical part.

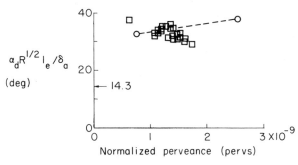

FIG. 45. Normalized deflection for translation vectoring data. ○, theory; □, experiment.

aperture effect) and (ii) nonlinear effects of finite beam diameter. The latter effect is probably small inasmuch as extrapolation of Nudd and Amboss solutions to zero perveance does not appear to approach the 14.3 deg value. This apparently leaves only the aperture effect to account for nearly a factor of 2 discrepancy. Perhaps the effect of substantial ion beams on the potential near the saddle point (discussed in connection with backstreaming in Section III,B,3) results in more beam deflection with aperture displacement than can be explained in terms of Laplace solutions.

The grid translation approach to vectoring thrust appears reliable and applicable to a wide range of thruster size. A normalized deflection of 30–40 deg appears to cover the likely range of sensitivity. A maximum deflection (without large direct interception) of 5 deg is apparently easy to obtain, and a 10 deg deflection reasonable with some development. The normalized perveances for experimental data appear to be higher than for electrostatic vectoring. But the maximum deflections were also less, so that a trade-off may exist between maximum angle and normalized perveance. Either the accelerator or screen can be translated, although in space applications the movement of the accelerator does not involve the ion chamber sealing problems associated with screen movement. Thermal effects have been used in both translation mechanisms discussed in the literature, and should be satisfactory if there are no adverse thermal interactions with the spacecraft and its environment. Other mechanisms may be desired for quicker motion in ground applications.

### F. Durability

The charge exchange of escaping neutral propellant atoms with the accelerated ions results in ion trajectories originating within the acceleration region. As indicated in Fig. 46, some of these charge-exchange ions follow

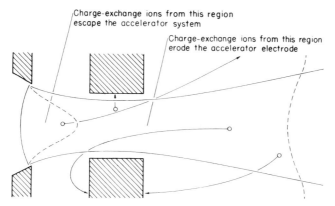

FIG. 46. Trajectories of charge-exchange ions.

FIG. 47. Typical erosion pattern for downstream side of accelerator. (Courtesy of NASA.)

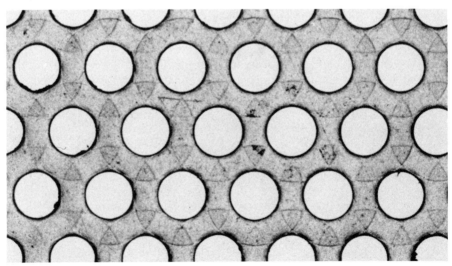

FIG. 48. Typical erosion pattern for upstream side of accelerator. Depth of pattern varies with ion current density. (Courtesy of NASA.)

escape trajectories. The rest impact on the accelerator electrode, causing erosion. The usual appearance of an eroded accelerator is shown in Fig. 47. The deep pits are caused by the focusing of the electric field downstream of the accelerator—with the location of the pits corresponding to those points on the closure surface that are farthest from the accelerator. There is also erosion on the upstream surface of the accelerator, as shown in Fig. 48.

The charge-exchange erosion was found by Kerslake (90) to be the major source of accelerator wear in normal operation. While this conclusion is still valid for the configurations studied by Kerslake, there is some doubt as to its validity for accelerator systems with smaller holes. For the charge-exchange portion of the following analysis, the approach of Kerslake will be roughly followed.

### 1. Charge-Exchange Current

The charge-exchange current is the product of charge exchange cross section, the neutral atom density, the total ion beam current, and an effective length that will be assumed to be $l_e$ for the acceleration portion of the charge exchange current.

$$J_a = \sigma_c n_0 J_b l_e . \tag{16}$$

The neutral density can in turn be expressed in terms of the total ion beam current, propellant utilization, and effective escape area for the accelerator grid. For an assumed neutral temperature of 400°C (the mean ion chamber wall temperature), this neutral density is

$$n_0 = 4.7 \times 10^{16}(1 - \eta_u)J_b/\eta_u A_e . \tag{17}$$

This equation follows the approach used by Kerslake in that only neutral velocities in the downstream direction were used to calculate $n_0$. This would be correct for very short interelectrode gaps and very thin electrodes. The actual $n_0$ would exceed the value given as larger gaps and thicker electrodes are used. The effective area $A_e$ can be defined in terms of the fraction open area of the accelerator (hole area over beam area), the total beam diameter, and the Clausing factor (63) for free molecular flow through the accelerator hole.

$$A_e = D_b^2 F_a K_c/4. \tag{18}$$

The Clausing factor $K_c$ is given in terms of the ratio of grid thickness to hole diameter in Table IX. The value of $\sigma_c$ was determined from Kushnir *et al.*

TABLE IX

CLAUSING FACTOR[a]

| $t/d$ | $K_c$ | $t/d$ | $K_c$ | $t/d$ | $K_c$ |
|-------|-------|-------|-------|-------|-------|
| 0.05 | 0.9524 | 0.55 | 0.6514 | 1.1 | 0.4914 |
| 0.10 | 0.9092 | 0.60 | 0.6320 | 1.2 | 0.4711 |
| 0.15 | 0.8699 | 0.65 | 0.6139 | 1.3 | 0.4527 |
| 0.20 | 0.8341 | 0.70 | 0.5970 | 1.4 | 0.4359 |
| 0.25 | 0.8013 | 0.75 | 0.5810 | 1.5 | 0.4205 |
| 0.30 | 0.7711 | 0.80 | 0.5659 | 1.6 | 0.4062 |
| 0.35 | 0.7434 | 0.85 | 0.5518 | 1.7 | 0.3931 |
| 0.40 | 0.7177 | 0.90 | 0.5384 | 1.8 | 0.3809 |
| 0.45 | 0.6940 | 0.95 | 0.5256 | 1.9 | 0.3695 |
| 0.50 | 0.6720 | 1.00 | 0.5136 | 2.0 | 0.3589 |

[a] From Clausing (63).

(119), Iovitsu and Ionescu-Pallas (120), and Zuccaro (121) to be about $6 \times 10^{-19}$ m$^2$ for $H_g - H_g^+$ at 1000 eV.* Using this value for $\sigma_c$, as well as substituting for $n_0$ and $A_e$, we can rewrite Eq. (16) as

$$J_a = 0.036(1 - \eta_u)J_b^2 l_e/\eta_u D_b^2 F_a K_c .$$   (19)

This, then, is the expression for the charge-exchange current originating within the acceleration portion of the accelerator system.

The deceleration region downstream of the accelerator also produces charge-exchange ions that reach the accelerator. Assuming for this region that the ion flow rate is uniform over the cross-sectional area of the total beam, the ratio of deceleration length to acceleration length is given by

$$l_d/l_e = [(1 + 3R^{1/2} - 4R^{3/2})/F_s(J/J_{cl})]^{1/2},$$   (20)

where $F_s$ is the fraction open area of the screen and $J/J_{cl}$ is the ratio of current density in the screen open area to Child's law current density. This ratio can be obtained for mercury ions by dividing normalized perveance (per hole) by $3.03 \times 10^{-9}$. Equation (20) can be obtained from Eq. (28) of Fay et al. (122), which has the boundary condition of zero electric field at the end of the deceleration region. This boundary condition is required to match the plasma sheath of the ion beam. The charge-exchange current for this region has a form similar to Eq. (19).

$$J_a = 0.036(1 - \eta_u)J_b^2 l_d/\eta_u D_b^2.$$   (21)

* The charge-exchange cross section increases slowly with decreasing velocity. It is about $8 \times 10^{-19}$ m$^2$ for $H_g - H_g^+$ at 100 eV.

There is another source of charge-exchange current that was not considered by Kerslake—that from the upstream end of the ion beam plasma. The electric field does not stop at the edge of the ion beam plasma but falls off exponentially within the plasma. The e-folding distance is the Debye shielding length

$$l_D = 7.43 \times 10^3 (T_e/n_e)^{1/2}, \tag{22}$$

where $T_e$ is the electron temperature in eV. In the ion beam, as in the ion chamber, Maxwellian electrons tend towards about 5 eV because of the excitation cross section of neutral mercury.* Using this temperature and an electron density $n_e$ equal to the ion density, one can obtain for the ion beam

$$l_D = 1.85 \times 10^{-4} D_b V_n^{1/4}/J_b^{1/2}. \tag{23}$$

The value of $l_D$ typically ranges from about 0.2 to 1.0 mm. If the ion beam were perfectly parallel, a small potential gradient would tend to return ions from an infinite distance away. Because of the divergence that actually exists, the ion density falls off with distance, and the potential in the plasma can be related to the density variation by the equation from Section V,A.

$$n_e = n'_e \exp(V/T_e). \tag{24}$$

Equation (24), with the decrease in density, gives a general decrease in potential in the downstream direction. This general decrease equals the rise in potential due to the plasma sheath in about 5 Debye shielding lengths. Because the fall-off of electric field at the upstream end of the ion beam is exponential, different configurations will have only a $\pm 20\%$ effect on this length. The charge exchange current over this distance is obtained by substituting $5l_D$ for $l_d$ in Eq. (22). For the configurations calculated, this distance of $5l_D$ was from 20 to 50% of $l_d$.

With the minor omission of the contribution from this $5l_D$ distance, Kerslake (90) calculated charge-exchange currents which showed reasonable agreement with experimental data from thrusters with diameters of 5, 10, and 20 cm. A particularly wide range of operating conditions was obtained with a 10 cm diam thruster that also satisfies the minimum gap used herein $\geq 1/60$ of beam diameter. A further limitation was made by using only data with $\eta_u \leq 0.6$ to avoid having a small value for $1 - \eta_u$. A small value of $1 - \eta_u$ could result in a large error due to double ionization from a 50 V

---

* There are other processes in the ion beam that can also affect electron temperature, as discussed in Section V,A. The 5 eV temperature, though, is close to the typical experimental value.

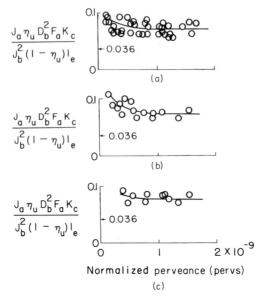

FIG. 49. Charge-exchange parameter. (a) $R = 0.8$. (b) $R = 0.6$. (c) $R = 0.4$. Data from Kerslake (90).

discharge voltage. These data are shown in Fig. 49 in terms of the charge-exchange parameter $J_a \eta_u D_b^2 F_a K_c / J_b^2 (1 - \eta_u) l_e$, which can be obtained from Eq. (19). If the charge-exchange ions from the acceleration region were the only major constituent of experimental accelerator current, then the data should make a near horizontal line at 0.036. The data are at about twice this value, indicating that the contribution downstream of the accelerator is on the average about as important as that upstream of the accelerator. (The reader should not draw any general conclusions about the ratio of total to acceleration region charge-exchange current from Fig. 49. Other configurations would have different ratios.) The effect of normalized perveance on the charge-exchange level appears small, except at the lowest values of perveance. There is a small trend towards slightly higher charge exchange with decreasing $R$. This small trend is in agreement with the qualitative trend shown by Brewer (15) on p. 211 of his book.

## 2. High Perveance Designs

The preceding description of the charge-exchange relation to accelerator current is in essential agreement with Kerslake. The problem mentioned earlier arises in more recent data in which smaller accelerator system holes were used. Four configurations were selected to demonstrate the size effect: those of Pawlik and Reader (*123*), Nakanishi *et al.* (*99*), King *et al.* (*111*), and Rawlin (*124*). The first and last were operated for a long enough time (2200 and 1500 hr) to erode away the portions of the accelerator subject to direct impingement at the edge of the beam. The second was a well aligned grid with a center support and small enough holes (0.75 mm) that the effect of plasma density variation across a hole diameter should be minimal. The third was a thrust-vector grid which could demonstrate substantial grid motion from the central position before the accelerator current increased significantly. In other words, there were good reasons for all of these accelerator systems to have minimal direct impingement. The ratio of actual accelerator current to calculated charge-exchange current (both upstream and downstream of the accelerator) for these configurations is shown in Fig. 50a.

The uncertainty in charge-exchange calculations is, at most, a factor of 2. Yet, the data show measured accelerator currents up to 4 and 5 times the calculated value. Although the points are few, there does seem to be a general trend to higher values of this current ratio for smaller holes. A hypothesis was made that the relatively larger sheath thickness in the ion chamber was the cause of defocusing for smaller holes, which led to increased direct impingement.* Such an effect would be assumed to vary as the ratio of Debye shielding length (in the ion chamber) to screen hole diameter. Using a 5 eV temperature for ion chamber Maxwellian electrons, together with the Bohm (*16*) criteria for directed ion velocity, the Debye length in the ion chamber was found to be

$$l_D = 2.32 \times 10^{-4} d_s N^{1/2} / J_b^{1/2}. \tag{25}$$

The magnitude for $l_D$ in the ion chamber typically ranges from about 0.02 to 0.1 mm. A reasonable fit was found with the assumption of an accelerator direct interception of

$$J_a = 0.04 J_b l_D / d_s. \tag{26}$$

The ratio of actual-to-calculated accelerator current, with the accelerator current including both the charge-exchange of Fig. 50a and the assumed direct impingement of Eq. (26), is shown in Fig. 50b. The agreement between actual and calculated values is clearly much improved.

---

* In the accelerator performance section, it was suggested that the maximum perveance might be less for a thicker ion chamber sheath. The further suggestion is made here that some high angle trajectories are introduced by the thicker sheath that still strike the accelerator at this reduced maximum perveance.

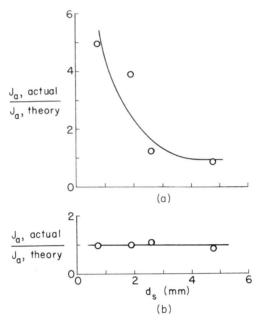

FIG. 50. Ratio of experimental to theoretical accelerator currents for operation with low direct impingement. (a) Theory includes only charge exchange. (b) Theory includes direct interception correction. Data from Nakanishi *et al.* (*99*), King *et al.* (*111*), Pawlik and Reader (*123*), and Rawlin (*124*).

The reader should keep in mind that the preceding is not intended as a proof of a sheath thickness effect in the ion chamber. But the results of Fig. 50b and the earlier apparent size effect in accelerator performance should both be considered by anyone who wishes to use very small spacings and hole diameters. The direct interception of high energy ions, instead of charge-exchange ions with a much lower average energy, would also have significant effects on accelerator lifetime.

## 3. *Accelerator Erosion*

The accelerators of mercury bombardment thrusters have been made almost exclusively of molybdenum. The advantages of molybdenum are high strength, low thermal expansion, and low sputtering yield to mercury ions. The sputtering yield for the range of interest is shown from data by Meyer and Güntherschulze (*125*), Wehner (*126*), Musket and Smith (*127*), and

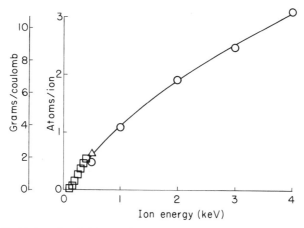

FIG. 51. Sputtering of molybdenum by normal incidence mercury ions. References: ○, (125); □, (126); △, (128).

Carter and Colligon (128) for normal incidence ions. (Oblique incidence generally gives higher yields.) The data for polycrystalline molybdenum are shown in Fig. 51, while the single-crystal data of Musket and Smith indicate that the sputtered atoms should roughly approximate a cosine distribution. As a comparison to Fig. 51, Pawlik and Reader (123) found wear rates from 2.7 to 3.6 g/A-hr of impingement current in long duration tests. The equivalent average ion energy for such erosion rates is less than 1000 eV despite the fact that $V_t = 5$ kV and $R = 0.6$. The use of lower voltages would be expected to drop the erosion per amp-hour roughly in proportion to the drop in yield shown in Fig. 51. Pawlik and Reader (123) estimated a minimum of 30% of the accelerator structure within the beam area could be eroded away before failure. Accelerator system electrodes should, of course, be stress-relieved to reduce the distortion when substantial amounts of structure are eroded away.

### 4. Other Propellants

The use of other propellants will change the preceding durability discussion to the extent of cross section, sputtering yield, and particle velocity changes. Data from a variety of sources are presented by Marino et al. (129) for $C_s$ charge-exchange cross sections. The mean value at 1000 eV is about $2 \times 10^{-18}$ m$^2$. While the accelerator current can be identified (within 10%)

with the impingement of mercury ions, no such identification can be made with cesium. For example, an 8000 hr test of a cesium bombardment thruster reported by Sohl *et al.* (*130*) had an average accelerator current of 5.5 mA with $V_t = 2500$ V and $R = 0.8$. Only 1–2 mA of this total can be identified as due to charge exchange. The remainder is due to secondary electron emission from ion collisions and electron emission from the accelerator.* The erosion rate measured by Sohl *et al.* during this test was 0.14g/A-hr of drain current, as based on the 5.5 mA figure. For charge-exchange ions alone it would be higher by a factor of several. Sohl *et al.* also estimated that a minimum of about 30% of the accelerator structure within the beam area could be eroded away before failure.

For the propellants xenon and argon, Kushnir *et al.* (*119*), Iovitsu and Ionescu-Pallis (*120*), and Rapp and Francis (*132*) gave results that averaged to about $4.5 \times 10^{-19}$ m$^2$ and $2.5 \times 10^{-19}$ m$^2$ at 1000 eV. The sputtering rates for these ions striking molybdenum are given by Laegreid and Wehner (*133*), Rosenberg and Wehner (*134*), and Carter and Colligon (*128*).

Materials other than molybdenum should be considered for ground applications. Where temperature and gap spacing are not critical, stainless steel should be considered. For very low thermal expansion, high resistance to warping, and low sputtering, carbon is a possible choice. Both Kerslake (*60*) and Kerslake and Pawlik (*61*) give some information on alternate choices for grid materials.

## IV. NEUTRALIZATION

The need for neutralization in space propulsion is twofold: (i) there is the need for *current neutralization* to avoid charge buildup on the spacecraft and (ii) there is the need for *charge neutralization* to avoid charge buildup within the ion beam. If current neutralization suddenly ceases, a spacecraft potential sufficient to stop ions from leaving will usually be reached in a matter of microseconds. An absence of charge neutralization will usually force the accelerator system into a high-impingement mode in less than one microsecond. Because of the speed with which these effects take place, it is customary to consider only self-balancing neutralization concepts. This means placing a more than ample source of electrons in a close enough

* Wilson (*131*) found aluminum oxide emits electrons in the presence of cesium vapor with an effective work function of about 1.4 eV, so that thin layers of oxide on the aluminum accelerator of this test could have contributed to the accelerator current. The lack of clear correspondence of accelerator current to impingement with cesium has resulted in the use of the term "drain current," as opposed to the usual "impingement current" with mercury.

proximity to the ion beam for the beam potential to establish the correct flow of electrons. The implementation of this principle is more difficult than it might seem.

The need for a neutralizer is less in a ground application than in space because charge neutralization can be obtained from secondary electrons at the target and current neutralization is often unnecessarily due to ground return circuits. The use of a neutralizer still offers the advantages of less electrical noise in the ion beam and less (and more predictable) beam spread. In the cases of sputtering insulators or operating with ion energies less than about 500–1000 eV, operation without a neutralizer is extremely difficult.

The width of the beam is significant in neutralization problems, with width defined relative to acceleration distance. A narrow beam (typically from a single-aperture acceleration system) is relatively tolerant of changes in charge neutralization. Poor charge neutralization results in enlargement of the beam, but is usually accompanied by no adverse effect within the accelerator system. A broad beam (from an accelerator system with many apertures) goes into a turnaround condition without charge neutralization, in which condition most of the ion current returns to impinge upon the accelerator. Most of this section is concerned with the problems of neutralizing broad beams.

The reader should not misinterpret the word "neutralization." The recombination cross section is small enough that radiative recombination of electrons and ions is a negligible process. Neutralization is concerned only with the net charge in a volume of ion beam and the net current in that beam.

### A. Theory

Early theoretical attempts were concerned with analytical space-charge flow solutions to the neutralization problem (*135–137*). Because no dissipative processes were included in these attempts, they concluded no solution was possible with a longitudinal electron velocity greater than twice the ion velocity. This velocity limit is well below the random motion from thermionic emission. Wells (*138*) proposed the "plasma bottle" concept to overcome this velocity limit. In this concept the electrons are assumed to have some large effectively random velocity relative to the ions. The initial escape of electrons would produce electric fields such that the remainder would be reflected by the boundaries of the beam, including the end boundary when the thruster is started. Thus, the electrons could be moving rapidly in all directions, but with an average velocity equal to that of the ions. The potential barrier produced by using "accel–decel" would prevent the electrons from escaping back through the accelerator system (see Section III,A for accel–decel description).

The approach of Wells was more realistic than the preceding closed-form solutions, but some pulsed-beam experiments by Sellen and Shelton (*139*) and Sellen and Kemp (*140*) soon showed that dynamic processes were also taking place, processes that could not be explained by the plasma bottle approach. Sellen and Kemp (*140*), in a classic electric propulsion experiment, demonstrated that an ion beam could be turned on rapidly and neutralized before the front of the beam reached the far wall of the vacuum facility. This experiment reassured workers in the field that neutralization could, indeed, be achieved in space. But a number of other phenomena that were observed in related experiments led to a more detailed look at dynamic processes.

The dynamic processes were studied by computer simulation of the electrons and the ion beam. One-dimensional simulations were used by Buneman and Kooyers (*141*) and Dunn and Ho (*142*). A two-dimensional simulation was then carried out by Wadwa *et al.* (*143*). These simulations showed "turbulence" when electrons were injected with large velocity, the turbulence serving effectively to randomize the electron motion. Inasmuch as ions are so much heavier, calculations long enough to show ion trajectories for realistic ion masses encountered time-accuracy problems. With the advances that have since taken place in computers and their use, it is not necessarily true that the same time and accuracy limitations exist today.

The most recent work on neutralization has been done by Wilhelm (*144*). Using analytic techniques, Wilhelm has shown the propagation of nonlinear electron waves into ion beams as part of the initial neutralization process. Wilhelm has also shown an analytic solution to the damping of longitudinal electron waves. This latter solution used Buneman's relaxation frequency (*145*), and showed the typical damping distance in an ion beam was several centimeters.

## B. Experimental Performance

There are three basic types of neutralizers. All three have current areas of application.

### 1. External Thermionic Emitter

The external thermionic emitter neutralizer is located near (but not in) the ion beam. The emitter should be located in a narrow strip around the beam close to the accelerator—close enough for the beam to be well defined, but not so close that the field from the accelerator makes a potential well around the neutralizer that stops emission. The coupling potential is approximated by treating the ion beam as a virtual anode for the electron current.* The coupling voltage should be low enough that charge-exchange ions from the beam will not cause excessive erosion of the emitter during the required lifetime.

Not only were early conceptual designs of neutralizers always of the external thermionic type, the first effective demonstration of a neutralizer by Brewer *et al.* (*146*) used this type. Effective applications of the external thermionic emitter neutralizer, however, have been limited to thin beams (*146, 147*). The reason for this limitation can be seen by considering the area of the strip around the ion beam that can be used for electron injection. The area of this strip increases linearly with beam diameter as accelerator holes are added. But, the ion current to be neutralized increases as the diameter squared. From Child's law, the coupling voltage to carry the required electron current must increase as the 3/4 power of beam diameter. In bombardment thrusters, thermionic emitters are presently considered as neutralizers only with accelerator systems of one hole or, at most, several holes or the equivalent (*47*). The thermionic emitter can, in such applications, give long life with moderate power losses.

## 2. Immersed Thermionic Emitter

The immersed thermionic emitter is placed within the ion beam. This location results in erosion by high energy beam ions, but it also permits excellent coupling to a broad beam. Because of sputtering, only tantalum or tungsten wires and strips have been normally used for immersed emitters.

The coupling voltage can be estimated by assuming a virtual anode at a distance from the cathode such that the densities of the ions and emitted electrons are equal. The conduction within the ion beam is quite good so that the immersed emitter can be used in a wide range of locations.

The first neutralizer used with an electron bombardment thruster was of this type (*1, 2*), as was also the neutralizer used on the first space test of an ion thruster—SERT I (*148*). The immersed emitter continues to be used in short ground tests of thrusters, and is the most common neutralizer used in ground applications of bombardment thrusters (such as producing a plasma to simulate the ionosphere or as a source for sputtering). The usual form is a tungsten or tantalum wire stretched across the ion beam. The end supports are located out of the beam or in the low density region at the edge of the beam. The lifetime, depending on ion energy and current density, can be from a few to over one hundred hours.

---

* If the calculated electron density exceeds the ion density, then the assumed location of the virtual anode must be moved back into the ion beam until the two densities are equal.

## 3. *Plasma Bridge*

The plasma bridge neutralizer uses a hollow cathode with a flow of propellant through a small orifice. The discharge in the orifice not only emits electrons, but also generates ions that are carried along by the neutral flow. These ions neutralize most of the space charge of the electrons. Moderate coupling voltages ($\leq 20$ V) can be obtained with large neutralizer currents and distances of several centimeters to the ion beam. This is opposed to, at most, a fraction of a centimeter that can be used with a thermionic emitter.

The plasma bridge neutralizer was used first with cesium propellant (*149*). It was used with mercury propellant in the first long-duration space test of an ion thruster—SERT II (*93*). The plasma bridge is at present the standard neutralizer for all space applications of broad-beam ion thrusters. It also offers greatly increased life over immersed emitters for ground applications.

The major design problem in using the plasma bridge neutralizer has been ions from the neutralizer causing erosion on the accelerator. This erosion can be substantial, as was found in the SERT II program. This flow of ions from the neutralizer to the accelerator has been since found to vary approximately as the inverse square of distance, as would be expected from Child's law. Bechtel (*150*) was able to decrease this ion current in a 30 cm thruster by more than a factor of 10 by simply moving the neutralizer farther from the accelerator, without a significant increase in coupling voltage. Two present examples can be given of neutralizer positions that have been selected to minimize direct ion impingement on the neutralizer, ion impingement from the neutralizer on the accelerator, and coupling voltage. A 5 cm thruster described by Nakanishi (*69*) and Nakanishi and Finke (*151*) has the neutralizer position given by Nakanishi and Finke (*152*). The position used for a 30 cm thruster is given by Bechtel (*153*). Both of these positions are indicated in Fig. 52. Note that the neutralizers are not just located away

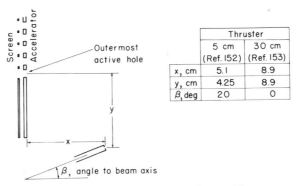

|  | Thruster | |
|---|---|---|
|  | 5 cm (Ref. 152) | 30 cm (Ref. 153) |
| $x$, cm | 5.1 | 8.9 |
| $y$, cm | 4.25 | 8.9 |
| $\beta$, deg | 20 | 0 |

FIG. 52. Hollow cathode neutralizer positions.

from the accelerator, they are also pointed away from the accelerator. From these two examples, one would expect that for a new design the neutralizer tip should be pointing nearly downstream and be located near a 45 deg angle from the last accelerator hole. The distance of the neutralizer from the accelerator should apparently be greater for larger neutralizer currents and neutralizer propellant flows. This would be consistent with the expected greater generation of ions at these conditions. The 45 deg angle results from avoiding a few high angle energetic-ion trajectories—the bulk of the beam is much closer to the axial direction.

## V. Thruster Efflux

The trajectories of beam ions are assumed known from either space-charge flow solutions or probe surveys of the ion beam.* There are other effluents from the thruster besides these beam ions. These other effluents and their effects are described in this section.

### A. Electrons

Electrons are emitted by the neutralizer and leave the thruster at the same rate as ions. Information on the distribution of these electrons is obtained from the analysis of emissive probe data, as described by Ogawa *et al.* (*155, 156*). In broad beams the electron density equals the local ion density. The electrons from the neutralizer are randomized rapidly into a Maxwellian distribution, presumably as a result of two-stream instability. The temperature of this distribution in electron volts is equal to about 0.3 of the injection (coupling) voltage. This temperature is nearly uniform throughout the ion beam.† The plasma potential and the electron (and ion) density are related by

$$n_e = n'_e \exp(V/T_e), \tag{27}$$

where $T_e$ is the electron temperature in electron volts and the potential $V$ is defined as zero at the reference electron density $n'_e$. Equation (27) has been found to agree with experimental results over a range in electron density of more than $100 : 1$.

---

* Probe surveys have generally shown more trajectories at large angles to the beam axis than can be explained by space-charge flow solutions. A survey with a collimating slit probe by Byers (*154*) indicates most of these large angle trajectories originate near the edge of the accelerator area. This result is consistent with the viewpoint that the plasma density gradient near the outside of the ion chamber is responsible for many of the observed large angle trajectories.

† A large drop in electron temperature in the downstream direction was measured in one test with a small ion source mounted in a 21 m long vacuum facility. Such a test is not representative of ground applications and may not even be representative of space applications.

Equation (27) would be consistent with an electron gas in equilibrium with its potential distribution, but the electron gas actually has a drift velocity equal to the ion velocity. There is the problem, therefore, of the energy source for the electrons to drift against the adverse potential gradient produced by the expanding beam. Over the distances likely to be encountered in ground vacuum facilities, Ogawa, Cole, and Sellen believe that this energy is conducted along the beam by the electron gas. Further, this flow of energy is consistent with the measurement of an electron temperature substantially below the injection energy.

Only the major aspects of electrons in ion beams have been presented here. For some of the more esoteric aspects, including the transition from an ion beam plasma to the ambient space plasma, the original references are recommended.

## B. Neutrals

Measurements of the neutral efflux from a thruster are consistent with a cosine distribution from the accelerator apertures (90). The neutral density falls off rapidly with distance from the thruster so that charge-exchange ion production is concentrated near the accelerator. The condensation of mercury propellant on spacecraft surfaces has been studied, but is not felt to be much of a problem (157). The lower vapor pressure of cesium requires greater care by a spacecraft designer so that cold surfaces do not "see" the accelerator area, or are even close enough to such surfaces to receive much reevaporated propellant.

## C. Charge-Exchange Ions

The primary ion trajectories together with the neutral density can be used to determine the distribution of charge-exchange ion production. The first systematic study of charge-exchange ions and their trajectories was made by Staggs et al. (158). They divided charge-exchange ions into four groups: (1) those leaving the thruster with angles of 0–20 deg to the beam axis, and considered to be within the primary ion beam, (2) those leaving the thruster at angles of 20–90 deg, (3) those returning to impinge on the accelerator electrode, and (4) those produced downstream of the neutralization plane and not returning to the thruster.

The first group presents no problem, while the third group has been already discussed in Section III,F. The total rate of escape for group 2 is shown in Fig. 53a. The corresponding energies are shown in Fig. 53b. These data were calculated for a thruster with a 15 cm diam beam operating at 0.25 A and 3000 V net with an 80% utilization of the mercury propellant.

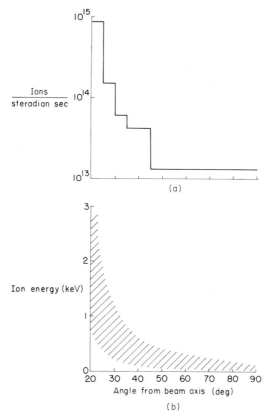

FIG. 53. Group 2 charge-exchange ions, produced between the screen and accelerator and leaving the thruster at angles of 20–90 deg with the beam axis. (a) Rate of escape. (b) Range of energy. From Staggs *et al.* (*158*).

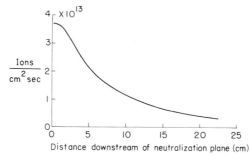

FIG. 54. Group 4 charge-exchange ions, produced downstream of the neutralization plane and leaving the ion beam in an approximately radial direction. The rate is for the surface of a cylinder surrounding the 15 cm diam beam. From Staggs *et al.* (*158*).

The interelectrode gap was 2.5 mm and an $R$ ratio of 0.6 was used. Scaling can be used for estimates at other operating conditions and other sizes, although the results depend somewhat on the exact electrode configuration used.

The production of group 4 charge-exchange ions is shown in Fig. 54. Due to the potential distribution of Eq. (27), the flow of these ions will be nearly radial in direction. Their velocity is low enough, however, that potentials of surrounding hardware can deflect their trajectories.

### D. Sputtered Particles

The sputtered efflux from thrusters has been studied in detail by Reynolds and Richley (157) and Reynolds (159). This efflux is believed to be directed downstream with somewhat more than a cosine distribution, both as a result of experimental deposition from a thruster (157) and from basic sputtering measurements (127). Sputtering data with either actual impingement measurements or the methods of Section III,F,1 can be used to estimate the total amount of material sputtered.

There is another group of particles leaving the thruster that leads to the deposition of opaque layers of accelerator material. Although Staggs et al. (158) only considered propellant atoms and ions in group 4 charge-exchange ions, it is also possible to have sputtered particles from the accelerator charge-exchange with beam ions. The upper limits of such cross sections were estimated by Dugan (160). These cross sections (for $Hg^+$ on Mo and $Cs^+$ on Al) can be used in the procedure of Reynolds (159) to show that, in some circumstances, these charge-exchange accelerator atoms can cause difficulty on a spacecraft.

In a ground application with an inert gas as propellant, particles sputtered directly from the accelerator and the target will probably be the only condensibles that need consideration. The calculation procedure should be essentially the determination of view factors for the accelerator, target, and (if there is one) deposition substrate. Using information such as that given by Musket and Smith (127), corrections can then be made for the noncosine portion of the sputtering.

### E. Light

Light is emitted from the ion chamber of a bombardment thruster, and can be used to analyze ion chamber processes, as shown by Milder and Sovey (65). The fraction of doubly ionized propellant atoms, the Maxwellian electron temperature, and the primary electron have all been obtained from spectrographic analysis of this light.

Measurable light can also be emitted from the beam downstream of the thruster, as was also shown by Milder and Sovey (*161*). Although faint, this light could offer interference with a spacecraft star tracker if a wide field of view through the beam was used and/or excessive neutralizer coupling voltage used. This light from the beam should not be confused with the pictures occasionally made that show well-defined operating ion beams. To obtain such a picture, the background pressure is usually raised to $\geq 10^{-5}$ Torr. The light then emitted is mostly from ion-neutral collisions.

## VI. ELECTRON EMISSION

A variety of cathode types have been used in bombardment thrusters. All of these types are described in this section, but the presentation emphasizes refractory metal and hollow cathodes. The refractory metal cathode is the simplest and most convenient to use in ground applications, while the hollow cathode is the best choice for long-duration space missions. The long life of the hollow cathode may also justify its use in some ground applications.

### A. Refractory Metal Cathodes

Tantalum and tungsten are the two refractory metals used for cathodes in bombardment thrusters. Tantalum has been used more often because it is easily worked and retains its ductility after use. The only shortcoming of tantalum is a tendency for cantilevered shapes with small cross sections to sag. Tungsten is more difficult to fabricate and becomes extremely brittle after use. Thoriated tungsten has been used, but its emissive advantage disappears after a few minutes of use due to ion bombardment. The pertinent properties of tantalum and tungsten at elevated temperatures are given in Table X. These data were obtained from Kohl (*162*).

### 1. Discharge Erosion Rates

The major lifetime limitation of refractory metal cathodes in ion chambers is erosion due to bombardment by discharge ions. A small amount of information on this erosion is given by Kerslake (*163*), but the most complete information is given by Milder and Kerslake (*164*). A 7.5 cm source was used by Milder and Kerslake to investigate tantalum ribbon cathodes that were 3.6 mm wide and 0.05 mm thick. The initial erosion rates for these cathodes are given in Fig. 55. The effect of propellant flow rate (and hence neutral density) is shown in Fig. 55a for two emission current densities. The

TABLE X

PROPERTIES OF TANTALUM AND TUNGSTEN EMITTERS[a]

| Temperature (deg K) | Emission (A/cm$^2$) | Radiation (W/cm$^2$) | Resistivity (ohm-cm) | Evaporation (cm/hr) |
|---|---|---|---|---|
| (a) TANTALUM | | | | |
| 293 | — | — | $15.5 \times 10^{-6}$ | — |
| 2000 | 0.00658 | 21.6 | $78.9 \times 10^{-6}$ | $3.47 \times 10^{-10}$ |
| 2100 | 0.0226 | 27.1 | $82.0 \times 10^{-6}$ | $2.99 \times 10^{-9}$ |
| 2200 | 0.0695 | 34.2 | $85.2 \times 10^{-6}$ | $2.12 \times 10^{-8}$ |
| 2300 | 0.195 | 42.2 | $88.3 \times 10^{-6}$ | $1.28 \times 10^{-7}$ |
| 2400 | 0.503 | 51.3 | $91.3 \times 10^{-6}$ | $6.59 \times 10^{-7}$ |
| 2500 | 1.21 | 62.4 | $94.4 \times 10^{-6}$ | $2.97 \times 10^{-6}$ |
| 2600 | 2.72 | 75.4 | $97.4 \times 10^{-6}$ | $1.20 \times 10^{-5}$ |
| 2700 | 5.79 | 89.9 | $100. \times 10^{-6}$ | $4.34 \times 10^{-5}$ |
| 2800 | 11.7 | 106. | $103. \times 10^{-6}$ | $1.43 \times 10^{-4}$ |
| 2900 | 22.6 | 123. | $106. \times 10^{-6}$ | $4.34 \times 10^{-4}$ |
| 3000 | 41.8 | 144. | $109. \times 10^{-6}$ | $1.26 \times 10^{-3}$ |
| (b) TUNGSTEN | | | | |
| 293 | — | — | $5.49 \times 10^{-6}$ | — |
| 2000 | 0.00100 | 23.6 | $55.7 \times 10^{-6}$ | $1.40 \times 10^{-11}$ |
| 2100 | 0.00392 | 29.8 | $59.0 \times 10^{-6}$ | $1.39 \times 10^{-10}$ |
| 2200 | 0.0133 | 37.2 | $62.4 \times 10^{-6}$ | $1.12 \times 10^{-9}$ |
| 2300 | 0.0407 | 45.7 | $65.8 \times 10^{-6}$ | $7.52 \times 10^{-9}$ |
| 2400 | 0.116 | 55.7 | $69.2 \times 10^{-6}$ | $4.31 \times 10^{-8}$ |
| 2500 | 0.298 | 67.2 | $72.7 \times 10^{-6}$ | $2.16 \times 10^{-7}$ |
| 2600 | 0.716 | 80.6 | $76.2 \times 10^{-6}$ | $9.46 \times 10^{-7}$ |
| 2700 | 1.63 | 95.6 | $79.7 \times 10^{-6}$ | $3.75 \times 10^{-6}$ |
| 2800 | 3.54 | 112. | $83.2 \times 10^{-6}$ | $1.34 \times 10^{-5}$ |
| 2900 | 7.31 | 132. | $86.8 \times 10^{-6}$ | $4.40 \times 10^{-5}$ |
| 3000 | 14.1 | 154. | $90.4 \times 10^{-6}$ | $1.33 \times 10^{-4}$ |

[a] From *Materials and Techniques for Vacuum Devices* by H. Kohl © 1967. Reprinted by permission of Van Nostrand-Reinhold Company.

increase in erosion rate at low neutral flow rates indicates that high utilization applications will have higher erosion rates. The effects of discharge voltage and emission current density are shown in Fig. 55b. The qualitative effect of voltage is roughly what would be expected from sputtering erosion. The erosion rate is nearly proportional to emission current density at a given voltage.

The data in Fig. 55 were obtained in a bell jar vacuum facility with only the discharge voltage difference to extract ions through the screen grid.

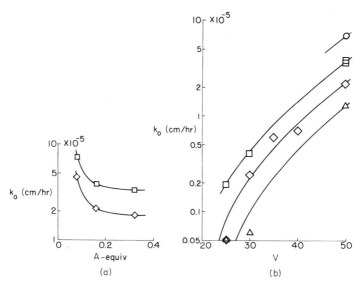

FIG. 55. Initial erosion rates for tantalum ribbon cathodes. (a) Propellant flow rate. (b) Discharge voltage. Emission, A/cm$^2$: $\bigcirc$, 0.95; $\square$, 0.55; $\diamondsuit$, 0.28; $\triangle$, 0.14. From Milder and Kerslake (*164*).

Check points on the effects of contaminants (probably oxygen and water vapor) in the vacuum system were also shown by Milder and Kerslake. These data were obtained with all operating parameters the same, except that no discharge was used. These points gave random erosion rates up to $6 \times 10^{-6}$ cm/hr for bell jar pressures up to $2 \times 10^{-6}$. It is likely that the ions from the discharge helped to pump contaminants, because the erosion rates in Fig. 55b appear reasonably self-consistent at rates below the $6 \times 10^{-6}$ cm/hr value. The presence of contaminants, though, may set a lower limit on erosion rates observed in a given environment.

Tests were also conducted by Milder and Kerslake using a 10 cm source operated in the conventional manner in a large vacuum facility. The initial erosion rates for the 0.25 mm tantalum wire cathodes used in this investigation are shown in Fig. 56. These data showed no effect of emission current density over the range investigated. The rise in erosion rates at low voltages is believed due indirectly to space-charge limitations. As the discharge was decreased below 40 V, the emission became space-charge limited. The emission of the desired level was obtained in this condition by overheating the cathode sufficiently for enough of the more energetic electrons to overcome the space-charge potential barrier. The higher erosion rates at low discharge voltages are therefore believed due to excessive operating temperatures. A check point using 0.25 mm wire in the 7.5 cm source operated in the bell jar facility gave agreement with Fig. 56 data within about a factor of 2.

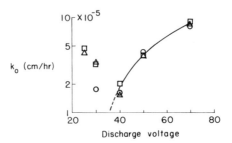

FIG. 56. Initial erosion rates for tantalum wire cathodes. Emission, A/cm$^2$: ○, 0.42; □, 0.63; △, 0.84. From Milder and Kerslake (*164*).

The data of Figs. 55 and 56 can be compared in an approximate manner with sputtering data. Stuart and Wehner (*165*) have obtained sputtering rates for mercury ions on tungsten at very low energies. Using their yield, an assumed ion density of $10^{17}/m^3$, and the Bohm sheath velocity for an electron temperature of 5 eV, the sputtering rate shown in Fig. 57 was obtained. Except for the space-charge limited region of Fig. 56, the agreement is within a factor of about 2–10. Some of the more obvious reasons for Figs. 55 and 56 being higher are doubly ionized mercury, a possible temperature effect of the sputtered surface (*166*), and the actual ion energy being higher than just the discharge voltage. (The plasma is usually several volts higher than the anode.) The fact that the sputtering data is tungsten instead of tantalum and the approximate nature of the ion density and electron temperature

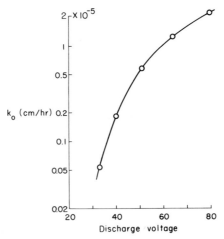

FIG. 57. Initial erosion rate predicted from the sputtering of tungsten by mercury ions. Based on sputtering data from Stuart and Wehner (*165*).

assumptions should keep us from making too detailed a comparison. The order-of-magnitude agreement, though, is reassuring.

Erosion rates such as shown in Figs. 55 and 56 are dependent on the specific configuration and operating conditions. Extrapolations to new applications can easily result in a factor of several in uncertainty. The erosion rate for tungsten is probably the same, within this uncertainty, as for tantalum.

The specific power loss for the data in Figs. 55 and 56 is 75–270 W per emitted ampere. The higher specific losses correspond to the lower emission current densities. These power losses are high for space applications, but reasonable for most ground applications. The electron emission is 5–10 times the ion beam current, so refractory metal cathode losses of 75–270 W/A can easily exceed discharge losses.

## 2. Lifetime in Ion Chamber

The lifetime of a refractory metal cathode can be estimated by assuming that the initial erosion rate of cathode mass continues until all cathode material is eroded away. This gives a maximum lifetime $T_{max}$ of

$$T_{max} = t/2k_0 \tag{28}$$

for a ribbon cathode of thickness $t$ and a lifetime of

$$T_{max} = d/4k_0 \tag{29}$$

for a wire cathode of diameter $d$. Milder and Kerslake found the actual lifetime was typically half of the value given by this simplified calculation, and it was occasionally as low as a quarter of that value. The existence of initial hot spots (in the uneroded cathode) has been found to further reduce the probable lifetime.

The variation in resistance with time can be predicted from the initial erosion rate and the initial resistance $R_0$. For ribbons this variation is

$$R = R_0/(1 - 2k_0 T/t) \tag{30}$$

and for wires it is

$$R = R_0/(1 - 4k_0 T/d). \tag{31}$$

These equations were found by Milder and Kerslake to predict the actual variation fairly accurately. The largest discrepancy is usually at the end of life, where the resistance increases above the predicted value as a hot spot develops progressively toward burnout. The range in resistance to expect in typical operation can be obtained by using a time $T$ equal to half the values given by Eqs. (28) and (29). This substitution in Eqs. (30) and (31) gives a final resistance of $2R_0$ for both ribbons and wires.

Assuming no major effect of contamination, initial hot spots, or over-heating due to a space-charge limit, the lifetime of a refractory metal cathode is fairly reproducible. The biggest uncertainty is in establishing the initial erosion rate for the particular design and operating conditions of interest.

The maximum lifetime obtained by Milder and Kerslake was about 1500 hr, which is far short of the needs of electric propulsion. This lifetime could probably be increased substantially if the much larger cathode power losses of more massive cathodes were acceptable. The combination of short demonstrated life and increased losses for further increases in life makes refractory metal cathodes unpromising for space applications. For ground applications, though, the lifetime and efficiency requirements are usually less severe. That, plus the simplicity of a refractory metal cathode, makes it an obvious choice for ground applications.

3. *Neutralizer*

Refractory metal can be used for durable neutralizers as long as direct impingement by beam ions is avoided. As discussed in Section IV,B,1, this is feasible only for thin beams (accelerator systems with not more than several holes). The information in Table X can be used in such an application, with only the level of erosion from charge-exchange ions to be determined by charge-exchange calculations or test.

A carburized surface on thoriated tungsten has been used successfully by Dulgeroff *et al.* (*147*) in a single-slit cesium contact thruster. The charge-exchange rate is usually lower for a contact thruster than for a bombard-ment thruster, but such an approach might be considered to obtain a longer effective life with thoriated tungsten in a thin-beam bombardment neutralizer.

An immersed-emitter neutralizer is basically limited by erosion from energetic beam ions. The sputtering curves shown in Fig. 58 can be used for a rough estimate of cathode life in such an application. The curves of Fig. 58 were obtained by fairing data from Meyer and Güntherschulze (*125*), Wehner (*126*), Carter and Colligon (*128*), Laegreid and Wehner (*133*), and Rosenberg and Wehner (*134*). One possible reason that the estimate will be only a rough one is that cathode operating temperatures are far above the temperatures at which these data were obtained. Another more certain reason is that Fig. 58 is for normal ion incidence which, in general, will not be the case for a neutralizer. That is, there is no sheath effect that tends to direct ions normal to the surface, such as is found in the ion chamber.

An example of immersed emitter lifetime was given by Kemp *et al.* (*167*). Both 2.5 mm by 0.05 mm ribbons and 0.25 mm diam wires of tantalum were used, with minimum lifetimes of 2–3 hr in a beam with 1.4 mA/cm$^2$ average

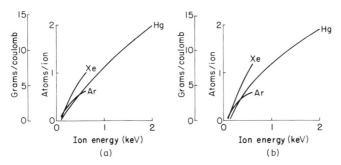

FIG. 58. Sputtering by normal incidence ions. (a) Tantalum. (b) Tungsten. Data from Meyer and Güntherschulze (125), Wehner (126), Carter and Colligon (128), Laegreid and Wehner (133), and Rosenberg and Wehner (134).

current density of 3000 eV mercury ions. The ribbon, mounted approximately edge-on to the ions, might have been expected to last much longer than the wire. What erosion did take place on the flat surfaces, though, was probably at a much higher yield due to the oblique angle.

Kemp *et al.* also tried heavy shields upstream of the neutralizer cathodes. This reduced the cathode erosion at the cost of increased coupling voltage. The total amount of sputtered material was also increased by the presence of the shield. The best approach for longer lifetimes in ground applications appears to be simply moving the neutralizer to less dense portions of the ion beam.

## B. Oxide Cathodes

A wide variety of oxide cathodes were evaluated for use in bombardment thrusters. Some of the types have been a thick oxide layer in a matrix of fine tungsten or tantalum wires (168), an oxide magazine type (169), and an oxide coated screen in a loose roll (170). Descriptions of early oxide cathode tests are included in a survey paper by Kerslake (163), while a later survey paper by Weigand and Nakanishi (171) covers all types.

Some types of oxide cathodes achieved lifetimes of over 4000 hr with moderate losses of 15–20 W per emitted ampere. This lifetime was sufficient for some space missions, while the losses were low enough for efficient performance. The basic problem was one of reliability. The cathode had to be exposed to atmosphere after completion of ground tests and prior to launch. (The only alternative is to launch a thruster in a heavy sealed tank, the weight of which will offset any planned advantage for electric propulsion.) The exposure to atmosphere meant that the oxide had to be removed and replaced with fresh (and hence untested) carbonate. The use of an unconditioned cathode was never found to have a probability of success over 90–95%, which is unacceptably low for a single component of a thruster.

The hollow cathode for mercury propellant became available during the SERT II program. This cathode used a small amount of oxide, but it could be exposed repeatedly to the atmosphere with no serious effects. A switch to the hollow cathode was made during the SERT II program, followed by similar choices for all subsequent bombardment-thruster programs.

There are no present plans to use oxide cathodes in space. In ground applications the extra problems of using oxide (including exposure to atmosphere) do not appear worth the small power saving relative to a refractory metal cathode. Even if a low power loss is important in a ground application, the hollow cathode is at least as efficient and offers far more reliability and tolerance to atmospheric exposure.

### C. Hollow Cathodes

The starting point in bombardment-thruster hollow cathodes was the cesium autocathode of Speiser and Branson (172). This was followed by a smaller version of the autocathode to operate as the neutralizer of a cesium contact thruster by Ernstene et al. (173). Sohl et al. (174) used the same neutralizer on a cesium bombardment thruster. A hollow cathode similar to this neutralizer was then operated on mercury propellant by Kemp and Sellen (175), with an internal oxide coating replacing the low-work function cesium. [The content of Kemp and Sellen (175) is included with additional information in Hall et al. (176).] The hollow cathode was developed into the SERT II neutralizer by Rawlin and Pawlik (177). Because of the problems with the original oxide cathode (discussed in the preceding section), the main cathode of the SERT II thruster was then switched to the hollow type—see Kerslake et al. (178). Since then, all bombardment thruster programs for space application have used hollow cathodes.

### 1. Theory and Typical Operation

A sketch of a typical hollow cathode used with mercury propellant is shown in Fig. 59. The vaporized propellant flows up the tantalum tube, past an oxide coated or impregnated insert, and through the orifice in a thoriated

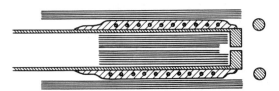

FIG. 59. Hollow cathode used with mercury propellant.

tungsten tip. The discharge is usually initiated by heating the cathode tip to ~ 1000°C and applying several hundred volts to the keeper electrode. The power supply to the keeper is such that the voltage drops rapidly as the keeper current increases. Normal operation for the keeper is in the range of 5–25 V and 0.1–2 A.

Typical performance at fixed levels of keeper and anode currents are shown in Fig. 60 for a plasma anode (in this case an ion beam). Keeper and

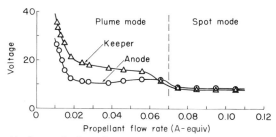

FIG. 60. Typical hollow cathode performance with plasma anode. Anode current is 0.25 A and keeper current is 0.13 A. From Rawlin and Pawlik (177).

anode voltages both decrease as propellant flow is increased, until they become nearly constant at high flows. As first observed by Kemp and Sellen (175) the performance can be divided into a plume or noisy mode and a spot or quiet mode. The plume mode, as indicated by the name, has a luminous plume extending up to several centimeters downstream of the orifice. At the higher propellant flows associated with the spot mode, the luminous region is restricted to a small spot in the orifice.

Performance curves with a metallic anode (a plate or screen one to several centimeters from the cathode) is shown in Fig. 61. Here we find the

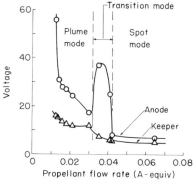

FIG. 61. Typical hollow cathode performance with metallic anode. Anode current is 0.25 A and keeper current is 0.16 A. From Rawlin and Pawlik (177).

additional transition mode—found by Rawlin and Pawlik (*177*)—between
the plume and spot modes. Except for the "hump" in anode voltage, the
general trends in anode and keeper voltages are the same as shown in
Fig. 60.* The relative distribution in luminosity for the three modes is indicated in Fig. 62. Rawlin and Pawlik showed the usual low noise level for
keeper and anode voltages in the spot mode and 1–2 V fluctuations for both
in the plume mode. For the transition mode they found a low noise level for
the keeper and 5–10 V fluctuations of the anode. Note that these noise
measurements form a consistent pattern with the luminosity distributions,
with the intermediate mode showing both electrical noise and light at the
anode and no noise and light at the keeper. The coupling or anode voltage in

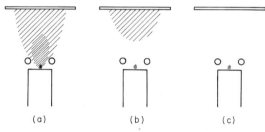

FIG. 62. Luminosity distributions for the different operating modes with a metallic anode.
(a) Plume. (b) Transition. (c) Spot.

the plume mode increases both with increasing anode distance and decreasing propellant flow rate—as described by Hall *et al.* The coupling voltage is
nearly constant in the spot mode, but an increase in anode distance can
result in a shift to the plume mode. This latter effect was found by Csiky
(*179*) and is shown in Fig. 63. The propellant necessary to remain in the spot
mode increases by a factor of several as the anode distance increases from 2
to 4 cm.

The discharge restricts the flow of propellant through the hole. Hall *et al.*
found qualitatively that the amount of restriction increased continuously
with discharge current. A quantitative comparison was made at a constant
propellant supply pressure and tip temperature, showing about 38% less
flow with a one-ampere discharge. A more recent investigation by Snyder
and Banks (*180*) showed that, for a constant propellant flow rate and tip
temperature, the discharge increased the thrust (of the cathode alone) by
15–30%. Inasmuch as ions were not extracted, this was taken to indicate an

---

* Some evidence for an intermediate or transition mode also exists with a plasma anode.
The steady-state data with a plasma anode, though, do not give the clear distinction of a
different mode shown in Fig. 61.

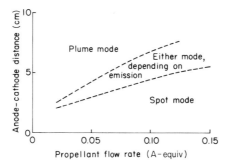

FIG. 63. Effect of anode distance and propellant flow rate on operating mode of hollow cathode. From Csiky (*179*).

increase in temperature of propellant atoms and ions. The flow restriction observed by Hall *et al.* could thus be a purely thermal effect of the discharge.

The potential of the plasma efflux from the hollow cathode was first measured by Hall *et al.* They found that most of the plasma between cathode and anode was at a nearly uniform level of potential. More detailed measurements were made by Csiky, who found that the uniform potential continued right to the anode in the spot made, but turned up in the plume mode—as shown in Fig. 64. An electron temperature difference was also found for this region by Csiky, 1.5–2.5 eV for the plume mode and $\sim 0.5$ for the spot.

A rough physical description can be assembled from the preceding observations of the plasma efflux. In the plume mode, the ion production within the efflux is the important process for neutralizing the electron space charge. This process was suggested by Hall *et al.*, with some calculations made later by Ward and King (*181*). Ionization within the plume mode efflux is supported both by the higher electron temperature and the light emission. In the spot mode, the ions to neutralize the electron space charge come from the cathode orifice. The higher propellant density and flow rate presumably carry the ions along.

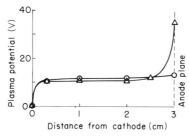

FIG. 64. Variation of plasma potential within hollow cathode efflux. Mode: $\bigcirc$, spot; $\triangle$, plume. From Csiky (*179*).

The preceding physical description is also supported by electron density measurements made by Csiky.* From plasma neutrality, of course, the electron density equals the ion density. If the ions were to flow out with the neutrals at constant velocity, then the electron (ion) density should fall off inversely as the square of distance. To compare the data with this simple picture, the electron densities given by Csiky are multiplied by the square of the distance from the cathode and plotted in Fig. 65. There is actually a small accelerating field for the electrons, and thus a tendency for ions to be attracted toward the cathode. For the spot mode, then, the ions propelled by the neutral atoms slow down as they leave the orifice, and thus $n_e l^2$ increases with distance instead of remaining constant. For the plume mode, with ions created approximately proportional to neutral density, the ions

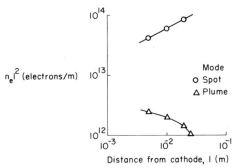

FIG. 65. Electron density parameter as a function of distance from cathode orifice. Data from Csiky (179).

from far locations accelerate toward the cathode and add their space charge to the ions created in the closer locations. This gives the increase in $n_e l^2$ as the cathode is approached. If both curves are extrapolated back to the orifice radius of 0.125 mm, a density of about $4 \times 10^{21}/m^3$ is obtained. This compares with an estimate by Kemp and Sellen of $\sim 10^{22}/m^3$ in the orifice with cesium propellant.

The transition mode with a metallic electrode can be described in similar terms. The presence of the metallic electrode results in the recombination of ions that reach it. If operation is between the transition and spot modes, this ion depletion results in inadequate electron neutralization near the anode. This depletion must be made up by ion production near the anode, as indicated by the emission of light in that location and the electrical noise at the anode.

---

*The variation of electron density with distance is of interest here, and is available only in the second portion of Csiky (179).

Processes are much less clear inside the hollow cathode. Fearn and Philip (*182*) suggest that the discharge motion with increased current (from plume to spot) is continued as the current is increased, resulting in the discharge moving into the cavity upstream of the orifice. [As shown clearly by Philip (*183*), the transition from plume to spot can be made either by increasing propellant flow or discharge current.] The steps in the high voltage portion of a voltage–current curve were associated with the steps in an orifice to support this conclusion. But stepped curves can be obtained with an unstepped orifice, as was shown by Sheheen and Finke (*184*). Further, as shown by Sheheen and Finke, Byers (*185*), and Byers and Snyder (*186*), the power supply characteristics can determine what shape the high voltage portion of the curve will have. In addition, Hall *et al.* saw no major change in internal illumination of the cathode in going from plume to spot mode.

The author prefers a single description of discharge changes for both propellant and emission effects. That is, the change from plume to spot caused by increasing propellant flow should be closely related to the change from plume to spot caused by increasing electron emission. Fearn and Philip found that increasing the propellant flow decreased both electron temperature and density in the upstream cavity. This is consistent with the luminous discharge being concentrated towards the orifice from both ends as the propellant flow is increased. Increasing emission may increase electron temperature and density in the upstream cavity, but similarity to propellant flow effects would give an even greater increase in electron temperature and density within the orifice. At the highest propellant flows, with only small mean path lengths, the densest portion of the discharge may be concentrated in only a short section of the orifice hole.

The source of emitted electrons in a hollow cathode is not clear at all for mercury propellant. Hall *et al.* feel that the emission mechanism may be similar to that of the liquid mercury cathode, which has been studied for many years without firm conclusions. Rawlin and Pawlik operated without oxide coatings or inserts of any kind, and found the biggest difference was hard starting with very high flows needed. Fearn and Philip obtained more detailed starting data—about a 400°C higher temperature was required to start at the same voltage without the oxide. Fearn and Philip also found an operating temperature about 200°C higher without the oxide. A series of curves obtained by Mirtich (*187*) while the oxide was being depleted give a more complete picture of the role of oxide on performance. The plume mode was affected the most, first increasing voltages for the same emission and propellant flow, and finally unstable operation. There was some shift to higher voltages with the spot mode and an increase of almost 50% in minimum flow required to obtain the spot mode. Thermionic emission appears most important for starting and plume mode and least important for spot mode. The fact that emission generally continues as the oxide is depleted, although at a higher voltage, indicates there is no single emission process. Instead, thermionic electrons can apparently be used if readily available, but a shift toward other processes can take place if they become less available, particularly in the spot mode.

Hall *et al.* suggested both one- and two-step ionization of neutrals by electrons in the plasma efflux, at least one of which appears certain for the plume mode. Similar processes are also a possibility for the orifice in the spot mode. In both cases, the circuit is completed by the ion going to the cathode and the electron to the anode. Hall *et al.* also suggested high field emission at the cathode surface. Rawlin and Kerslake (*188*) suggested the formation of a cathode spot similar to that encountered in cold cathode arcs. Philip suggests electron emission from the impact of excited atoms on the cathode surfaces.

The author has no intention of giving a personal order of preference for these different processes. Instead, he wishes to point out a similarity of vacuum-arc data with that of hollow cathodes. According to Guile (*189*), Gilmour and Lockwood (*190*), and Lockwood (*191*), vacuum-arc data tend toward an approximate cathode erosion limit of $10^{-7}$ kg/coulomb in the absence of sputtered droplets and fragments. The reason given for this erosion limit is that it represents the amount of gas required for cathode operation. This approximate limit corresponds in mercury propellant to 21 emitted electrons per atom of propellant. Examining a wide range of tests with low erosion rates, a maximum of about three times this electron-atom ratio was found. The minimum propellant flow rate for a hollow cathode is thus in approximate agreement with vacuum-arc data. This agreement carries the implication that the discharge may erode the cathode rapidly if it does not have this minimum gas flow rate. All this does not solve the problem of where the electrons come from, but it does suggest a closer comparison to vacuum-arc data with high current, low-propellant-flow operation.

Operation with cesium is substantially different as far as electron emission is concerned. Tantalum is used for the emitting surface which, with a cesium coating, is a good emitter at 600°C (*46, 173*). Because the operating temperature is consistent with thermionic emission, it is believed that most electrons are supplied in that manner with cesium propellant. The lower temperature of operation also results in main cathode construction being larger than for mercury (*46*).

Operation of mercury-type hollow cathodes with other propellants is limited (*50, 51*). Operation with xenon is similar enough to mercury that similar lifetimes might be expected. Argon performance is less similar, but no obvious problems were encountered in the short tests conducted.

## 2. Experimental Configurations and Performance

Thoriated tungsten was found to be more resistant to tip damage than several other materials in extensive tests by Rawlin and Pawlik (*177*). Thoriated tungsten was used in place of tungsten for its higher recrystallization temperature. The sputtering of bombardment ions is believed to prevent any emissive advantage of thoriated tungsten. The tip is electron-beam welded to a tantalum tube. The most reliable heater design has been tungsten–rhenium wire imbedded in flame-sprayed alumina (*188*). A corrosion barrier of tungsten is flame-sprayed on the tantalum tube prior to depositing the alumina. A radiation shield of tantalum foil is usually wrapped around the outside of the alumina, with another flame-sprayed layer of tungsten applied first.

The optimum keeper design is 1–2 mm from the tip and of a large enough diameter that it just makes contact with the dense efflux region (about a 45 deg half-angle cone). Smaller diameters may facilitate starting, but they result in increased ion recombination on the keeper surface (tantalum). In fact, added keeper surface outside of the minimum required for durability usually results in increased recombination losses.

Low-emission-current designs use an enclosed keeper approach that makes all the propellant pass through the keeper opening (see Fig. 66). The enclosed keeper design was initially used by Hall *et al.* (*176*), and then developed further by Reader *et al.* (*58*) and Hyman (*192*). The keeper hole diameter in the enclosed design usually corresponds to a 20–30 deg efflux half-angle, probably because the pressure between the tip and the keeper contains the efflux better than a nonenclosed design.

The insert used most often is a strip of tantalum foil, coated with barium carbonate (later barium and strontium carbonates), and rolled into a tight spiral (*177*). The position of the insert was originally tight against the inside of the tip, except for a small on-axis cavity behind the orifice. Positive electrical contact of the insert with the rest of the cathode is required to avoid erratic operation. A more recent investigation has used a recessed insert location for slower depletion of oxide and to decouple tip temperature

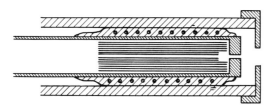

FIG. 66. Hollow cathode with enclosed keeper used with mercury propellant.

from insert temperature for high emissions, as described by Mirtich (187). Hudson and Weigand (193) have investigated barium oxide impregnated porous tungsten for inserts, which offers more precision fabrication than the handmade foil inserts. Hudson and Weigand also tested an impregnated cathode tip [first used by Kemp and Sellen (175)]. The use of an impregnated porous tungsten tip without an insert may be a useful approach where temperature and lifetime requirements are moderate.

The lifetime of a hollow cathode is normally limited by ion erosion, starting with a rounding or chamfering at the downstream end of the orifice. Operation at excessive temperature leaves a conical scoured appearance instead of the fairly smooth erosion of normal operation. For the very long lifetimes desired in space propulsion ($\sim 10,000$ hr), depletion of oxide can also be a serious problem. In ground applications, repeated exposures to atmosphere could have even more effect on the oxide than total accumulated lifetime.

The neutralizer of the 5 cm thruster is at the low emission end of mercury hollow cathodes. This cathode has operated with a little over 2 mA equivalent propellant flow and an emission of 25 mA. This low flow, however, resulted in a high coupling voltage of up to $\sim 20$ V, which in turn produced erosion that limited life to $\sim 10,000$ hr (151). A 3.2 mm tip diameter was used with a 1.3 mm tip thickness. But a deep machined conical chamfer reduced the cylindrical portion of the orifice to an initial length only equal to the diameter ($\sim 0.25$ mm).

The SERT II cathode was operated in a bell jar test at about 2.5 A emission with a wear rate estimated to give a 5000 hr life. To obtain a 10,000 hr life, the emission should be dropped to about 1.9 A (188). This design used a hole about 0.2 mm in diameter in a tip 1 mm thick and 3.2 mm in diameter.

The 30 cm cathodes are usually made with tips of 1.2 mm thickness. Mirtich (187) found holes of 0.38, 0.75, and 1.52 mm diameter had little or no wear with emission currents of 2, 8, and 15 A. This trend of higher maximum currents with larger tip orifice diameters is generally found in all cathode investigations.

The foregoing should serve as examples of typical hollow cathode performance, although some additional values are given in Section VI,C,3. All the preceding cathode tips were attached to tantalum tubes of equal diameter, with the remainder of the construction as described earlier. In most of these designs, operation without heater power was possible (after starting) at some operating conditions.

A problem that must be considered with hollow cathodes is the sensing of propellant flow rate for a closed loop control. In space applications this is normally done by sensing some operating voltage. For the neutralizer, the keeper voltage is monitored at a constant keeper current. Both the small 5 cm thruster and the 15 cm SERT II thruster have neutralizers that operate in the plume mode. This means that an increase in propellant flow generally gave a decrease in keeper voltage. If some ambiguity is encountered in the voltage flow curve, it can usually be resolved by using a different fixed keeper current. In the 30 cm thruster the neutralizer is the spot mode at high beam current, and hence is in the region of little voltage change for a changing propellant flow.

King *et al.* (*194*) found that higher keeper currents could be used to reestablish a voltage effect of flow rate while in the spot mode. The 30 cm thruster presently uses a keeper current of 1.5 A with maximum beam current [see Bechtel (*153*)].

The main cathode flow is normally regulated to maintain a fixed discharge voltage. The main cathode is typically in the spot mode and prefers to emit into a plasma $\sim 10$ V positive to the cathode, although the control method does not require the spot mode. This low voltage would give excessive discharge losses if used as the discharge voltage. On the other hand, throttling the cathode flow to put it at 30–40 V in the plume mode would result in rapid failure by ion erosion. A baffle is used to decouple the cathode plasma efflux from the ion chamber plasma (*178*). (This use of a baffle was adapted from the earlier work by King *et al.* (*195*) that used a baffle for the same purpose with a liquid metal cathode.) With the baffle, the cathode maintains its $\sim 10$ V environment, and the bulk of the 30–40 V discharge is dropped across the baffle restriction. The effect of cathode propellant flow is felt at this restriction, more neutral density permitting more rapid electron flow (mostly across magnetic field lines). The discharge voltage is thus used to sense and control main cathode propellant flow. Some cut-and-try is usually required to obtain a baffle that gives the right flow level in initial prototype testing.

Bechtel (*196*) found that if the initial cathode–plasma voltage drop was subtracted from the discharge voltage, a better correlation of experimental discharge losses was obtained. This indicates that the initial cathode–plasma drop is essentially wasted, as far as producing beam ions is concerned. Wells (*197*) in an analytical study concluded that the electrons crossed the baffle restriction by anomalous diffusion, which heated the electrons but did not prevent this energy from being used to produce beam ions. This result is in agreement with the previous one of Bechtel that the use of a baffle does not cause any losses beyond the initial cathode–plasma drop. An experimental investigation of Serafini and Terdan (*198*) supports the conclusion that the diffusion was anomalous at the baffle restriction, in at least some operating conditions. Serafini and Terdan used the variable magnetic baffle of Poeschel (*199*) and Bechtel (*200*). This device is useful for electrically changing the baffle restriction to permit a broader range of beam current with efficient operation. Under some circumstances the variable baffle field can be put in series with the discharge chamber, so that no additional circuits are required.

One more aspect of hollow cathode operation should be mentioned— that of starting. Many starting tests have been conducted with hollow cathodes without showing adverse effects of repeated starts (193, 201). Perhaps the most interesting tests have been conducted by Kerslake and Finke (202) using the still orbiting SERT II vehicle in space. Another recent development in hollow cathodes has been the rapid turn-on tests of Wintucky (203) using high voltage pulses. This pulse technique uses a small additional electrode near the keeper.

## 3. Performance Correlations

The theoretical knowledge of hollow cathodes is far from complete, as should be evident from Section VI,C,1. Some approaches to performance correlation can still be used despite the lack of theoretical knowledge. This is especially true of the spot mode, where the effects of an oxide coating or impregnation are minimized.

One aspect of interest is the prediction of the conditions at which the spot mode is encountered. From experimental performance, the most important parameters are propellant flow $J_0$, electron emission $J_e$, orifice diameter $d_0$, and plasma bridge length $l_p$. To obtain a clear definition of both plasma bridge length $l_p$ and the existence of the spot mode, bell jar data with metallic collector electrodes are used. Dimensionless groupings can be assembled out of the pertinent parameters, with $J_e/J_0$ and $l_p/d_0$ the obvious choices. But $J_0$ is not independent of all length measurements and another ratio will be required.* The mean path length $l_m$ for an atom or molecule of a given propellant obeys the proportionality

$$l_m \propto 1/n_0 . \tag{32}$$

The neutral density in the orifice $n_0$ can in turn be related to the neutral "current."

$$n_0 \propto J_0/d_0^2 . \tag{33}$$

---

* According to the $\Pi$ theorem of dimensional analysis, the number of dimensionless ratios required to describe a problem is the number of variables less the number of independent dimensions. If $J_0$ (or $J_e$) was truly independent of length, two dimensionless ratios would suffice for the problem. But since $J_0$ (or $J_e$) is a rate that affects directly the mean free path for neutrals (or electrons) the $\pi$ theorem requires three dimensionless ratios to describe the problem.

(The temperature effect in the preceding proportionality is omitted because it enters as the square root and does not vary over a wide range.) The ratio of orifice diameter to mean path length can therefore be expressed as

$$d_0/l_m \propto J_0/d_0 .\tag{34}$$

Because $J_0/d_0$ is far more convenient to calculate and use than $d_0/l_m$, $J_0/d$ will be used despite its dimensional form. It should be kept in mind, however, that the justification for the use of $J_0/d_0$ is in this last proportionality.

Experimental data for the junction of the transition and spot modes were obtained from Byers and Snyder (186), Mirtich (187), Rawlin and Kerslake (188), Hudson and Weigand (193), and Rawlin (204). These data are a strong function of $J_0/d_0$, as shown in Fig. 67. The range of data in Fig. 67 can be expressed as

$$J_0/d_0 = 0.27 \pm 0.13 \text{ A-equiv/mm.}\tag{35}$$

Because these data are for the junction of transition and spot modes, the spot mode corresponds to an $J_0/d_0$ equal to, or exceeding, the Eq. (35) value.

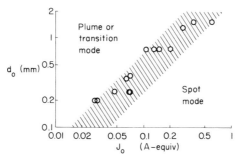

FIG. 67. Junction of transition and spot modes as a function of propellant flow rate and orifice diameter. Data from Byers and Snyder (186), Mirtich (187), Rawlin and Kerslake (188), Hudson and Weigand (193), and Rawlin (204).

The data of Fig. 67 were found to be weak functions of $J_e/J_0$ and $l_p/d_0$. The scatter of Eq. (35) could be reduced by about 35% by using the correlation parameter $(J_0/d_0)(J_e/J_0)^{0.2}(l_p/d_0)^{0.2}$ obtained from log–log plots of the data. But the added complexity was not felt to be worth the added accuracy. Also, there is something misleading about the correlation parameter obtained. An increase in $J_e/J_0$ in this parameter moves operation toward the spot mode. This agrees with experimental observation because, for an increase in $J_e/J_0$ without changing other ratios in the parameter, $J_e$ must be increased at constant $J_0$. An increase in the ratio $l_p/d_0$ also moves towards the spot mode, according to the correlation parameter. This trend is the opposite of the experimentally observed trend shown in Fig. 63. After obtaining this unexpected result, the nature of the collector electrodes was checked. All the data obtained at the shortest plasma bridge lengths of 1–1.3 cm were for wire screen collectors, while all the longer data at about 2 cm were for solid plate collectors. The wire screen was dense, but would still be expected to let some neutrals through and, therefore, reduce local neutral density. The power relationship obtained for $l_p/d_0$ was felt to be anomalous, and probably a result of the collector configurations. The effect of $l_p/d_0$ is probably not great in the 1–2 cm range (or $l_p/d_0$ range of 11–84, if you prefer), but the trend of Fig. 63 should be considered for longer distances. The range of $l_p/d_0$ for the 18 combinations of cathodes and operating conditions that were investigated was, as indicated, 11–84. The range of $J_e/J_0$ was 11–79. In view of the vacuum-arc data discussed in Section VI,C,1, there is no assurance that the higher values of $J_e/J_0$ correspond to long cathode life.

For cathode lifetime, data were obtained from Bechtel (153), Rawlin and Pawlik (177), Philip (183), Mirtich (187), Rawlin and Kerslake (188), Hudson and Weigand (193), and Rawlin (204). Some of the data were simply in the form of statements of acceptable or unacceptable current for a given size orifice. The majority of the data, though, included measured wear rates. For the latter data a linear extrapolation of 10,000 hr, or more, was considered an acceptable erosion rate on tip thickness. These data are plotted in Fig. 68. The data of Fig. 68 for acceptable erosion rates can be described by the approximate expression

$$J_e/d_0 \leq 12 \text{ A/mm.} \tag{36}$$

Note that a value of current divided by orifice diameter is obtained again. Ignoring the possible dimensional analysis interpretation, a simple thermal

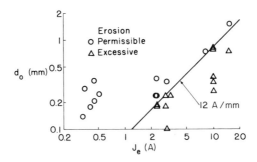

Fig. 68. Permissible emission for lifetimes of $\sim$ 10,000 hr using cathode tips of 1–1.3 mm thickness. Data from Bechtel (153), Rawlin and Pawlik (177), Philip (183), Rawlin and Kerslake (188), Hudson and Weigand (193), and Rawlin (204).

explanation can be given. The erosion and plasma heating in the spot mode [the mode of data near the limit of expression (36)] affects the perimeter of orifice. This perimeter is proportional to $d_0$, and a limiting value of $J_e/d_0$ simply means that erosion and/or heating is proportional to $J_e$.

The more detailed erosion data are shown in Fig. 69. Making use of the results of Fig. 68, the wear rate data of Fig. 69 are plotted against $J_e/d_0$. The data at more than 9 A/mm are all for the spot mode with $< 20$ V coupling. A conservative extrapolation of spot-mode data to lower values of $J_e/d_0$ is shown by the dotted line. This wear rate for lower values of $J_e/d_0$ is based on the assumption of wear being proportional to emission ampere-hours, which may be an overly conservative extrapolation. The data for less than 3 A/mm are all for the plume mode with 25–30 V coupling. The plume mode data are included for completeness and should not be compared to the dotted line.

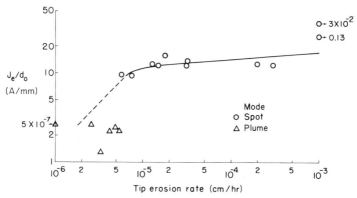

FIG. 69. Tip erosion rate as a function of emission diameter ratio. Data from Bechtel (153), Rawlin and Kerslake (188), and Rawlin (204).

The wear rates of Fig. 69 were obtained in many cases from tests of only several hundred hours duration. As described by King et al. (194), the initial erosion rate near the downstream edge of the orifice is more rapid during initial operation. This means that many of the data in Fig. 69 may be conservative. It also means, as pointed out by King et al., that the initial rapid erosion rate and changes in orifice shape can be avoided by starting out with a rounded or chamfered orifice at the downstream end.

## D. Liquid Metal Cathodes

The liquid metal cathode has been studied by King et al. (195), Eckhardt et al. (205, 206), and Hyman et al. (207). The emission is from a liquid mercury surface in the shape of either a small disc or annulus. Performance and durability of the liquid metal cathode are both excellent. It has the advantage of using no oxide so that it cannot be poisoned by atmospheric exposure. It has a drawback for space applications of requiring a heat sink at or below 300 C. The feed system requirements also constitute a drawback for space applications. As described in Section VII,A, the isolator prevents a high voltage short in one thruster from shorting all the thrusters in an array. An isolator that injects periodic bubbles of hydrogen into an insulated mercury feedline is described by Hyman *et al.* Such a device appears more difficult to develop than the simple vapor phase isolator used with the hollow cathode. The requirement for the precise control of liquid mercury, relative to the simple vaporizer used with the hollow cathode, is also a shortcoming. The most serious shortcoming for ground applications is that the liquid metal cathode has been used only with mercury.

## VII. ADDITIONAL FUNCTIONS

There are several additional functions that are clearly subsidiary to the primary functions discussed in the preceding sections, but are still important enough to mention sources of information.

### A. High Voltage Isolation

The propellant feed lines are potential conductors that could parallel a high voltage short in one thruster across all the other thrusters using that propellant system. There are several alternative solutions, but the most practical has been to use an isolator in the vapor phase portion of the propellant food system.

The first isolator was operated by Nakanishi and Pawlik (*208*). The initial device was large and cumbersome, but established basic feasibility. Smaller, more practical devices have been studied since then, as described by Nakanishi (*209, 210*) and Mantenieks (*211*). Isolators for cesium vapor have also been investigated, but so far without success.

The reliability with multiple thrusters is not an important reason for using isolators in ground applications. Instead, the convenience and safety of a grounded propellant feed system would be far more important. The propellants used in most ground applications will not encounter the condensation problem studied for space propulsion and mercury propellant. Also, the need for extreme compactness and ruggedness should not apply on the ground. In short, the use of the concepts and principles described in the cited references should permit quick, efficient design for most ground applications.

## B. Propellant Vaporization

The need for controlled feed of mercury propellant in the space environment has resulted in the development of small lightweight vaporizers. As described by Kerslake (212), these devices use porous tungsten with several-micron pore size to keep the nonwetting liquid mercury on one side and let the vapor pass through. Liquid pressures of several atmospheres can be easily contained by surface tension forces in the small porous tungsten passages.

Cesium wets most materials, so a vaporizer and feed system based on this tendency is used for cesium (46). Most ground applications use gaseous propellant so that both mercury and cesium vaporizer technology will not be of interest.

## C. Control

Many aspects of thruster control have been discussed in this paper. More extensive and systematic discussions of control problems are given by Pawlik et al. (37), Nakanishi et al. (91), Pawlik and Nakanishi (213), Bechtel (214), and Terdan and Bechtel (215).

## D. Mechanical Design

Some of the more troublesome thruster problems involve mechanical design. The author is only aware of one publication devoted solely to these problems, that of Zavesky and Hurst (216).

## VIII. CONCLUDING REMARKS

The electric propulsion program has been (and is) one of the most concentrated efforts ever mounted on charged particle sources and accelerators. Because the end product has been space propulsion, this technology has been followed primarily by those interested in space propulsion.

The source and accelerator technology developed in this program—particularly that of bombardment thrusters—has applications in many areas unrelated to space propulsion. These other applications will result in devices that bear little or no resemblance to space-propulsion thrusters, but the basic knowledge required will be the same. The intent of this paper, then, has been not just a review of bombardment-thruster technology for workers in that field, but to present the information in the general form required for a broad range of potential applications.

ACKNOWLEDGMENT

The author wishes to give special thanks to David C. Byers for his assistance in both discussions and a critical review.

REFERENCES

1. H. R. Kaufman and P. D. Reader, *Amer. Rocket Soc. Pap.* No. 1374-60 (1960); also *in* Langmuir and Stuhlinger, Ref. 12, pp. 3–20.
2. H. R. Kaufman, *NASA Tech. Note* **TN D-585** (1961).
3. C. E. S. Phillips, *Proc. Roy. Soc., Ser. A* **64**, 172–176 (1898).
4. F. M. Penning and J. H. A. Moubis, *Physica (Utrecht)* **4**, 1190–1199 (1937).
5. M. S. Livingston, M. G. Holloway, and C. P. Baker, *Rev. Sci. Instrum.* **10**, 63–73 (1939).
6. A. T. Finkelstein, *Rev. Sci. Instrum.* **11**, 94–97 (1940).
7. J. Backus, *in* "The Characteristics of Electrical Discharges in Magnetic Fields" (A. Guthrie and R. K. Wakerling, eds.), pp. 345–369, McGraw-Hill, New York, 1949.
8. A. Guthrie and R. K. Wakerling, TID-5218. Radiat. Lab., Univ. of California, Berkeley, 1949.
9. M. Von Ardenne, *U.K. At. Energy Res. Estab., Lib. Transl.* No. 758 (1957).
10. R. G. Meyerand, Jr. and S. C. Brown, *Rev. Sci. Instrum.* **30**, 110–111 (1959).
11. C. D. Moak, H. E. Banta, J. N. Thurston, J. W. Johnson, and R. F. King, *Rev. Sci. Instrum.* **30**, 694–699 (1959).
12. D. B. Langmuir and E. Stuhlinger, eds., "Electrostatic Propulsion," Progress in Astronautics and Rocketry, Vol. 5. Academic Press, New York, 1961.
13. E. Stuhlinger, "Ion Propulsion for Space Flight." McGraw-Hill, New York, 1964.
14. G. F. Au, "Elektrische Antriebe von Raumfahrzeugen." Verlag G. Braun, Karlsruhe, 1968.
15. G. R. Brewer, "Ion Propulsion." Gordon & Breach, New York, 1970.
16. D. Bohm, *in* "The Characteristics of Electrical Discharges in Magnetic Fields" (A. Guthrie and R. K. Wakerling, eds.), pp. 77–86. McGraw-Hill, New York, 1949.
17. R. G. Jahn, "Physics of Electric Propulsion." Gordon & Breach, New York, 1970.
18. P. D. Reader, *ARS J.* **32**, 711–714 (1962).
19. H. R. Kaufman, *NASA Tech. Note* **TN D-3041** (1965).
20. R. D. Evans, "The Atomic Nucleus," p. 659. McGraw-Hill, New York, 1955.
21. H. J. King and I. Kohlberg, *NASA Contract. Rep.* **CR-52440** (1963).
22. I. Kohlberg and S. Nablo, *in* "Physics and Technology of Ion Motors" (F. E. Marble and J. Surugue, eds.), pp. 155–206. Gordon & Breach, New York, 1966.
23. A. J. Cohen, *NASA Tech. Note* **TN D-3731** (1966).
24. E. K. Shaw, *NASA Contract. Rep.* **CR-54665** (1966).
25. T. D. Masek, *NASA Contract. Rep.* **CR-94554** (1968).
26. W. Knauer, R. L. Poeschel, H. J. King, and J. W. Ward, *NASA Contract. Rep.* **CR-72440** (1968).
27. W. Knauer, *AIAA Pap.* No. 70-177 (1970).
28. N. L. Milder, *J. Spacecr. Rockets* **7**, 641–649 (1970).
29. T. D. Masek, *AIAA Pap.* No. 69-256 (1969).
30. T. D. Masek, *AIAA J.* **9**, 205–212 (1971).
31. W. B. Strickfaden and K. L. Geiler, *AIAA Pap.* No. 63056 (1963).
32. A. R. Martin, *J. Spacecr. Rockets* **8**, 548–550 (1971). A similar paper was presented by A. R. Martin, *EUROMECH 18 Colloq. Advan. Instrum. Meas. Tech. Hypersonic Flow, Univ. Southampton, 1970.*

33. R. C. Speiser, *in* "Physics and Technology of Ion Motors" (F. E. Marble and J. Surugue, eds.), pp. 255–271. Gordon & Breach, New York, 1966.
34. H. R. Kaufman, *J. Spacecr. Rockets* 9, 511–517 (1972).
35. D. A. Dunn and S. A. Self, *J. Appl. Phys.* 35, 113–122 (1964).
36. P. D. Reader, *NASA Tech. Note* TN D-1163 (1962).
37. E. V. Pawlik, S. Nakanishi, and H. R. Algeri, *NASA Tech. Note* TN D-4204 (1967).
38. R. T. Bechtel, *J. Spacecr. Rockets* 5, 795–800 (1968).
39. W. Knauer, R. L. Poeschel, and J. W. Ward, *AIAA Pap.* No. 69-259 (1969).
40. R. D. Moore, *AIAA Pap.* No. 69-260 (1969).
41. W. D. Ramsey, *J. Spacecr. Rockets* 9, 318–321 (1972).
42. P. D. Reader, *AIAA Pap.* No. 64-689 (1964).
43. H. R. Kaufman and A. J. Cohen, *Proc. Symp. Ion Sourc. Form. Ion Beams, Brookhaven Nat. Lab.* pp. 61–68 (1971).
44. W. R. Hudson and B. A. Banks, *AIAA Pap.* No. 73-1131 (1973).
45. V. K. Rawlin, *AIAA Pap.* No. 73-1053 (1973).
46. G. Sohl, G. C. Reid, and R. C. Speiser, *J. Spacecr. Rockets* 3, 1093–1098 (1966).
47. G. Sohl, V. V. Fosnight, S. J. Goldner, and R. C. Speiser, *J. Spacecr. Rockets* 4, 1180–1183 (1967).
48. G. Sohl, R. H. Vernon, K. G. Wood, and T. R. Dillon, *J. Spacecr. Rockets* 5, 165–167 (1968).
49. P. D. Reader, *Int. Conf. Electron Ion Beam Sci., Technol.* [*Pap.*], *1st, Toronto, 1964* (1965).
50. R. J. Schertler, *AIAA Pap.* No. 71-157 (1971).
51. D. C. Byers and P. D. Reader, *Symp. Electron, Ion, Laser Beam Technol., 11th, Boulder, Colo.* (1971).
52. D. C. Byers and P. D. Reader, *NASA Tech. Note* TN D-6620 (1971).
53. D. C. Byers, W. R. Kerslake, and J. Grobman, *NASA Tech. Note* TN D-2401 (1964).
54. J. V. Dugan, Jr., *NASA Tech. Note* TN D-1185 (1964).
55. N. L. Milder, *NASA Tech. Note* TN D-2592 (1965).
56. R. L. Poeschel, J. W. Ward, and W. Knauer, *AIAA Pap.* No. 69-257 (1969).
57. P. D. Reader, *NASA Tech. Note* TN D-2587 (1965).
58. P. D. Reader, S. Nakanishi, W. C. Lathem, and B. A. Banks, *J. Spacecr. Rockets* 7, 1287–1292 (1970).
59. D. C. Byers, *J. Electrochem. Soc.* 116, 9–17 (1969).
60. W. R. Kerslake, *NASA Tech. Note* TN D-1168 (1962).
61. W. R. Kerslake and E. V. Pawlik, *NASA Tech. Note* TN D-1411 (1963).
62. B. A. Free and W. R. Mickelsen, *AIAA Pap.* No. 66-598 (1966).
63. P. Clausing, *Ann. Phys. (Leipzig)*, 12, 961–989 (1932).
64. N. L. Milder, *NASA Tech. Note* TN D-1219 (1962).
65. N. L. Milder and J. S. Sovey, *NASA Tech. Note* TN D-6565 (1971).
66. E. V. Pawlik, R. Goldstein, D. J. Fitzgerald, and R. W. Adams, *AIAA Pap.* No. 72-475 (1972).
67. P. J. Wilbur, NASA Grant NGR 06-002-112, Mo. Rep. (Sept. 1973). The cross sections of this reference are also included in P. J. Wilbur, *NASA Contract. Rep.* CR-121038 (1972).
68. M. Gryziński, *Phys. Rev. A* 138, 336–358 (1965).
69. S. Nakanishi, *AIAA Pap.* No. 72-1151 (1972).
70. J. L. Power, *AIAA Pap.* No. 73-1109 (1973).
71. C. D. Child, *Phys. Rev.* 32, 492–511 (1911).
72. I. Langmuir, *Phys. Rev.* 2, 450–486 (1913).
73. W. Schottky, *Phys. Z.* 15, 526–528 (1914).

74. P. C. Thoneman and E. R. Harrison, *U.K. At. Energy Auth., Res. Group, Rep.* **AER GP/R 1190** (1955).

75. J. R. Pierce, *J. Appl. Phys.* **11**, 548–554 (1940).

76. D. L. Hollway, *Aust. J. Phys.* **8**, 74–89 (1955).

77. G. R. Brewer, *J. Appl. Phys.* **28**, 7–15 (1957).

78. T. Van Duzer and G. R. Brewer, *J. Appl. Phys.* **30**, 291–301 (1959).

79. J. Hyman, Jr., W. O. Eckhardt, R. C. Knechtli, and C. R. Buckey, *AIAA J.* **2**, 1739–1748 (1964).

80. V. Hamza and E. A. Richley, *NASA Tech. Note* **TN D-1323** (1962).

81. V. Hamza, *NASA Tech. Note* **TN D-1711** (1963).

82. C. D. Bogart and E. A. Richley, *NASA Tech. Note* **TN D-3394** (1966).

83. F. V. Pawlik, P. M. Margosion, and J. F. Staggs, *NASA Tech. Note* **TN D-2804** (1965).

84. N. B. Kramer and H. J. King, *J. Appl. Phys.* **38**, 4019–4023 (1967).

85. W. C. Lathem, *J. Spacecr. Rockets* **6**, 1237–1242 (1969).

86. W. C. Lathem, personal communication (1973).

87. G. R. Nudd and K. Amboss, *AIAA J.* **8**, 649–656 (1970).

88. K. R. Spangenberg, "Vacuum Tubes," p. 348. McGraw-Hill, New York, 1948.

89. G. Sohl, G. C. Reid, F. A. Barcatta, S. Zafran, and R. C. Speiser, *NASA Contract. Rep.* **CR-54323** (1965).

90. W. R. Kerslake, *NASA Tech. Note* **TN D-1657** (1963).

91. S. Nakanishi, E. V. Pawlik, and C. W. Baur, *NASA Tech. Note* **TN D-2171** (1964).

92. V. K. Rawlin, *AIAA Pap.* No. 73-1086 (1973).

93. W. R. Kerslake, R. G. Goldman, and W. C. Nieberding, *J. Spacecr. Rockets* **8**, 213–224 (1971).

94. F. M. Charbonnier, C. J. Bennette, and L. W. Swanson, *J. Appl. Phys.* **38**, 627–633 (1967).

95. C. J. Bennette, L. W. Swanson, and F. M. Charbonnier, *J. Appl. Phys.* **38**, 634–640 (1967).

96. P. M. Margosian, *NASA Tech. Memo.* **TM X-1342** (1967).

97. B. Banks, *Int. Conf. Electron Ion Beam Sci. Technol., 3rd, Boston, Mass.* (1968).

98. R. T. Bechtel, B. A. Banks, and T. W. Reynolds, *AIAA Pap.* No. 71-156 (1971).

99. S. Nakanishi, E. A. Richley, and B. A. Banks, *J. Spacecr. Rockets* **5**, 356–358 (1968).

100. D. C. Byers, *NASA Tech. Memo.* **TM X-67842** (1971).

101. C. R. Collett, NASA Contract NAS3-15523, Mo. Rep. 11 (May 1972).

102. V. K. Rawlin, B. A. Banks, and D. C. Byers, *J. Spacecr. Rockets* **10**, 29–35 (1973).

103. V. K. Rawlin, B. A. Banks, and D. C. Byers, *AIAA Pap.* No. 72-486 (1972).

104. R. L. Danilowicz, V. K. Rawlin, B. A. Banks, and E. G. Wintucky, *AIAA Pap.* No. 73-1051 (1973).

105. D. C. Byers and B. A. Banks, *NASA Tech. Memo.* **TM X-67922** (1971).

106. P. LeVaguerèse and D. Pigache, *Rev. Phys. Appl.* **6**, 325–327 (1971).

107. D. Pigache, *AIAA J.* **11**, 129–130 (1973).

108. C. R. Collett, H. J. King, and D. E. Schnelker, *AIAA Pap.* No. 71-691 (1971).

109. W. C. Lathem, *NASA Tech. Memo.* **TM X-68133** (1972).

110. H. J. King, D. Schnelker, J. W. Ward, C. Dulgeroff, and R. Vahrenkamp, *NASA Contract. Rep.* **CR-121142** (1972).

111. H. J. King, C. R. Collett, and D. E. Schnelker, *NASA Contract. Rep.* **CR-72677** (1971).

112. W. R. Hudson, W. C. Lathem, J. L. Power, and B. A. Banks, *NASA Tech. Memo.* **TM X-68096** (1972).

113. W. C. Lathem, *J. Spacecr. Rockets* **5**, 735–737 (1968).

114. G. Sohl and V. V. Fosnight, *J. Spacecr. Rockets* **6**, 143–147 (1969).

115. W. C. Lathem and W. B. Adam, *NASA Tech. Memo.* **TM X-67911** (1971).

116. K. Amboss, *IEEE Trans. Electron Devices* **12**, 313–321 (1965).

117. C. J. Davisson and C. J. Calbick, *Phys. Rev.* **38**, 585 (1931); *Phys. Rev.* **42**, 580 (1932).
118. V. V. Fosnight, T. R. Dillon, and G. Sohl, *J. Spacecr. Rockets* **7**, 266–270 (1970).
119. R. M. Kushnir, B. M. Palyukh, and L. A. Sena, *Bull. Acad. Sci. USSR, Phys. Ser.* **23**, 995–999 (1959).
120. I. P. Iovitsu and N. Ionescu-Pallas, *Sov. Phys.—Tech. Phys.* **4**, 781–791 (1960).
121. D. Zuccaro, *NASA Contract. Rep.* **CR-72398** (1968).
122. C. E. Fay, A. L. Samuel, and W. Shockley, *Bell Syst. Tech. J.* **17**, 49–79 (1938).
123. E. V. Pawlik and P. D. Reader, *NASA Tech. Note* **TN D-4054** (1967).
124. V. K. Rawlin, personal communication (1973).
125. K. Meyer and A. Güntherschulze, *Z. Phys.* **71**, 279–290 (1931).
126. G. K. Wehner, *Phys. Rev.* **108**, 35–45 (1957).
127. R. G. Musket and H. P. Smith, Jr., *J. Appl. Phys.* **39**, 3579–3586 (1968).
128. G. Carter and J. S. Colligon, "Ion Bombardment of Solids," pp. 323–324. Amer. Elsevier, New York, 1968.
129. L. L. Marino, A. C. H. Smith, and E. Caplinger, *Phys. Rev.* **128**, 2243–2250 (1962).
130. G. Sohl, V. V. Fosnight, and S. J. Goldner, *NASA Contract. Rep.* **CR-54711** (1967).
131. R. G. Wilson, *Surface Sci.* **38**, 261–264 (1973).
132. D. Rapp and W. E. Francis, *J. Chem. Phys.* **37**, 2631–2645 (1962).
133. N. Laegreid and G. K. Wehner, *J. Appl. Phys.* **32**, 365–369 (1961).
134. D. Rosenberg and G. K. Wehner, *J. Appl. Phys.* **33**, 1842–1845 (1962).
135. H. R. Kaufman, *NASA Tech. Note* **TN D-261** (1960).
136. H. Mirels and B. M. Rosenbaum, *NASA Tech. Note* **TN D-266** (1960).
137. Staff of the Ramo-Wooldridge Res. Lab., *Proc. IRE* **48**, 477–491 (1960).
138. W. H. Wells, *JPL Tech. Release* No. 34–118, Pasadena, Calif. (Oct. 1960).
139. J. M. Sellen and H. Shelton, *Amer. Rocket Soc. Pap.* No. 1379-60 (1960).
140. J. M. Sellen and R. F. Kemp, *Amer. Rocket Soc. Pap.* No. 61-84-1778 (1961).
141. O. Buneman and G. P. Kooyers, *AIAA Pap.* No. 63042 (1963).
142. D. A. Dunn and I. T. Ho, *AIAA J.* **1**, 2770–2777 (1963).
143. R. P. Wadwa, D. F. Brauch, and O. Bunemen, *AIAA Pap.* No. 64-698 (1964).
144. H. E. Wilhelm, NASA Grant NGR-06-002-147, Annu. Rep. (June 1973). (One part published in *J. Appl. Phys.* **44**, 4562–4566 (1973); another part to be published in *Phys. Fluids* (1974).)
145. O. Buneman, *Phys. Rev.* **115**, 503–517 (1959).
146. G. R. Brewer, J. E. Etter, and J. R. Anderson, *AIAA Pap.* No. 1125-60 (1960).
147. C. R. Dulgeroff, D. E. Zuccaro, S. Kami, D. E. Schnelker, and J. W. Ward, *AIAA Pap.* No. 72-494 (1972).
148. R. J. Cybulski, D. J. Shellhammer, R. R. Lovell, E. J. Domino, and J. T. Kotnik, *NASA Tech. Note* **TN D-2718** (1965); also *in* "Advanced Propulsion Concepts" (A. T. Forrester and G. Kuskevics, eds.), pp. 21–64. Gordon & Breach, New York, 1965.
149. M. P. Ernstene, E. L. James, G. W. Purmal, R. M. Worlock, and A. T. Forrester, *J. Spacecr. Rockets* **3**, 744–747 (1966); also *in* "Ion Propulsion" (A. T. Forrester and G. Kuskevics, eds.), pp. 42–45. AIAA Sel. Repr., New York, 1968.
150. R. T. Bechtel, *NASA Tech. Memo.* **TM X-67926** (1971).
151. S. Nakanishi and R. C. Finke, *AIAA Pap.* No. 73-1111 (1973).
152. S. Nakanishi and R. C. Finke, *NASA Tech. Memo.* **TM X-68155** (1972).
153. R. T. Bechtel, *AIAA Pap.* No. 73-1052 (1973).
154. D. C. Byers, *NASA Tech. Note* **TN D-5844** (1970).
155. H. S. Ogawa, R. K. Cole, and J. M. Sellen, *AIAA Pap.* No. 69-263 (1969).
156. H. S. Ogawa, R. K. Cole, and J. M. Sellen, *AIAA Pap.* No. 70-1142 (1970).
157. T. W. Reynolds and E. A. Richley, *NASA Tech. Note* **TN D-7038** (1971).

158. J. F. Staggs, W. P. Gula, and W. R. Kerslake, *J. Spacecr. Rockets* **5**, 159–164 (1968).
159. T. W. Reynolds, *NASA Tech. Memo.* **TN X-68043** (1972).
160. J. V. Dugan, Jr., *NASA Tech. Memo.* **TM X-2527** (1972).
161. N. L. Milder and J. S. Sovey, *AIAA Pap.* No. 72-441 (1972).
162. W. H. Kohl, "Materials and Techniques for Vacuum Devices," pp. 261, 301, 303, 478, 482. Van Nostrand-Reinhold, Princeton, New Jersey, 1967.
163. W. R. Kerslake, *AIAA Pap.* No. 64-683 (1964).
164. N. L. Milder and W. R. Kerslake, *NASA Tech. Note* **TN D-2173** (1964).
165. R. V. Stuart and G. K. Wehner, *J. Appl. Phys.* **33**, 2345–2352 (1962).
166. G. K. Wehner, *Advan. Electron. Electron Phys.* **7**, 239–298 (1955).
167. R. F. Kemp, J. M. Sellen, Jr., and E. V. Pawlik, *Amer. Rocket Soc. Pap.* No. 2663-62 (1962).
168. W. R. Kerslake, *NASA Tech. Note* **TN D-3818** (1967).
169. D. C. Byers, *NASA Tech. Note* **TN D-5074** (1969).
170. H. E. Gallagher and W. Knauer, *J. Spacecr. Rockets* **5**, 730–732 (1968).
171. A. J. Weigand and S. Nakanishi, *Proc. Symp. Ion Sourc. Form. Ion Beams, Brookhaven Nat. Lab.* pp. 93–101 (1971).
172. R. C. Speiser and L. K. Branson, *Amer. Rocket Soc. Pap.* No. 2664-62 (1962).
173. M. P. Ernstene, A. T. Forrester, E. L. James, G. W. Purmal, and R. M. Worlock, *AIAA Pap.* No. 65-375 (1965); also M. P. Ernstene, E. L. James, G. W. Purmal, R. M. Worlock, and A. T. Forrester, *J. Spacecr. Rockets* **3**, 744–747 (1966).
174. G. Sohl, F. A. Barcatta, and S. Zafran, *NASA Contract. Rep.* **CR-54416** (1965).
175. R. F. Kemp and J. M. Sellen, Jr., *NASA Contract. Rep.* **CR-54692** (1966).
176. D. F. Hall, R. F. Kemp, and H. Shelton, *AIAA Pap.* No. 67-669 (1967).
177. V. K. Rawlin and E. V. Pawlik, *AIAA Pap.* No. 67-670 (1967).
178. W. R. Kerslake, D. C. Byers, and J. F. Staggs, *AIAA Pap.* No. 67-700 (1967).
179. G. A. Csiky, *J. Spacecr. Rockets* **7**, 474–475 (1970). A more complete version appears in *NASA Tech. Note* **TN D-4966** (1969).
180. A. Snyder and B. A. Banks, *NASA Tech. Note* **TN D-6705** (1972).
181. J. W. Ward and H. J. King, *J. Spacecr. Rockets* **5**, 1161–1164 (1968).
182. D. G. Fearn and C. M. Philip, *AIAA Pap.* No. 72-416 (1972).
183. C. M. Philip, *AIAA J.* **9**, 2191–2196 (1971).
184. T. W. Sheheen and R. C. Finke, *NASA Tech. Memo.* **TM X-2799** (1973).
185. D. C. Byers, *NASA Tech. Memo.* **TM X-52543** (1969).
186. D. C. Byers and A. Snyder, *AIAA Pap.* No. 70-1090 (1970).
187. M. J. Mirtich, *AIAA Pap.* No. 73-1138 (1973).
188. V. K. Rawlin and W. R. Kerslake, *J. Spacecr. Rockets* **7**, 14–20 (1970). More tabulated data are contained in the original, *AIAA Pap.* No. 69-304 (1969).
189. A. E. Guile, *Proc. Inst. Elec. Eng.* **118**, 1131–1154 (1971).
190. A. S. Gilmour, Jr., and D. L. Lockwood, *Proc. IEEE* **60**, 977–991 (1972).
191. D. L. Lockwood, Ph.D. Thesis, State Univ. of New York at Buffalo, 1973.
192. J. Hyman, Jr., *AIAA Pap.* No. 72-492 (1972).
193. W. R. Hudson and A. J. Weigand, *AIAA Pap.* No. 73-1142 (1973).
194. H. J. King, R. L. Poeschel, D. E. Schnelker, B. G. Herron, and C. R. Collett, *NASA Contract. Rep.* **CR-120919** (1972).
195. H. J. King, W. O. Eckhardt, J. W. Ward, and R. C. Knechtli, *J. Spacecr. Rockets* **4**, 599–602 (1967).
196. R. T. Bechtel, *AIAA Pap.* No. 70-1100 (1970).
197. A. A. Wells, *AIAA Pap.* No. 72-418 (1972).
198. J. S. Serafini and F. F. Terdan, *AIAA Pap.* No. 73-1056 (1973).
199. R. L. Poeschel, *AIAA Pap.* No. 72-488 (1972).

*200.* R. T. Bechtel, *AIAA Pap.* No. 72-489 (1972).

*201.* W. R. Kerslake, *Elec. Prop. Conf., Inst. Elec. Eng., Culham, Eng., 1973.*

*202.* W. R. Kerslake and R. C. Finke, *AIAA Pap.* No. 73-1136 (1973).

*203.* E. G. Wintucky, *AIAA Pap.* No. 73-1140 (1973).

*204.* V. K. Rawlin, *NASA Tech. Memo.* **TM X-2785** (1973).

*205.* W. O. Eckhardt, H. J. King, J. Snyder, and R. C. Knechtli, *AIAA Pap.* No. 66-245 (1966).

*206.* W. O. Eckhardt, K. W. Arnold, G. Hagen, J. Hyman, Jr., J. Snyder, and R. C. Knechtli, *AIAA Pap.* No. 67-667 (1967).

*207.* J. Hyman, Jr., J. R. Bayless, D. E. Schnelker, J. W. Ward, and J. Simpkins, *J. Spacecr. Rockets* **8**, 717–721 (1971).

*208.* S. Nakanishi and E. V. Pawlik, *NASA Tech. Memo.* **TM X-1026** (1964).

*209.* S. Nakanishi, *NASA Tech. Memo.* **TM X-1579** (1968).

*210.* S. Nakanishi, *NASA Tech. Note* **TN D-3535** (1966).

*211.* M. A. Mantenieks, *AIAA Pap.* No. 73-1088 (1973).

*212.* W. R. Kerslake, *AIAA Pap.* No. 72-484 (1972); also *NASA Tech. Note* **TN D-6782** (1972).

*213.* E. V. Pawlik and S. Nakanishi, *NASA Tech. Note* **TN D-2470** (1964).

*214.* R. T. Bechtel, *AIAA Pap.* No. 69-238 (1969).

*215.* F. F. Terdan and R. T. Bechtel, *AIAA Pap.* No. 73-1079 (1973).

*216.* R. J. Zavesky and E. B. Hurst, *NASA Tech. Memo.* **TM X-2518** (1972).

# Author Index

Numbers in parentheses are reference numbers and indicate that an author's work is referred to although his name is not cited in the text. Numbers in italics show the page on which the complete reference is listed.

Domino, E. J., 339(148), *371*
Donskaya, N. P., 59, 86, 88, *149*
Dryer, M., 41, *53*
Dugan, Jr., J. V., 278, 344, *369, 372*
Dulgeroff, C. R., 321(110), 322(110), 339(147), 350, *370, 371*
Dunn, D. A., 272(35), 338, *369, 371*

**E**

Eckhardt, W. O., 292(79), 361(195), 365(195), *370, 372, 373*
Effemey, H. G., 178(23), *193*
Ekers, R. D., 38, *53*
Ernstene, M. P., 340(149), 352, 358(173), *371, 372*
Etter, J. E., 339(146), *371*
Evans, R. D., 270(20), *368*
Evans, R. G., 89, 90, 116, *149*
Eviatar, A., 13, 17, 18, 27, 33, 42, *53, 55*
Eyni, M., 13, *53*

**F**

Fairfield, D. H., 42, *53*
Faisal, F. H. M., 134, 140, *149*
Farkas, G., 125, *149*
Fay, C. E., 330, *371*
Fearn, D. G., 357, *372*
Feyman, R. P., 119, *149*
Finke, R. C., 340, 357, 362, *371, 372, 373*
Finkelstein, A. T., 266(6), *368*
Finley, L. T., 9, 10, 11, *55*
Fisk, L. A., 42, *53*
Fitzgerald, D. J., 283(66), *369*
Fornaca, G., 143, *149*
Forrester, A. T., 340(149), 352(173), 358(173), *371, 372*
Forslund, D. W., 17, 32, *53*
Fosnight, V. V., 277(47), 324, 325, 326(114, 118), 336(130), 339(47), *369, 370, 371*
Fox, R. A., 59, 60, 89, 140, *149, 150*
Francis, W. E., 336, *371*
Franken, P. A., 59, *149*
Frazier, E. N., 35, *55*
Fredricks, R. W., 31, *53*
Free, B. A., 282, *369*

**G**

Gallagher, H. E., 351(170), *372*
Garrett, C. G. B., 59, *150*
Gatti, E., 143, *148*
Geiler, K. L., 270, *368*
Geltman, S., 59, 96, 97, 98, *149, 152*
Ghanekar, K. M., 171(14), *193*
Gilmour, Jr., A. S., 358, *372*
Giordmaine, J. A., 59, *150*
Glauber, R. J., 142, *150*
Godart, J. L., 60, 98, 99, *148*
Göppert-Mayer, M., 58, 62, *150*
Gold, A., 63, 66, 68, 74, 78, 95, 99, 125, *148, 150*
Goldberger, M. L., 12, 14, 15, 48, *53*
Goldman, R. G., 309(93), 340(93), *370*
Goldner, S. J., 277(47), 336(130), 369(47), *369, 371*
Goldsborough, J. P., 59, *151*
Goldstein, M., 42, *53*
Goldstein, R., 283(66), *369*
Gontier, Y., 72, 73, 74, 95, 117, 118, 119, 120, 121, *150*
Gorbunkov, V. M., 59, 77, 79, 83, 99, 108, *152*
Gordon, W., 70, *150*
Gosling, J. T., 10, *53*
Griffel, D. H., 16, *53*
Gringauz, K. I., 3, *53*
Grobman, J., 278(53), *369*
Grutzmann, S., 175, *193*
Gryziński, M., 284, *369*
Guccione-Gush, R., 143, *150*
Güntherschulze, A., 334, 350, 351, *371*
Guile, A. E., 358, *372*
Gula, W. P., 342(158), 344(158), *372*
Gurevich, L. E., 190(45), *194*
Gush, H. P., 143, *150*
Guthrie, A., 266(8), *368*

**H**

Hagen, G., 365(206), *373*
Hall, D. F., 352, 359, *372*
Hall, J. L., 59, 96, *150*
Hampshire, M. J., 191(66), *194*
Hamza, V., 293, *370*
Harmuth, H. F., 200(1), 202, *264*

# Subject Index